Praise for *The Stuff of Thought* by Steven Pinker

"Pinker brings an engaging and witty style to the study of subject matter that—were it not as important to us as it is complex—might otherwise be off-putting. . . . An inviting and important book. Everyone with an interest in language and how it gets to be how it is—that is, everyone interested in how we get to be human and do our human business—should read *The Stuff of Thought.*" —Robin Lakoff, *Science*

"Packed with information, clear, witty, attractively written, and generally persuasive . . . [Pinker] is unfailingly engaging to read, with his aptly chosen cartoons, his amusing examples, and his bracing theoretical rigor." —Colin McGinn, *The New York Review of Books*

"Engaging and provocative . . . filled with humor and fun. It's good to have a mind as lively and limpid as his bringing the ideas of cognitive science to the public while clarifying them for his scientific colleagues." —Douglas Hofstadter, *Los Angeles Times*

"Pinker is not only wonderfully clear; he is also blessedly witty. There's plenty of stuff to think about in *The Stuff of Thought,* but a lot of fun stuff too." —George Scialabba, *The Boston Globe*

"An excellent window not only into human nature but into Pinker's nature: curious, inventive, fearless, naughty." —William Saletan, *The New York Times Book Review*

"[Pinker] is the cognitive philosopher of our generation, and his work on language and mind has implications for anybody interested in human expression and experience. . . . [He] has changed the way we understand where we have come from and where we are going." —Seth Lerer, *The New York Sun*

"A fascinating look at how language provides a window into the deepest functioning of the human brain." —Josie Glausiusz, *Wired*

"A perceptive, amusing and intelligent book." —Douglas Johnstone, *The Times* (London)

"This is Steven Pinker at his best—theoretical insight combined with clear illustration and elegant research summary, presented throughout with an endearing wit and linguistic creativity which has become his hallmark. Metaphor, he says, with typical Pinkerian panache, 'provides us with a way to eff the ineffable.' The book requires steady concentration, but despite the abstract character of its subject matter it is not difficult to read. That is Pinker's genius. He effs like no other." —David Crystal, *Financial Times*

"Immensely readable and stimulating."
—David Papineau, *The Independent on Sunday*

"Illuminating and astonishingly readable."
—Robert Hanks, *Sunday Telegraph* (London)

"*The Stuff of Thought* delivers the same rewards as Pinker's earlier books for a general audience. He has a very good eye for the apt example, the memorable quote, and the joke that nails the point; he is lucid in explanation and vigorous in argument. . . . *The Stuff of Thought* [has] the two most important qualities in a good popular science book: it makes the subject accessible, and it makes its readers think." —Deborah Camerson, *The Guardian* (London)

"The pleasure of Pinker's book is in watching the careful skill with which he peels back the linguistic layers that clothe those models. The whole performance brought to my mind (very Pinkerishly, I now see) those elaborate colored diagrams in anatomy textbooks, in which you can leaf through successive transparencies to remove the skin, musculature, and organs to reveal at last the skeleton. . . . Like [Pinker's other books], it breathes the spirit of good-natured, rational, humane inquiry." —John Derbyshire, *American Conservative*

"[A] brilliant book." —Emma Garman, *Huffington Post*

"A cracking read." —Shane Hegarty, *The Irish Times*

"I recommend the book as highly as I can recommend any book, without reservation. *Buy it.* And read it. You'll find yourself educated and entertained at the same time." —S. Abbas Raza, *3 Quarks Daily*

"A spicy stew." —Chris Scott, *The Globe and Mail* (Toronto)

"Its sheer range is astonishing. If you wish to know why metaphors are both inescapable and inadequate, why and how people swear, how English expresses concepts of space and time, or why we often avoid saying what we mean, I find it hard to imagine a better guide. As always, Pinker displays an apparently effortless talent for illuminating complex ideas with pointed, witty examples. . . . He has fun with ideas and draws ideas from fun. An impressive achievement, all in all, on many levels."   —Mark Abley, *Montreal Gazette*

"[An] awesome combination of analytical and imaginative thinking . . . Pinker writes lucidly and elegantly, and leavens the text with scores of perfectly judged anecdotes, jokes, cartoons, and illustrations."   —Rita Carter, *Daily Mail*

"Pinker is fascinating, authoritative, intense. His book is packed with ideas that have been fully thought out and carefully rendered to prompt us each to marvel at the determinants of human nature."
                        —Anne Brataas, *Star Tribune* (Minneapolis)

"A fascinating explanation of how we think and why we do what we do. . . . While you might have to wrap your brain around tenses, Extreme Nativism, and polysemy before you can figure out why you're constantly swearing like a drunken sailor, it's abso-fucking-lutely worth it."
                        —Courtney Ferguson, *The Portland Mercury*

"*The Stuff of Thought* is an excellent book . . . easily his most accessible and fun book to read . . . [and] on a scientific level, the book does something quite amazing: it bridges the chasm that many academics have over language itself."
                        —Daniel Schneider, *Monsters and Critics*

"[A] stimulating volume . . . From politics to poetry, children's wonderful malapropisms to slang, Pinker's fluency in the nuances of words and syntax serves as proof of his faith in language as 'a window into human nature.'"
                        —Donna Seamon, *Booklist*

"A book on semantics may not sound especially enticing, but with Pinker as your guide, pondering what the meaning of 'is' is can be mesmerizing."
                                        —*Details*

STEVEN PINKER

The
STUFF
of
THOUGHT

*Language as a Window
into Human Nature*

PENGUIN BOOKS

For Rebecca

PENGUIN BOOKS

Published by the Penguin Group

Penguin Group (USA) Inc., 375 Hudson Street, New York, New York 10014, U.S.A.
Penguin Group (Canada), 90 Eglinton Avenue East, Suite 700, Toronto,
Ontario, Canada M4P 2Y3 (a division of Pearson Penguin Canada Inc.)
Penguin Books Ltd, 80 Strand, London WC2R 0RL, England
Penguin Ireland, 25 St Stephen's Green, Dublin 2, Ireland (a division of Penguin Books Ltd)
Penguin Group (Australia), 250 Camberwell Road, Camberwell,
Victoria 3124, Australia (a division of Pearson Australia Group Pty Ltd)
Penguin Books India Pvt Ltd, 11 Community Centre, Panchsheel Park, New Delhi – 110 017, India
Penguin Group (NZ), 67 Apollo Drive, Rosedale, North Shore 0632,
New Zealand (a division of Pearson New Zealand Ltd)
Penguin Books (South Africa) (Pty) Ltd, 24 Sturdee Avenue, Rosebank, Johannesburg 2196, South Africa

Penguin Books Ltd, Registered Offices:
80 Strand, London WC2R 0RL, England

First published in the United States of America by Viking Penguin,
a member of Penguin Group (USA) Inc. 2007
Published in Penguin Books 2008

1 3 5 7 9 10 8 6 4 2

Grateful acknowledgment is made for permission to reprint excerpts from the following copyrighted works:
"This Be the Verse" from *Collected Poems* by Philip Larkin. Copyright © 1988, 2003 by the Estate of Philip
Larkin. Reprinted by permission of Farrar, Straus and Giroux, LLC and Faber and Faber Ltd.
The Nurture Assumption: Why Children Turn Out the Way They Do by Judith Rich Harris (Free Press).
Copyright © 1998 by Judith Rich Harris. Reprinted with permission.

THE LIBRARY OF CONGRESS HAS CATALOGED THE HARDCOVER EDITION AS FOLLOWS:
Pinker, Steven, 1954–
The stuff of thought : language as a window into human nature / Steven Pinker.
p. cm.
Includes bibliographical references and index.
ISBN 978-0-670-06327-7 (hc.)
ISBN 978-0-14-311424-6 (pbk.)
1. Language and languages—Philosophy.   2. Thought and thinking.   I. Title.
P107.P548 2007
401—dc22   2007026601

Printed in the United States of America

PREFACE

There is a theory of space and time embedded in the way we use words. There is a theory of matter and a theory of causality, too. Our language has a model of sex in it (actually, two models), and conceptions of intimacy and power and fairness. Divinity, degradation, and danger are also ingrained in our mother tongue, together with a conception of well-being and a philosophy of free will. These conceptions vary in their details from language to language, but their overall logic is the same. They add up to a distinctively human model of reality, which differs in major ways from the objective understanding of reality eked out by our best science and logic. Though these ideas are woven into language, their roots are deeper than language itself. They lay out the ground rules for how we understand our surroundings, how we assign credit and blame to our fellows, and how we negotiate our relationships with them. A close look at our speech—our conversations, our jokes, our curses, our legal disputes, the names we give our babies—can therefore give us insight into who we are.

That is the premise of the book you are holding, the third in a trilogy written for a wide audience of readers who are interested in language and mind. The first, *The Language Instinct*, was an overview of the language faculty: everything you always wanted to know about language but were afraid to ask. A language is a way of connecting sound and meaning, and the other two books turn toward each of those spheres. *Words and Rules* was about the units of language, how they are stored in memory, and how they are assembled into the vast number of combinations that give language its expressive power. *The Stuff of Thought* is about the other side of the linkage, meaning. Its vistas include the meanings of words and constructions and the way that language is used in social settings, the topics that linguists call semantics and pragmatics.

At the same time, this volume rounds out another trilogy: three books on human nature. *How the Mind Works* tried to reverse-engineer the psyche in the light of cognitive science and evolutionary psychology. *The Blank Slate* explored the concept of human nature and its moral, emotional, and political colorings. This one broaches the topic in still another way: what we can learn about our makeup from the way people put their thoughts and feelings in words.

As in my other books on language, the early chapters occasionally dip into technical topics. But I have worked hard to make them transparent, and I am confident that my subject will engage anyone with an interest in what makes us tick. Language is entwined with human life. We use it to inform and persuade, but also to threaten, to seduce, and of course to swear. It reflects the way we grasp reality, and also the image of ourselves we try to project to others, and the bonds that tie us to them. It is, I hope to convince you, a window into human nature.

In writing this book I have enjoyed the advice and support of many people, beginning with my editors, Wendy Wolf, Stefan McGrath, and Will Goodlad, and my agent, John Brockman. I have benefited tremendously from the wisdom of generous readers who reviewed the entire manuscript—Rebecca Newberger Goldstein, David Haig, David Kemmerer, Roslyn Pinker, and Barbara Spellman—and from the mavens who commented on chapters in their areas of expertise: Linda Abarbanell, Ned Block, Paul Bloom, Kate Burridge, Herbert Clark, Alan Dershowitz, Bruce Fraser, Marc Hauser, Ray Jackendoff, James Lee, Beth Levin, Peggy Li, Charles Parsons, James Pustejovsky, Lisa Randall, Harvey Silverglate, Alison Simmons, Donald Symons, J. D. Trout, Michael Ullman, Edda Weigand, and Phillip Wolff. Thanks, too, to those who answered my queries or offered suggestions: Max Bazerman, Iris Berent, Joan Bresnan, Daniel Casasanto, Susan Carey, Gennaro Chierchia, Helena Cronin, Matt Denio, Daniel Donoghue, Nicholas Epley, Michael Faber, David Feinberg, Daniel Fessler, Alan Fiske, Daniel Gilbert, Lila Gleitman, Douglas Jones, Marcy Kahan, Robert Kurzban, Gary Marcus, George Miller, Martin Nowak, Anna Papafragou, Geoffrey Pullum,

S. Abbas Raza, Laurie Santos, Anne Senghas, G. Richard Tucker, Daniel Wegner, Caroline Whiting, and Angela Yu. This is the sixth book of mine that Katya Rice has agreed to copyedit, and like the others it has benefited from her style, precision, and curiosity.

I thank Ilavenil Subbiah for the many examples of subtle semantic phenomena she recorded from everyday speech, for designing the chapter ornament, and for much else besides. Thanks also to my parents, Harry and Roslyn, and to my family: Susan, Martin, Eva, Carl, Eric, Rob, Kris, Jack, David, Yael, Gabe, and Danielle. Most of all, I thank Rebecca Newberger Goldstein, my *bashert,* to whom this book is dedicated.

The research for this book was supported by NIH Grant HD-18381 and by the Johnstone Family Chair at Harvard University.

CONTENTS

I

WORDS AND WORLDS

On September 11, 2001, at 8:46 A.M., a hijacked airliner crashed into the north tower of the World Trade Center in New York. At 9:03 A.M. a second plane crashed into the south tower. The resulting infernos caused the buildings to collapse, the south tower after burning for an hour and two minutes, the north tower twenty-three minutes after that. The attacks were masterminded by Osama bin Laden, leader of the Al Qaeda terrorist organization, who hoped to intimidate the United States into ending its military presence in Saudi Arabia and its support for Israel and to unite Muslims in preparation for a restoration of the caliphate.

9/11, as the happenings of that day are now called, stands as the most significant political and intellectual event of the twenty-first century so far. It has set off debates on a vast array of topics: how best to memorialize the dead and revitalize lower Manhattan; whether the attacks are rooted in ancient Islamic fundamentalism or modern revolutionary agitation; the role of the United States on the world stage before the attacks and in response to them; how best to balance protection against terrorism with respect for civil liberties.

But I would like to explore a lesser-known debate triggered by 9/11. Exactly how many events took place in New York on that morning in September?

It could be argued that the answer is one. The attacks on the buildings were part of a single plan conceived in the mind of one man in service of a single agenda. They unfolded within a few minutes and yards of each other, targeting the parts of a complex with a single name, design, and owner. And they launched a single chain of military and political events in their aftermath.

Or it could be argued that the answer is two. The north tower and the south tower were distinct collections of glass and steel separated by an expanse of space, and they were hit at different times and went out of existence at different times. The amateur video that showed the second plane closing in on the south tower as the north tower billowed with smoke makes the twoness unmistakable: in those horrifying moments, one event was frozen in the past, the other loomed in the future. And another occurrence on that day—a passenger mutiny that brought down a third hijacked plane before it reached its target in Washington—presents to the imagination the possibility that one tower or the other might have been spared. In each of those possible worlds a distinct event took place, so in our *actual* world, one might argue, there must be a pair of events as surely as one plus one equals two.

The gravity of 9/11 would seem to make this entire discussion frivolous to the point of impudence. It's a matter of mere "semantics," as we say, with its implication of picking nits, splitting hairs, and debating the number of angels that can dance on the head of a pin. But this book is about semantics, and I would not make a claim on your attention if I did not think that the relation of language to our inner and outer worlds was a matter of intellectual fascination and real-world importance.

Though "importance" is often hard to quantify, in this case I can put an exact value on it: three and a half billion dollars. That was the sum in dispute in a set of trials determining the insurance payout to Larry Silverstein, the leaseholder of the World Trade Center site. Silverstein held insurance policies that stipulated a maximum reimbursement for each destructive "event." If 9/11 comprised a single event, he stood to receive three and a half billion dollars. If it comprised two events, he stood to receive seven billion. In the trials, the attorneys disputed the applicable meaning of the term *event*. The lawyers for the leaseholder defined it in physical terms (two collapses); those for the insurance companies defined it in mental terms (one plot). There is nothing "mere" about semantics!

Nor is the topic intellectually trifling. The 9/11 cardinality debate is not about the facts, that is, the physical events and human actions that took place that day. Admittedly, those have been contested as well: according to various conspiracy theories, the buildings were targeted by American missiles, or demolished by a controlled implosion, in a plot conceived by American neoconservatives, Israeli spies, or a cabal of psychiatrists. But aside from the kooks, most people agree on the facts. Where they differ is in the *construal* of those facts: how the intricate swirl of matter in space ought to be conceptualized by human minds. As we shall see, the categories in this dispute permeate the meanings of words in our language because they permeate the way we represent reality in our heads.

Semantics is about the relation of words to thoughts, but it is also about the relation of words to other human concerns. Semantics is about the relation of words to reality—the way that speakers commit themselves to a shared understanding of the truth, and the way their thoughts are anchored to things and situations in the world. It is about the relation of words to a community—how a new word, which arises in an act of creation by a single speaker, comes to evoke the same idea in the rest of a population, so people can understand one another when they use it. It is about the relation of words to emotions: the way in which words don't just point to things but are saturated with feelings, which can endow the words with a sense of magic, taboo, and sin. And it is about words and social relations—how people use language not just to transfer ideas from head to head but to negotiate the kind of relationship they wish to have with their conversational partner.

A feature of the mind that we will repeatedly encounter in these pages is that even our most abstract concepts are understood in terms of concrete scenarios. That applies in full force to the subject matter of the book itself. In this introductory chapter I will preview some of the book's topics with vignettes from newspapers and the Internet that can be understood only through the lens of semantics. They come from each of the worlds that connect to our words—the worlds of thought, reality, community, emotions, and social relations.

WORDS AND THOUGHTS

Let's look at the bone of contention in the world's most expensive debate in semantics, the three-and-a-half-billion-dollar argument over the meaning of "event." What, exactly, is an event? An event is a stretch of time, and time, according to physicists, is a continuous variable—an inexorable cosmic flow, in Newton's world, or a fourth dimension in a seamless hyper-space, in Einstein's. But the human mind carves this fabric into the discrete swatches we call events. Where does the mind place the incisions? Some-times, as the lawyers for the World Trade Center leaseholder pointed out, the cut encircles the change of state of an object, such as the collapse of a building. And sometimes, as the lawyers for the insurers pointed out, it en-circles the goal of a human actor, such as a plot being executed. Most often the circles coincide: an actor intends to cause an object to change, the intent of the actor and the fate of the object are tracked along a single time line, and the moment of change marks the consummation of the intent.

The conceptual content behind the disputed language is itself like a lan-guage (an idea I will expand in chapters 2 and 3). It represents an analogue reality by digital, word-sized units (such as "event"), and it combines them into assemblies with a syntactic structure rather than tossing them together like rags in a bag. It's essential to our understanding of 9/11, for example, not only that bin Laden acted to harm the United States, and that the World Trade Center was destroyed around that time, but that it was bin Laden's act that *caused* the destruction. It's the causal link between the intention of a particular man and a change in a particular object that distinguishes the mainstream understanding of 9/11 from the conspiracy theories. Linguists call the inventory of concepts and the schemes that combine them "con-ceptual semantics."[1] Conceptual semantics—the language of thought— must be distinct from language itself, or we would have nothing to go on when we debate what our words mean.

The fact that rival construals of a single occurrence can trigger an ex-travagant court case tells us that the nature of reality does not dictate the way that reality is represented in people's minds. The language of thought allows us to frame a situation in different and incompatible ways. The un-folding of history on the morning of September 11 in New York can be thought of as one event or two events depending on how we mentally de-scribe it to ourselves, which in turn depends on what we choose to focus on and what we choose to ignore. And the ability to frame an event in alterna-tive ways is not just a reason to go to court but also the source of the rich-ness of human intellectual life. As we shall see, it provides the materials for scientific and literary creativity, for humor and wordplay, and for the dra-mas of social life. And it sets the stage in countless arenas of human dispu-tation. Does stem-cell research destroy a ball of cells or an incipient human?

Is the American military incursion into Iraq a case of invading a country or of liberating a country? Does abortion consist of ending a pregnancy or of killing a child? Are high tax rates a way to redistribute wealth or to confiscate earnings? Is socialized medicine a program to protect citizens' health or to expand government power? In all these debates, two ways of framing an event are pitted against each other, and the disputants struggle to show that their framing is more apt (a criterion we will explore in chapter 5). In the past decade prominent linguists have been advising American Democrats on how the Republican Party has outframed them in recent elections and on how they might regain control of the semantics of political debate by reframing, for example, *taxes* as *membership fees* and *activist judges* as *freedom judges*.[2]

The 9/11 cardinality debate highlights another curious fact about the language of thought. In puzzling over how to count the events of that day, it asks us to treat them as if they were objects that can be tallied, like poker chips in a pile. The debate over whether there was one event or two in New York that day is like a disagreement over whether there is one item or two at an express checkout lane, such as a pair of butter sticks taken out of a box of four, or a pair of grapefruits selling at two for a dollar. The similar ambiguity in tallying objects and tallying events is one of the many ways in which space and time are treated equivalently in the human mind, well before Einstein depicted them as equivalent in reality.

As we shall see in chapter 4, the mind categorizes matter into discrete things (like *a sausage*) and continuous stuff (like *meat*), and it similarly categorizes time into discrete events (like *to cross the street*) and continuous activities (like *to stroll*). With both space and time, the same mental zoom lens that allows us to count objects or events also allows us to zoom in even closer on what each one is made of. In space, we can focus on the material making up an object (as when we say *I got sausage all over my shirt*); in time, we can focus on an activity making up an event (as when we say *She was crossing the street*). This cognitive zoom lens also lets us pan out in space and see a collection of objects as an aggregate (as in the difference between *a pebble* and *gravel*), and it allows us to pan out in time and see a collection of events as an iteration (as in the difference between *hit the nail* and *pound the nail*). And in time, as in space, we mentally place an entity at a location and then shunt it around: we can *move a meeting from 3:00 to 4:00* in the same way that we move a car from one end of the block to the other. And

speaking of an *end*, even some of the fine points of our mental geometry
carry over from space to time. The *end of a string* is technically a point, but
we can say *Herb cut off the end of the string*, showing that an end can be
construed as including a snippet of the matter adjacent to it. The same is
true in time: the *end of a lecture* is technically an instant, but we can say *I'm
going to give the end of my lecture now*, construing the culmination of an
event as including a small stretch of time adjacent to it.[3]

As we shall see, language is saturated with implicit metaphors like
EVENTS ARE OBJECTS and TIME IS SPACE. Indeed, space turns out to be a
conceptual vehicle not just for time but for many kinds of states and circumstances. Just as a meeting can be moved from 3:00 to 4:00, a traffic light
can go from green to red, a person can go from flipping burgers to running
a corporation, and the economy can go from bad to worse. Metaphor is so
widespread in language that it's hard to find expressions for abstract ideas
that are *not* metaphorical. What does the concreteness of language say
about human thought? Does it imply that even our wispiest concepts are
represented in the mind as hunks of matter that we move around on a mental stage? Does it say that rival claims about the world can never be true or
false but can only be alternative metaphors that frame a situation in different ways? Those are the obsessions of chapter 5.

WORDS AND REALITY

The aftermath of 9/11 spawned another semantic debate, one with consequences even weightier than the billions of dollars at stake in how to count
the events on that day. This one involves a war that has cost far more money
and lives than 9/11 itself and that may affect the course of history for the
rest of the century. The debate hinges on the meaning of another set of
words—sixteen of them, to be exact:

> The British government has learned that Saddam Hussein re-
> cently sought significant quantities of uranium from Africa.

This sentence appeared in George W. Bush's State of the Union address in
January 2003. It referred to intelligence reports suggesting that Saddam
may have tried to buy five hundred tons of a kind of uranium ore called yellowcake from sources in Niger in West Africa. For many Americans and

Britons the possibility that Saddam was assembling nuclear weapons was the only defensible reason to invade Iraq and depose Saddam. The United States led the invasion in the spring of that year, the most despised American foreign policy initiative since the war in Vietnam. During the occupation it became clear that Saddam had had no facilities in place to manufacture nuclear weapons, and probably had never explored the possibility of buying yellowcake from Niger. In the words of placards and headlines all over the world, "Bush Lied."

Did he? The answer is not as straightforward as partisans on both sides might think. Investigations by the British Parliament and the U.S. Senate have established that British intelligence did believe that Saddam was trying to buy yellowcake. They showed that the evidence for the British intelligence officers' belief at the time was not completely unreasonable but that it was far short of conclusive. And they revealed that the American intelligence experts had doubts that the report was true. Given these facts, how are we to determine whether Bush lied? It isn't a question of whether he was unwise in putting credence in British intelligence, or of whether he made a calculated risk based on uncertain information. It's a question of whether he was dishonest in how he conveyed this part of his rationale for the invasion to the world. And this question hinges on the semantics of one of those sixteen words, the verb *learn*.[4]

Learn is what linguists call a factive verb; it entails that the belief attributed to the subject is true. In that way it is like the verb *know* and unlike the verb *think*. Say I have a friend Mitch who mistakenly believes that Thomas Dewey defeated Harry Truman in the 1948 presidential election. I could truthfully say *Mitch thinks that Dewey defeated Truman,* but I couldn't say *Mitch knows that Dewey defeated Truman,* because Dewey did not, in fact, defeat Truman. Mitch may think he did, but you and I know he didn't. For the same reason I couldn't honestly say that Mitch has *admitted, discovered, observed, remembered, showed,* or, crucially, *learned* that Dewey defeated Truman. There is, to be sure, a different sense of *learn,* roughly "be taught that," which is not factive; I can say *When I was in graduate school, we learned that there were four kinds of taste buds,* though I now know, thanks to a recent discovery, that there are five. But the usual sense, especially in the perfect tense with *have,* is factive; it means "acquire true information."

People, then, are "realists" in the philosophers' sense. They are tacitly committed, in their everyday use of language, to certain propositions' *being* true or false, independent of whether the person being discussed *believes* them to be true or false. Factive verbs entail something a speaker assumes to be indisputably true, not just something in which he or she has high confidence: it is not a contradiction to say *I'm very, very confident that Oswald shot Kennedy, but I don't* know *that he did*. For this reason factive verbs have a whiff of paradox about them. No one can be certain of the truth, and most of us know we can never be certain, yet we honestly use factive verbs like *know* and *learn* and *remember* all the time. We must have an intuition of a degree of certitude that is so high, and so warranted by standards we share with our audience, that we can vouch for the certainty of a particular belief, while realizing that in general (though presumably not this time) we can be mistaken in what we say. Mark Twain exploited the semantics of factive verbs when he wrote, "The trouble with the world is not that people know too little, but that they know so many things that aren't so."[5] (He also allegedly wrote, "When I was younger, I could remember anything, whether it had happened or not; but my faculties are decaying now, and soon . . . I will remember [only] the things that never happened.")

So did Bush lie? A strong case could be made that he did. When Bush said that the British government had "learned" that Saddam had sought uranium, he was committing himself to the proposition that the uranium seeking *actually* took place, not that the British government *believed* it did. If he had reason to doubt it at the time—and the American intelligence community had made its skepticism known to his administration—the sixteen words did contain a known untruth. Defense Secretary Donald Rumsfeld, speaking in Bush's defense, said that the statement was "technically accurate," and National Security Advisor Condoleezza Rice added that "the British have said that." But note the switch of verbs: Bush didn't state that the British had *said* that Saddam sought yellowcake, which would be true regardless of what Saddam did; he stated that they had *learned* it, which could be true only if Saddam had in fact gone shopping. The logic of factivity, then, is what Bush's critics implicitly appeal to when they accused him of lying.

Lying is an impeachable offense for a president, especially when it comes to the *casus belli* of a terrible war. Could semantics really be that consequential in political history? Is it plausible that the fate of an American president could ever hinge on fine points of a verb? We shall return to that question in chapter 4, where we will see that it depends upon what the meaning of the word *is* is.

Words are tied to reality when their meanings depend, as factive verbs do, on a speaker's commitments about the truth. But there is a way in which words are tied to reality even more directly. They are not just about facts about the world stored in a person's head but are woven into the causal fabric of the world itself.

Certainly a word meaning depends on *something* inside the head. The other day I came across the word *sidereal* and had to ask a literate companion what it meant. Now I can understand and use it when the companion is not around (it means "pertaining to the stars," as in *a sidereal day,* the time it takes for the Earth to make a complete rotation relative to a star). Something in my brain must have changed at the moment I learned the word, and someday cognitive neuroscientists might be able to tell us what that change is. Of course most of the time we don't learn a word by looking it up or asking someone to define it but by hearing it in context. But however a word is learned, it must leave some trace in the brain. The meaning of a word, then, seems to consist of information stored in the heads of the people who know the word: the elementary concepts that define it and, for a concrete word, an image of what it refers to.

But as we will see in chapter 6, a word must be more than a shared definition and image. The easiest way to discover this is to consider the semantics of names.[6] What is the meaning of a name, such as *William Shakespeare*? If you were to look it up in a dictionary, you might find something like this:

> **Shakespeare, William** (1564–1616), n.: English poet and dramatist considered one of the greatest English writers. His plays, many of which were performed at the Globe Theatre in London, include historical works, such as *Richard II,* comedies, including *Much Ado about Nothing* and *As You Like It,* and tragedies, such as *Hamlet, Othello,* and *King Lear.* He also composed 154 sonnets. [Syn.: Shakespeare, Shakspeare, William Shakspere, the bard]

And the definition would typically be accompanied by the famous engraving of a doe-eyed balding man with a very small mustache and a very big ruff. Presumably that is not too far from your understanding of the name.

But is that what *William Shakespeare* really means? Historians agree that there was a man named William Shakespeare who lived in Stratford-on-Avon and London in the late sixteenth and early seventeenth centuries. But for 250 years there have been doubts as to whether that man composed the plays we attribute to him. This might sound like the theory that the CIA imploded the World Trade Center, but it has been taken seriously by Walt Whitman, Mark Twain, Henry James, and many modern-day scholars, and it rests on a number of damning facts. Shakespeare's plays were not published as serious literature in his lifetime, and authorship in those days was not recorded as carefully as it is today. The man himself was relatively uneducated, never traveled, had illiterate children, was known in his hometown as a businessman, was not eulogized at his death, and left no books or manuscripts in his will. Even the famous portraits were not painted in his lifetime, and we have no reason to believe that they resembled the man himself. Because writing plays was a disreputable occupation in those days, the real author, identified by various theories as Francis Bacon, Edward de Vere, Christopher Marlowe, and even Queen Elizabeth, may have wanted to keep his or her identity a secret.

My point isn't to persuade you that William Shakespeare was not the great English poet and dramatist who wrote *Hamlet, As You Like It,* and 154 sonnets. (Mainstream scholars say he was, and I believe them.) My point is to get you to think about the possibility that he wasn't, and to understand the implications for the idea that the meanings of words are in the head. For the sake of argument, imagine that forensic evidence proved beyond doubt that the Shakespearean oeuvre was written by someone else. Now, if the meaning of *William Shakespeare* were something like the dictionary entry stored in the head, we would have to conclude either that the meaning of the term *William Shakespeare* had changed or that the real author of *Hamlet* should be posthumously christened William Shakespeare, even though no one knew him by that name in his lifetime. (We would also have to give full marks to the hapless student who wrote in an exam, "Shakespeare's plays were written by William Shakespeare or another man of that name.") Actually, it's even worse than that. We would not have been able to ask "Did Shakespeare write *Hamlet*?" in the first place, because he did by definition.

It would be like asking "Is a bachelor unmarried?" or "Who's buried in Grant's Tomb?" or "Who sang 'Hey, Hey, We're the Monkees'?" And the conclusion, "William Shakespeare did not in fact write *Hamlet*," would be self-contradictory.

But these implications are bizarre. In fact we *are* speaking sensibly when we ask whether Shakespeare wrote *Hamlet;* we would not be contradicting ourselves if we were to conclude that he did not; and we would still feel that *William Shakespeare* means what it always meant—some guy who lived in England way back when—while admitting that we were mistaken about the man's accomplishments. Even if *every* biographical fact we knew about Shakespeare were overturned—if it turned out, for example, that he was born in 1565 rather than 1564, or came from Warwick rather than Stratford—we would still have a sense that the name refers to the same guy, the one we've been talking about all along.

So what exactly does *William Shakespeare* mean, if not "great writer, author of *Hamlet*," and so on? A name really *has* no definition in terms of other words, concepts, or pictures. Instead it *points* to an entity in the world, because at some instant in time the entity was dubbed with the name and the name stuck. *William Shakespeare,* then, points to the individual who was christened William by Mr. and Mrs. Shakespeare around the time he was born. The name is connected to that guy, whatever he went on to do, and however much or little we know about him. A name points to a person in the world in the same way that I can point to a rock in front of me right now. The name is meaningful to us because of an unbroken chain of word of mouth (or word of pen) that links the word we now use to the original act of christening. We will see that it's not just names, but words for many kinds of things, that are rigidly yoked to the world by acts of pointing, dubbing, and sticking rather than being stipulated in a definition.

The tethering of words to reality helps allay the worry that language ensnares us in a self-contained web of symbols. In this worry, the meanings of words are ultimately circular, each defined in terms of the others. As one semanticist observed, a typical dictionary plays this game when it tells the user that "*to order* means *to command,* that *to direct* and *instruct* 'are not so strong as *command* or *order,*' that *command* means 'to direct, with the right to be obeyed,' that *direct* means 'to order,' that *instruct* means 'to give orders'; or that *to request* means 'to demand politely,' *to demand* [means] 'to claim as if by right,' *to claim* [means] 'to ask for or demand,' *to ask* [means] 'to make a request,' and so on."[7] This cat's cradle is dreaded by those who crave certainty in words, embraced by adherents of deconstructionism and postmodernism, and exploited by the writer of a dictionary of computer jargon:

> **endless loop,** n. See loop, endless.
> **loop, endless,** n. See endless loop.

The logic of names, and of other words that are connected to events of dubbing, allay these concerns by anchoring the web of meanings to real events and objects in the world.

The connectedness of words to real people and things, and not just to *information* about those people and things, has a practical application that is very much in the news. The fastest-growing crime in the beginning of this century is identity theft. An identity thief uses information connected with your name, such as your social security number or the number and password of your credit card or bank account, to commit fraud or steal your assets. Victims of identity theft may lose out on jobs, loans, and college admissions, can be turned away at airport security checkpoints, and can even get arrested for a crime committed by the thief. They can spend many years and much money reclaiming their identity.

Put yourself in the shoes of someone who has lost his wallet, or inadvertently divulged information on his computer, and now has a doppelgänger using his name (say, Murray Klepfish) to borrow money or make purchases. Now you have to convince a bureaucrat that *you*, not the impostor, are the real Murray Klepfish. How would you do it? As with *William Shakespeare*, it comes down to what the words *Murray Klepfish* mean. You could say, " 'Murray Klepfish' means an owner of a chain of discount tire stores who was born in Brooklyn, lives in Piscataway, has a checking account at Acme Bank, is married with two sons, and spends his summers on the Jersey Shore." But they would reply, "As far as we are concerned, 'Murray Klepfish' means a personal trainer who was born in Delray Beach, gets his mail at a post office box in Albuquerque, charged a recent divorce to a storefront in Reno, and spends his summers on Maui. We do agree with you about the bank account, which, by the way, is severely overdrawn."

So how would you prove that you are the real referent of the name *Murray Klepfish*? You could provide any information you wanted—social

security number, license number, mother's maiden name—and the imper-
sonator can either duplicate it (if he stole that, too) or contest it (if he augmented the stolen identity with his own particulars, including a photograph).
As with picking out the real Shakespeare after his familiar biographical particulars had been cast into doubt, ultimately you would have to point to a
causal chain that links your name as it is used today to the moment your
parents hailed your arrival. *Your* credit card was obtained from a bank account, which was obtained with a driver's license, which was obtained with
a birth certificate, which was vouched for by a hospital official, who was
in touch with your parents around the time of your birth and heard from
their lips that you are the Klepfish they were naming Murray. In the case of
your impostor, the chain of testimony peters out in the recent past, well
short of the moment of dubbing. The measures designed to foil identity
theft depend upon the logic of names and the connection of words to reality:
they are ways to identify unbroken chains of person-to-person transmission through time, anchored to a specific event of dubbing in the past.

WORDS AND COMMUNITY

Naming a child is the only opportunity that most people get to anoint an entity in the world with a word of their choosing. Apart from creative artists
like Frank Zappa, who named his children *Moon Unit* and *Dweezil,* traditionally most people select a prefabricated forename like *John* or *Mary* rather
than a sound they concoct from scratch. In theory a forename is an arbitrary
label with no inherent meaning, and people interpret it as simply pointing to
the individual who was dubbed with it. But in practice names take on a
meaning by association with the generation and class of people who bear
them. Most American readers, knowing nothing else about a man other than
that his name is Murray, would guess that he is over sixty, middle-class, and
probably Jewish. (When a drunken Mel Gibson let loose with an anti-
Semitic tirade in 2006, the editor Leon Wieseltier commented, "Mad Max is
making Max mad, and Murray, and Irving, and Mort, and Marty, and
Abe.")[8] That is because of another curiosity of names we will explore in
chapter 6. Names follow cycles of fashion, like the widths of ties and the
lengths of skirts, so people's first names may give away their generational
cohort. In its heyday in the 1930s, *Murray* had an aura of Anglo-Saxon respectability, together with names like *Irving, Sidney, Maxwell, Sheldon,* and

Herbert. They seemed to stand apart from the Yiddish names of the previous generation, such as *Moishe, Mendel,* and *Ruven,* which made their bearers sound as if they had a foot in the old country. But when the Murrays and Sids and their wives launched the baby boom, they gave their sons blander names like *David, Brian,* and *Michael,* who in their turn begat biblically inspired *Adams, Joshuas,* and *Jacobs.* Many of these Old Testament namesakes are now completing the circle with sons named *Max, Ruben,* and *Saul.*

Names follow trends because people in a community have uncannily similar reactions to the ones in the namepool (as parents often find when they take a child to school and discover that their unique choice of a name was also the unique choice of many of their neighbors). A name's coloring comes in part from the sounds that go into it and in part from a stereotype of the adults who currently bear it. For this reason, the faux-British names of first-generation Americans became victims of their own middle-class respectability a generation later. In a scene from *When Harry Met Sally* set in the 1970s, a pair of baby boomers get into an argument about Sally's sexual experience:

> HARRY: With whom did you have this great sex?
> SALLY: I'm not going to tell you that!
> HARRY: Fine. Don't tell me.
> SALLY: Shel Gordon.
> HARRY: Shel. Sheldon? No, no. You did not have great sex with
>   Sheldon.
> SALLY: I did too.
> HARRY: No, you didn't. A Sheldon can do your income taxes. If
>   you need a root canal, Sheldon's your man. But humpin'
>   and pumpin' is not Sheldon's strong suit. It's the name.
>   "Do it to me, Sheldon." "You're an animal, Sheldon."
>   "Ride me, big Sheldon." It doesn't work.

Though postwar parents probably didn't have great sex in mind, they must have recoiled from the name's nebbishy connotation even then: beginning in the 1940s, *Sheldon,* like *Murray,* sank like a stone and never recovered.[9] The reaction to the name is now so uniform across the English-speaking world that humorists can depend on it. The playwright Marcy Kahan, who recently adapted Nora Ephron's screenplay to the British stage, notes,

"I included the Sheldon joke in the stage play, and all three actors playing Harry got a huge laugh of recognition from it, every night, without fail."[10]

The dynamics of baby naming have become a talking point in newspapers and conversation now that the fashion cycles have accelerated. One of the most popular American names for baby girls in 2006 was unheard of only five years before: *Nevaeh,* or "heaven" spelled backwards. At the other end of the curve, people are seeing their own names, and the names of their friends and relations, becoming stodgy more quickly.[11] I don't think I ever felt so old as when a student told me that *Barbara, Susan, Deborah,* and *Linda,* some of the most popular names for girls of my generation, made her think of middle-aged women.

In naming a baby, parents have free rein. Obviously they are affected by the pool of names in circulation, but once they pick one, the child and the community usually stick with it. But in naming everything else, the community has a say in whether the new name takes. The social nature of words is illustrated in Calvin's presumably ill-fated attempt to pass a physics exam:

Calvin and Hobbes © 1995 Watterson. Dist. by Universal Press Syndicate. Reprinted with permission. All rights reserved.

The way in which we understand "your own words"—as referring only to how you combine them, not to what they are—shows that words are owned by a community rather than an individual. If a word isn't known to everyone around you, you might as well not use it, because no one will know what you're talking about. Nonetheless, every word in a language must have been minted at some point by a single speaker. With some coinages, the rest of the community gradually agrees to use the word to point to the same thing, tipping the first domino in the chain that makes the word available to subsequent generations. But as we shall see, how this tacit agreement is forged across a community is mysterious.

In some cases necessity is the mother of invention. Computer users, for instance, needed a term for bulk e-mail in the 1990s, and *spam* stepped into the breach. But many other breaches stay stubbornly unstepped into. Since the sexual revolution of the 1960s we have needed a term for the members of an unmarried heterosexual couple, and none of the popular suggestions has caught on—*paramour* is too romantic, *roommate* not romantic enough, *partner* too gay, and the suggestions of journalists too facetious (like *POSSLQ*, from the census designation "persons of opposite sex sharing living quarters," and *umfriend,* from "This is my, um, friend"). And speaking of decades, we are more than halfway through the first one of the twenty-first century, and no one yet knows what to call it. *The zeroes? The aughts? The nought-noughts? The naughties?*

Traditional etymology is of limited help in figuring out what ushers a word into existence and whether it will catch on. Etymologists can trace most words back for centuries or more, but the trail goes cold well before they reach the actual moment at which a primordial wordsmith first dubbed a concept with a sound of his or her choosing. With recent coinages, though, we can follow the twisted path to wordhood in real time.

Spam is not, as some people believe, an acronym for Short, Pointless, and Annoying Messages. The word *is* related to the name of the luncheon meat sold by Hormel since 1937, a portmanteau from SPiced hAM. But how did it come to refer to e-mailed invitations to enlarge the male member and share the ill-gotten gains of deposed African despots? Many people assume that the route was metaphor. Like the luncheon meat, the e-mail is cheap, plentiful, and unwanted, and in one variant of this folk etymology, *spamming* is what happens when you dump Spam in a fan. Though these intuitions may have helped make the word contagious, its origin is very different. It was inspired by a sketch from Monty Python's Flying Circus in which a couple enter a café and ask the waitress (a Python in drag) what's available. She answers:

> Well, there's egg and bacon; egg sausage and bacon; egg and spam; egg bacon and spam; egg bacon sausage and spam; spam bacon sausage and spam; spam egg spam spam bacon and spam; spam sausage spam spam bacon spam tomato and spam; spam

spam spam egg and spam; spam spam spam spam spam spam baked beans spam spam spam, or Lobster Thermidor: a Crevette with a mornay sauce served in a Provençale manner with shallots and aubergines garnished with truffle pâté, brandy and with a fried egg on top and spam.

You are probably thinking, "This sketch must be stopped—it's too silly." But it did change the English language. The mindless repetition of the word *spam* inspired late-1980s hackers to use it as a verb for flooding newsgroups with identical messages, and a decade later it spread from their subculture to the populace at large.[12]

Though it may seem incredible that such a whimsical and circuitous coinage would catch on, we shall see that it was not the first time that silliness left its mark on the lexicon. The verb *gerrymander* comes from a nineteenth-century American cartoon showing a political district that had been crafted by a Governor Elbridge Gerry into a tortuous shape resembling a salamander in an effort to concentrate his opponent's voters into a single seat. But most silly coinages go nowhere, such as *bushlips* for "insincere political rhetoric" (after George H. W. Bush's 1988 campaign slogan "Read my lips: No new taxes"), or *teledildonics* for computer-controlled sex toys. Every year the American Dialect Society selects a "word most likely to succeed." But the members of the society are the first to admit that their track record is abysmal. Does anyone remember the *information superhighway*, or the *Infobahn*?[13] And could anyone have predicted that *to blog, to google,* and *to blackberry* would quickly become part of everyone's language?

The dynamics of taking from the wordpool when naming babies and giving back to it when naming concepts are stubbornly chaotic. And as we shall see, this unpredictability holds a lesson for our understanding of culture more generally. Like the words in a language, the practices in a culture—every fashion, every ritual, every common belief—must originate with an innovator, must then appeal to the innovator's acquaintances and then to the acquaintance's acquaintances, and so on, until it becomes endemic to a community. The caprice in the rise and fall of names, which are the most easily tracked bits of culture, suggests we should be skeptical of most explanations for the life cycles of other mores and customs, from why men stopped wearing hats to why neighborhoods become segregated. But it also points to the patterns of individual choice and social contagion that might someday make sense of them.

WORDS AND EMOTIONS

The shifting associations to the name for a person are an example of the power of a word to soak up emotional coloring—to have a *connotation* as well as a *denotation*. The concept of a connotation is often explained by the conjugational formula devised by Bertrand Russell in a 1950s radio interview: I am firm; you are obstinate; he is pigheaded. The formula was turned into a word game in a radio show and newspaper feature and elicited hundreds of triplets. I am slim; you are thin; he is scrawny. I am a perfectionist; you are anal; he is a control freak. I am exploring my sexuality; you are promiscuous; she is a slut. In each triplet the literal meaning of the words is held constant, but the emotional meaning ranges from attractive to neutral to offensive.

The affective saturation of words is especially apparent in the strange phenomena surrounding profanity, the topic of chapter 7. It is a real puzzle for the science of mind why, when an unpleasant event befalls us—we slice our thumb along with the bagel, or knock a glass of beer into our lap—the topic of our conversation turns abruptly to sexuality, excretion, or religion. It is also a strange feature of our makeup that when an adversary infringes on our rights—say, by slipping into parking space we have been waiting for, or firing up a leaf blower at seven o'clock on a Sunday morning—we are apt to extend him advice in the manner of Woody Allen, who recounted, "I told him to be fruitful and multiply, but not in those words."

These outbursts seem to emerge from a deep and ancient part of the brain, like the yelp of a dog when someone steps on its tail, or its snarl when it is trying to intimidate an adversary. They can surface in the involuntary tics of a Tourette's patient, or in the surviving utterances of a neurological patient who is otherwise bereft of language. But despite the seemingly atavistic roots of cursing, the sounds themselves are composed of English words and are pronounced in full conformity with the sound pattern of the language. It is as though the human brain were wired in the course of human evolution so that the output of an old system for calls and cries were patched into the input of the new system for articulate speech.

Not only do we turn to certain words for sexuality, excretion, and religion when we are in an excitable state, but we are wary of such words when we are in any other state. Many epithets and imprecations are not just unpleasant but taboo: the very act of uttering them is an affront to listeners, even when the concepts have synonyms whose use is unexceptionable. The tendency of words to take on awesome powers may be found in the taboos and word magic in cultures all over the world. In Orthodox Judaism, the name of God, transcribed as *YHVH* and conventionally pronounced *Yahweh,* may never be spoken, except by high priests in the ancient temple on Yom Kippur in the "holy of holies," the chamber housing the ark of the covenant. In everyday conversation observant Jews use a word to refer to the word, referring to God as *hashem,* "the name."

While taboo language is an affront to common sensibilities, the *phenomenon* of taboo language is an affront to common sense. Excretion is an activity that every incarnate being must engage in daily, yet all the English words for it are indecent, juvenile, or clinical. The elegant lexicon of Anglo-Saxon monosyllables that give the English language its rhythmic vigor turns up empty-handed just when it comes to an activity that no one can avoid. Also conspicuous by its absence is a polite transitive verb for sex—a word that would fit into the frame *Adam verbed Eve* or *Eve verbed Adam.* The simple transitive verbs for sexual relations are either obscene or disrespectful, and the most common ones are among the seven words you can't say on television.

Or at least, the words you couldn't say in 1973, when the comedian George Carlin delivered his historic monologue arguing against the ban of those words in broadcast media. In a conundrum that reminds us of the rationale for unfettered free speech, a radio network that had broadcasted the monologue was punished by the Federal Communications Commission (in a case that ultimately reached the Supreme Court) for allowing Carlin to mention on the radio exactly those words that he was arguing ought to be allowed to be mentioned on the radio. We have a law that in effect forbids criticism of itself, a paradox worthy of Russell and other connoisseurs of self-referential statements. The paradox of identifying taboo words without using them has always infected attempts to regulate speech about sexuality. In several states, the drafters of the statute against bestiality could not bring themselves to name it and therefore outlawed "the abominable and detestable crime against nature," until the statutes were challenged for being void for vagueness. To avoid this trap, a New Jersey obscenity statute stipulated exactly which kinds of words and images would be deemed obscene. But the wording of the statute was so pornographic that some law libraries tore the page out of every copy of the statute books.[14]

Taboos on language are still very much in the news. While sexual and scatological language is more available than ever on cable, satellite, and the Internet, the American government, prodded by cultural conservatives, is trying to crack down on it, especially within the dwindling bailiwick of broadcast media. Legislation such as the "Clean Airwaves Act" and the "Broadcast Decency Enforcement Act" imposes draconian fines on broadcast stations that fail to censor their guests when they use the words on Carlin's list. And in an unscripted event that shows the unavoidable hypocrisy of linguistic taboos, the Broadcast Decency Enforcement Act was passed on the day in 2004 that Vice President Dick Cheney got into an argument with Senator Patrick Leahy on the Senate floor and Cheney told the senator to be fruitful and multiply, but not in those words.

No curious person can fail to be puzzled by the illogic and hypocrisy of linguistic taboos. Why should certain words, but not their homonyms or synonyms, be credited with a dreadful moral power? At the same time, no matter how illogical it may seem, everyone respects taboos on at least some words. Everyone? Yes, everyone. Suppose I told you there was an obscenity so shocking that decent people dare not mention it even in casual conversation. Like observant Jews referring to God, they must speak of it at one degree of separation by using a word that refers to the word. An elect circle of people are granted a special dispensation to use it, but everyone else risks grave consequences, including legally justifiable violence.[15] What is this obscenity? It is the word *nigger*—or, as it is referred to in respectable forums, the *n-word*—which may be uttered only by African Americans to express camaraderie and solidarity in settings of their choosing. The shocked reaction that other uses evoke, even among people who support free speech and wonder why there is such a fuss about words for sex, suggests that the psychology of word magic is not just a pathology of censorious bluenoses but a constituent of our emotional and linguistic makeup.

WORDS AND SOCIAL RELATIONS

In recent years the Internet has become a laboratory for the study of language. It not only provides a gigantic corpus of real language used by real people, but also acts as a superefficient vector for the transmission of infectious ideas, and can thereby highlight examples of language that people find intriguing enough to pass along to others. Let me introduce the last major topic of this book with a story that circulated widely by e-mail in 1998:

> During the final days at Denver's Stapleton airport, a crowded United flight was canceled. A single agent was rebooking a long line of inconvenienced travelers. Suddenly an angry passenger pushed his way to the desk and slapped his ticket down on the counter, saying, "I HAVE to be on this flight, and it HAS to be first class." The agent replied, "I'm sorry, sir. I'll be happy to try to help you, but I've got to help these folks first, and I'm sure we'll be able to work something out." The passenger was unimpressed. He asked loudly, so that the passengers behind him could hear, "Do you have any idea who I am?" Without hesitating, the gate agent smiled and grabbed her public address microphone. "May I have your attention, please?" she began, her voice bellowing through the terminal. "We have a passenger here at the gate WHO DOES NOT KNOW WHO HE IS. If anyone can help him find his identity, please come to the gate." With the folks behind him in line laughing hysterically, the man glared at the agent, gritted his teeth, and swore, "[Expletive] you!" Without flinching, she smiled and said, "I'm sorry, sir, but you'll have to stand in line for that too."

The story seems too good to be true, and is probably an urban legend.[16] But its two punch lines make a nice teaser for oddities of language that we will explore in later chapters. I have already touched on the puzzle behind the second punch line, namely that certain words for sex are also used in aggressive imprecations (chapter 7). But the first punch line introduces the final world that I wish to connect to words, the world of social relations (chapter 8).

The agent's comeback to "Do you have any idea who I am?" springs from a mismatch between the sense in which the passenger intended his rhetorical question—a demand for recognition of his higher status—and the sense in which she pretended to take it—a literal request for information. And the payoff to the onlookers (and the e-mail audience) comes from understanding the exchange at a third level—that the agent's feigned misunderstanding was a tactic to reverse the dominance relation and demote the arrogant passenger to well-deserved ignominy.

Language is understood at multiple levels, rather than as a direct parse of the content of the sentence.[17] In everyday life we anticipate our interlocutor's ability to listen between the lines and slip in requests and offers that we feel we can't blurt out directly. In the film *Fargo,* two kidnappers with a hostage hidden in the back seat are pulled over by a policeman because their car is missing its plates. The kidnapper at the wheel is asked to produce his driver's license, and he extends his wallet with a fifty-dollar bill protruding from it, saying, "So maybe the best thing would be to take care of that here in Brainerd." The statement, of course, is intended as a bribe, not as a comment on the relative convenience of different venues for paying the fine. Many other kinds of speech are interpreted in ways that differ from their literal meaning:

> If you could pass the guacamole, that would be awesome.
> We're counting on you to show leadership in our Campaign
>   for the Future.
> Would you like to come up and see my etchings?
> Nice store you got there. Would be a real shame if something
>   happened to it.

These are clearly intended as a request, a solicitation for money, a sexual come-on, and a threat. But why don't people just say what they mean—"If you let me drive off without further ado, I'll give you fifty bucks," "Gimme the guacamole," and so on?

With the veiled bribe and the veiled threat, one might guess that the technicalities of plausible deniability are applicable: bribery and extortion are crimes, and by avoiding an explicit proposition, the speaker could make a charge harder to prove in court. But the veil is so transparent that it is hard to believe it could foil a prosecutor or fool a jury—as the lawyers say, it wouldn't pass the giggle test. Yet we all take part in these charades, while knowing that no one is fooled. (Well, almost no one. In an episode of *Seinfeld,* George is asked by his date if he would like to come up for coffee. He declines, explaining that caffeine keeps him up at night. Later he slaps his forehead and realizes, " 'Coffee' doesn't mean coffee! 'Coffee' means sex!" And of course this can go too far in the other direction. In a joke recounted by Freud in *Jokes and Their Relation to the Unconscious,* a businessman meets a rival at a train station and asks him where he's going. The second businessman says he's going to Minsk. The first one replies, "You're telling me you're going to Minsk because you want me to think you're going to Pinsk. But I happen to know that you *are* going to Minsk. So why are you lying to me?")

If a speaker and a listener were ever to work through the tacit propositions that underlie their conversation, the depth of the recursively embedded mental states would be dizzying. The driver offers a bribe; the officer knows that the driver is offering him a bribe; the driver knows that the officer knows; the officer knows that the driver knows that the officer knows; and so on. So why don't they just blurt it out? Why do a speaker and a hearer willingly take on parts in a dainty comedy of manners?

The polite dinnertime request—what linguists call a whimperative— offers a clue. When you issue a request, you are presupposing that the hearer will comply. But apart from employees or intimates, you can't just boss people around like that. Still, you do want the damn guacamole. The way out of this dilemma is to couch your request as a stupid question ("Can you . . . ?"), a pointless rumination ("I was wondering if . . ."), a gross overstatement ("It would be great if you could . . ."), or some other blather that is so incongruous the hearer can't take it at face value. She does some quick intuitive psychology to infer your real intent, and at the same time she senses that you have made an effort not to treat her as a factotum. A stealth imperative allows you to do two things at once—communicate your request, and signal your understanding of the relationship.

As we shall see in chapter 8, ordinary conversation is like a session of tête-à-tête diplomacy, in which the parties explore ways of saving face, offering an "out," and maintaining plausible deniability as they negotiate the mix of power, sex, intimacy, and fairness that makes up their relationship. As with real diplomacy, communiqués that are too subtle, or not subtle enough, can ignite a firestorm. In 1991, the nomination of Clarence Thomas to the U.S. Supreme Court was nearly derailed by accusations that he had made sexual overtures to a subordinate, the lawyer Anita Hill. In one of the stranger episodes in the history of the Senate's exercise of its power of advice and consent, senators had to decide what Thomas meant when he spoke to Hill about a porn star named Long Dong Silver and when he asked the rhetorical question "Who has put pubic hair on my Coke?" It's presumably not what the Framers had in mind when they formulated the doctrine of the separation of powers, but this kind of question has become a part of our national discourse. Ever since the Thomas-Hill case put sexual harassment on the national stage, the adjudication of claims of harassment has been a major headache for universities, businesses, and government agencies, particularly when a putative come-on is conveyed by innuendo rather than a bald proposition.

These tidbits from the news and from the net show some of the ways in which our words connect to our thoughts, our communities, our emotions, our relationships, and to reality itself. It isn't surprising that language supplies so many of the hot potatoes of our public and private life. We are verbivores, a species that lives on words, and the meaning and use of language are bound to be among the major things we ponder, share, and dispute.

At the same time, it would be a mistake to think that these deliberations are really about language itself. As I will show in chapter 3, language is above all a medium in which we express our thoughts and feelings, and it mustn't be confused with the thoughts and feelings themselves. Yet another phenomenon of language, the symbolism in sound (chapter 6), offers a hint at this conclusion. Without a substrate of thoughts to underlie our words, we do not truly speak but only babble, blabber, blather, chatter, gibber, jabber, natter, patter, prattle, rattle, yammer, or yadda, yadda—an onomatopoeic lexicon for empty speech that makes plain the expectation that the sounds coming out of our mouths are ordinarily *about* something.

The rest of this book is about that something: the ideas, feelings, and attachments that are visible through our language and that make up our nature. Our words and constructions disclose conceptions of physical reality and human social life that are similar in all cultures but different from the products of our science and scholarship. They are rooted in our development as individuals, but also in the history of our language community, and in the evolution of our species. Our ability to combine them into bigger assemblies and to extend them to new domains by metaphorical leaps goes a long way toward explaining what makes us smart. But they can also clash with the nature of things, and when they do, the result can be paradox, folly, and even tragedy. For these reasons I hope to convince you that the three and a half billion dollars at stake in the interpretation of an "event" is just part of the value of understanding the worlds of words.

2

DOWN THE RABBIT HOLE

 The discovery of a world hidden in a nook or cranny of everyday life is an enduring device in children's fiction. The best-known example is Alice stumbling down a rabbit hole to find a surreal underworld, and the formula continues to enchant in endless variations: the wardrobe passageway to Narnia, the wrinkle in time, the subtle knife, Whoville in a speck of dust.[1]

In nonfiction as well, the revelation of a microcosm is a recurring source of fascination. In 1968, the designers Charles and Ray Eames made a film called *Powers of Ten,* which began with a view of galaxy clusters a billion light-years across, and zoomed by tenfold leaps to reveal our galaxy, solar system, planet, and so on, down to a picnicker asleep in a park, to his hand, his cells, his DNA, a carbon atom, and finally the atomic nucleus and its particles sixteen orders of magnitude smaller. This magnificent unfurling of physical reality can be seen in a companion book by the film's scientific consultants, Philip and Phyllis Morrison, and the idea has recently been adapted to one of the most enjoyable ways to waste time on the Web: zooming smoothly from a photograph of the Earth taken from space through seven orders of magnitude of satellite photographs down to a pigeon's-eye view of your street and house.

This chapter is about my own stumbling upon a microcosm—the world of basic human ideas and their connections—in the course of trying to solve what I thought was a mundane problem in psycholinguistics. It is a hidden world that I had glimpsed not by training a telescope on its whereabouts from the start but because it kept peeking out from under the phenomena I thought I was studying. By taking you through the layers of mental organization that must be exploded to make sense of the problem, I hope to offer you a view of this inner world.[2]

The rabbit hole that leads to this microcosm is the verb system of English—what verbs mean, how they are used in sentences, and how children figure it all out. This chapter will try to show you how cracking these problems led to epiphanies about the contents of cognition that serve as leitmotifs of this book. Why leap into the world of the mind through this particular opening? One reason, I confess, is personal: I simply find verbs fascinating. (A colleague once remarked, "They really are your little friends, aren't they?") But as every enthusiast knows, other people can't be counted on to share one's passion, and I like to think I have a better reason to introduce you to my little friends.

Science proceeds by studying particulars. No one has ever gotten a grant to study "the human mind." One has to study something more tractable, and when fortune smiles, a general law may reveal itself in the process. In the first chapter I introduced four ideas:

- The human mind can construe a particular scenario in multiple ways.
- Each construal is built around a few basic ideas, like "event," "cause," "change," and "intend."
- These ideas can be extended metaphorically to other domains, as when we count events as if they were objects or when we use space as a metaphor for time.
- Each idea has distinctively human quirks that make it useful for reasoning about certain things but that can lead to fallacies and confusions when we try to apply it more broadly.

These claims may strike you as reasonable enough, but not particularly meaty—just four out of hundreds of platitudes that could be listed as true of our thought processes. In this chapter I hope to show that they are more than that. In solving the problem of how children learn verbs, each of these hypotheses served as a puzzle piece that took a long time to find but then fit perfectly into its slot, together completing an attractive picture of the whole.

This offers some confidence that the themes of this book are real discoveries about the mind, not just innocuous comments about it.

My plan is as follows. First I will take you on a plunge from the intergalactic perspective to the quark's-eye view, showing how a general curiosity about how the mind works can lead to an interest in verbs and how children learn them.

Then we will bump up against a paradox—a case in which children seem to learn the unlearnable. Isaac Asimov once wrote, "The most exciting phrase to hear in science, the one that heralds new discoveries, is not 'Eureka!' (I found it!) but 'That's funny. . . .'" The following section presents a discovery—the mind's ability to flip between frames—that was the crucial opening to solve the paradox.

The remaining parts of the solution bring us face to face with two of the basic concepts in our mental inventory, moving and changing. The same line of reasoning, applied to other verbs, illuminates the other major elements our thoughts are built from: the concepts of having, knowing, and helping, and the concepts of acting, intending, and causing.

From there we step back up to reflect on what it all means. We will consider whether the signs of intelligent design in the English language imply a corresponding intelligence in every English speaker—a question that will recur throughout the book as we try to use language as a window into human nature. I will then suggest an inventory of basic human thoughts, ones that will be unpacked in later chapters. Finally, I will show how design quirks in these basic thoughts give rise to fallacies, follies, and foibles in the way that people reason about the conundrums of modern life.

POWERS OF TEN

Let me now take you, in a few turns of the zoom lens, from a wide concern with human nature to a close-up look at how children learn verbs.

The first, galaxy-wide view is of the human mind and its remarkable powers. It's easy for us humans, safe inside our well-functioning minds, to be jaded about the mundane activities of cognition and to attend instead to the extraordinary and the lurid. But the science of mind begins with a recognition that ordinary mental activities—seeing, hearing, remembering, moving, planning, reasoning, speaking—require our brains to solve fractious engineering problems.³ Despite the immense hazard and cost of manned space flight, most plans for planetary exploration still envision blasting people into the solar system. Partly it's because of the drama of following an intrepid astronaut in exploring strange new worlds rather than a silicon chip, but mainly it's because no foreseeable robot can match an ordinary person's ability to recognize unexpected objects and situations, decide what to do about them, and manipulate things in unanticipated ways, all while exchanging information with humans back home. Understanding how these faculties of mind work is a frontier of modern science.

Among these magnificent faculties, pride of place must go to language—ubiquitous across the species, unique in the animal kingdom, inextricable from social life and from the mastery of civilization and technology, devastating when lost or impaired.⁴

Language figures in human life in many ways. We inform, we request, we persuade, we interrogate, we orate, and sometimes we just schmooze. But the most remarkable thing we do with language is learn it in the first place.⁵ Babies are born into the world not knowing a word of the language being spoken around them. Yet in just three years, without the benefit of lessons, most of them will be talking a blue streak, with a vocabulary of thousands of words, a command of the grammar of the spoken vernacular, and a proficiency with the sound pattern (what tourist isn't momentarily amazed at how well the little children in France speak French!). Children deploy the code of syntax unswervingly enough to understand improbable events like a cow jumping over the moon and a dish running away with the spoon, or to share their childlike aperçus like "I think the wind wants to get in out of the rain" or "I often wonder when people pass me by do they wonder about me."⁶

To become so fluent in a language, children must have *analyzed* the speech around them, not just memorized it. We see this clearly when children say things that sound wrong to adult ears but that reveal acute hypotheses about how the ingredients of language may combine. When children make errors like "All the animals are wake-upped," "Don't tickle me; I'm laughable," or "Mommy, why did he dis it appear?" they could not have been imitating their parents. They must have extracted the mental equivalent of grammatical rules that add suffixes to words and arrange verbs and particles in phrases.

The triumph of language acquisition is even more impressive when we consider that a talking child has solved a knotty instance of the problem of induction: observing a finite sample of events and framing a generalization that embraces the infinite set from which the events are drawn.[7] Scientists engage in induction when they go beyond their data and put forward laws that make predictions about cases they haven't observed, such as that gas under pressure will be absorbed by a liquid, or that warm-blooded animals have larger body sizes at higher latitudes. Philosophers of science call induction a "scandal" because there are an infinite number of generalizations that are consistent with any set of observations, and no strictly logical basis for choosing among them.[8] There is no guarantee that a law discovered this year will continue to hold next year, no limit to the number of smooth curves that can connect a set of points on a graph, and, upon glimpsing a black sheep in Scotland, no strictly logical reason for choosing among the conclusions that all sheep in Scotland are black, that at least one sheep in Scotland is black, and that at least one sheep in Scotland is black on at least one side. As Mark Twain wrote, science is fascinating because "one gets such wholesale returns on conjecture out of such a trifling investment in fact." Yet the returns keep coming. Philosophers of science argue that theories are not just peeled off the data but constrained beforehand by reasonable assumptions about the way the universe works, such as that nature is lawful and that simpler theories that fit the data are more likely to be true than complex ones.

As children learn their mother tongue, they, too, are solving an induction problem. When listening to their parents and siblings, they can't just file away every sentence and draw on that list in the future, or they would be as mindless as parrots. Nor can they throw together all the words they have found in any order they please. They have to extract a set of rules that will allow them to understand and express new thoughts, and do it in a way that is consistent with the speech patterns used by those around them. The induction problem arises because ambient speech offers countless opportunities for the child to leap on seductive yet false generalizations. For instance, as children learn how to ask questions, they should be able to go from *He ate the green eggs with ham* to *What did he eat?* and *What did he eat the green eggs with?* But from *He ate the green eggs and ham* they should *not* be able to ask *What did he eat the green eggs and?* To take another example:

the sentences *Harriet appeared to Sam to be strong* and *Harriet appealed to Sam to be strong* differ by only the curl of the tongue in a single consonant. Yet their meanings (in particular, who is supposed to be the strong one) are completely different. A child hearing one sentence should not generalize its interpretation to the other just because they sound so similar.

In cracking the code of language, then, children's minds must be constrained to pick out just the right kinds of generalizations from the speech around them. They can't get sidetracked by how sentences sound but must dig into the grammatical structure hidden in the words and their arrangement. It is this line of reasoning that led the linguist Noam Chomsky to propose that language acquisition in children is the key to understanding the nature of language, and that children must be equipped with an innate Universal Grammar: a set of plans for the grammatical machinery that powers all human languages.[9] This idea sounds more controversial than it is (or at least more controversial than it should be) because the logic of induction mandates that children make *some* assumptions about how language works in order for them to succeed at learning a language at all.[10] The only real controversy is what these assumptions consist of: a blueprint for a specific kind of rule system, a set of abstract principles, or a mechanism for finding simple patterns (which might also be used in learning things other than language).[11] The scientific study of language acquisition aims to characterize the child's built-in analyzers for language, whatever they turn out to be.

Language itself is not a single system but a contraption with many components. To understand how children learn a language, it's helpful to focus on one of these components rather than try to explain everything at once. There are components that assemble sounds into words, and words into phrases and sentences. And each of these components must interface with brain systems driving the mouth, the ear, one's memory for words and concepts, one's plans for what to say, and the mental resources for updating one's knowledge as speech comes in.

The component that organizes words into sentences and determines what they mean is called syntax. Syntax itself encompasses several mechanisms, which are tapped to different extents by different languages. They include putting words in the right order, enforcing agreement between elements like the subject and the verb, and keeping track of special words that have their fingers in two places in the sentence at once (such as the *what* in

What do you want?—it serves as both the element being questioned and the thing that is wanted).

One of the key phenomena of syntax is the way that sentences are built around their verbs. The phenomenon goes by many technical names (including subcategorization, diathesis, predicate-argument structure, valence, adicity, arity, case structure, and theta-role assignment), but I'll refer to it using the traditional term *verb constructions*.[12]

Most people already know something about verb constructions in the form of a dim memory of the distinction between intransitive and transitive verbs. Intransitive verbs like *snore* appear without a direct object, as in *Max snored;* it sounds odd to say *Max snored a racket.* Transitive verbs like *sprain* require a direct object, as in *Shirley sprained her ankle;* it sounds odd to say *Shirley sprained.* The transitive and intransitive constructions are the tip of an iceberg. English also has verbs that require an oblique object (an object introduced by a preposition), as in *The swallow darted into a cave,* verbs that require an object and an oblique object, as in *They funneled rum into the jugs,* and verbs that require a sentence complement, as in *She realized that she would have to get rid of her wolverines.* A book by the linguist Beth Levin classifies three thousand English verbs into about eighty-five classes based on the constructions they appear in; its subtitle is *A Preliminary Investigation.*

A verb, then, is not just a word that refers to an action or state but the chassis of the sentence. It is a framework with receptacles for the other parts—the subject, the object, and various oblique objects and subordinate clauses—to be bolted onto. Then a simple sentence held together by a verb can be inserted into a more inclusive sentence, which can be inserted into a still more inclusive sentence, and so on without limit (as in the old sign "I know that you believe you understand what you think I said, but I am not sure you realize that what you heard is not what I meant").

The information packed into a verb not only organizes the nucleus of the sentence but goes a long way toward determining its meaning. We see this most clearly in sentences that differ only in their choice of verb, like *Barbara caused an injury* and *Barbara sustained an injury,* where Barbara is involved in the event in completely different ways. The same is true for Norm in *Norm gave a pashmina* and *Norm received a pashmina.* You can't figure out what a sentence means by guessing that the subject is the doer and the object is the done-to; you also have to check with the verb. The entry for the verb *give* in the mental dictionary indicates in some way that its subject is the giver and its object the gift. The entry for *receive* says in some way that its subject is the *recipient* and its object the gift. The difference between Harriet *appearing to Sam to be brave* and *appealing to Sam to be brave* shows that the different schemes for casting actors into roles can be quite intricate.

A good way to appreciate the role of verb constructions in language is to ponder jokes that hinge on an ambiguity between them: same words, different constructions. An old example is this exchange: "Call me a taxi." "OK, you're a taxi."[13] According to a frequently e-mailed list of badly translated hotel signs, a Norwegian cocktail lounge sported the notice "Ladies are requested not to have children in the bar." In *The Silence of the Lambs,* Hannibal Lecter (a.k.a. Hannibal the Cannibal) taunts his pursuer by saying, "I do wish we could chat longer, but I'm having an old friend for dinner." And in his autobiography the comedian Dick Gregory recounts an episode from the 1960s: "Last time I was down South I walked into this restaurant and this white waitress came up to me and said, We don't serve colored people here. I said, That's all right. I don't eat colored people. Bring me a whole fried chicken."[14]

The constructions that a verb may appear in depend in part on its meaning. It's no coincidence that *snore* is intransitive, snoring being an activity that one accomplishes without anyone's help, and that *kiss* is transitive, since a kiss ordinarily requires both a kisser and a kissee. According to a long-standing assumption in linguistics (accepted both in Chomsky's theory and in some of its rivals, like Charles Fillmore's Case Grammar), the way that the meaning of a verb affects the constructions it appears in is by specifying a small number of roles that the nouns can play.[15] (These roles go by many names, including semantic roles, case roles, semantic relations, thematic relations, and theta roles.) A verb with just an actor (like the snorer in *snore*) likes to be intransitive, naturally enough, with the actor as the subject. A verb with an agent and an acted-upon entity (like a kisser and a kissee) likes to be transitive, with the agent as the subject and the acted-upon as the object. And verbs that talk about things moving from place to place (like the verb *move* itself) also take one or more oblique objects, like a *from*-phrase for the source of the movement and a *to*-phrase for its goal.

Nonetheless, it has long been known that the fit between the scenario behind a verb and the constructions it may appear in is highly inexact.

Ultimately it's the verb itself, not the underlying concept, that has the final say. For instance, a given concept like "eating" can underlie both a transitive verb, as in *devour the pâté* (you can't say *Olga devoured*), and an intransitive one, as in *dine* (you can't say *Olga dined the pâté*). And in thousands of cases a verb refuses to appear in constructions that would seem to make perfect sense, given the verb's meaning. Based on meaning alone, one would expect that it would be natural to say *Sal rumored that Flo would quit,* or *The city destroyed,* or *Boris arranged Maria to come.* But while these sentences are perfectly understandable, they sound odd to an English speaker's ears.

In order for children to *acquire* an English speaker's ears, they must somehow learn this whole system: what each verb means, which constructions it naturally appears in, and which roles are played by the various nouns that accompany it in a sentence. This is the rabbit hole that I invite you to explore—one that leads to the world of human ideas and the dramas they engage in.

Before we descend into this world, I owe you an explanation of what it means to claim that "you can't say this" or "such-and-such is ungrammatical." These judgments are the most commonly used empirical data in linguistics: a sentence under a certain interpretation and in a certain context is classified as grammatical, ungrammatical, or having various degrees of iffiness.[16] These judgments aren't meant to accredit a sentence as being correct or incorrect in some objective sense (whatever that would mean), nor are they legislated by some council of immortals like the Académie Française. Designating a sentence as "ungrammatical" simply means that native speakers tend to avoid the sentence, cringe when they hear it, and judge it as sounding odd.

Note too that when a sentence is deemed ungrammatical, it might still be used in certain circumstances. There are special constructions, for example, in which English speakers use transitive verbs intransitively, as when a parent says to a child *Justin bites—I don't want you to bite.* There are also circumstances in which we can use intransitive verbs transitively, as when we say *Jesus died a long, painful death.* And we all stretch the language a bit when we paint ourselves into a syntactic corner or can't find any other way to say what we mean, as in *I would demur that Kepler deserves second place after Newton,* or *That really threatened the fear of God into the radio people.* Calling a sentence ungrammatical means that it sounds odd "all things being equal"—that is, in a neutral context, under its conventional meaning, and with no special circumstances in force.

Some people raise an eyebrow at linguists' practice of treating their own sentence judgments as objective empirical data. The danger is that a linguist's pet theory could unconsciously warp his or her judgments. It's a legitimate worry, but in practice linguistic judgments can go a long way. One of the perquisites of research on basic cognitive processes is that you always have easy access to a specimen of the species you study, namely, yourself. When I was a student in a perception lab I asked my advisor when we would stop generating tones to listen to and start doing the research. He corrected me: listening to the tones *was* research, as far as he was concerned, since he was confident that if a sequence sounded a certain way to him, it would sound that way to every other normal member of the species. As a sanity check (and to satisfy journal referees) we would eventually pay students to listen to the sounds and press buttons according to what they heard, but the results always ratified what we could hear with our own ears. I've followed the same strategy in psycholinguistics, and in dozens of studies I've found that the average ratings from volunteers have always lined up with the original subjective judgments of the linguists.[17]

A PARADOX IN BABY TALK

Put yourself in the booties of a child who is in the midst of figuring out how to speak the language as it is spoken by parents, friends, and siblings. You have learned a few thousand words, and have an inkling (not conscious, of course) of the difference between subjects, verbs, objects, and oblique objects. The verbs keep coming in, and as you learn them you have to figure out how you can use them. Just knowing what a verb means isn't enough, because, as we saw, verbs with similar meanings can appear in different constructions (like *dine* and *devour,* or *hinted* and *rumored*); you have to pay attention to which participants accompany the verb in the sentence.

For instance, say you've heard *load* in a sentence for the first time, such as *Hal is loading hay into the wagon.* Say you have an idea of what the words mean, and from watching what's going on, you can see that Hal is pitching hay into a wagon. A safe bet is to file away the information that *load* can appear in a sentence with a subject, which expresses the loader (Hal); an object, which expresses the contents being moved (the hay); and an object of *into,* which expresses the container (the wagon). You can now say or understand new examples with the same verb in the same construction, like *May loaded some compost into the wheelbarrow.* (Linguists call this the content-locative construction, because the contents being moved are focused upon in the object of the sentence.) But that's as far as you go—you don't venture into saying *May loaded* (meaning she loaded something into something else), or *May loaded into the wheelbarrow.*

So far so good. In a little while you hear *load* in a new construction, like *Hal loaded the wagon with hay.* Once again hay is being pitched into the wagon, and as far as you can see, the sentence has the same meaning as the familiar sentence *Hal loaded hay into the wagon.* You can add an addendum in your mental dictionary to the entry for *load:* the verb can also appear in a construction with a subject (the loader), an object (the container, such as a wagon), and an object of *with* (the contents, such as the hay). Linguists call this the container-locative construction, because now it's the container that's being focused upon.

As you continue to hoover up verbs over the months and years, you encounter other verbs that behave like *load:* they appear in two synonymous constructions but differ in whether it is the content or the container that shows up as the direct object:

> Jared sprayed water on the roses.
> Jared sprayed the roses with water.
>
> Betsy splashed paint onto the wall.
> Betsy splashed the wall with paint.
>
> Jeremy rubbed oil into the wood.
> Jeremy rubbed the wood with oil.

This is starting to look like a pattern (what linguists call an alternation), and now you face a critical choice. Do you keep accumulating these pairs of verbs, filing them away pair by pair? Or do you make a leap of faith and assume that any verb that appears in one of these constructions can appear in the other one? That generalization could be put to work by coining a rule that more or less says, "If a verb can appear in a content-locative construction, then it can also appear in a container-locative construction, and vice versa." With this rule (which we can call the locative rule) in hand, you could hear someone say *brush paint onto the fence* and then surmise that *brush the fence with paint* is fine, without having actually heard it. Likewise, if you hear *Babs stuffed the turkey with breadcrumbs,* you can assume that *Babs stuffed breadcrumbs into the turkey* is also OK.

It's a small step toward mastering the language, but a step in the right direction. English is crawling with families of constructions that admit verbs interchangeably, and if children can dig out the patterns and extend them to new verbs, they can multiply their learning speed by the average number of constructions per verb. This could be an important path to becoming a fluent and open-ended speaker of the language, as opposed to one who simply regurgitates a small number of formulas.

There is only one problem. When the locative rule is applied willy-nilly, it cranks out many errors. For example, if you apply it to *Amy poured water into the glass,* you get *Amy poured the glass with water,* which English speakers reject (as I've verified in questionnaires).[18] You can also get into trouble when you apply it in the other direction, to verbs like *fill:* though the input, *Bobby filled the glass with water,* is fine, the output, *Bobby filled water into the glass,* is not (again, a survey bears this out).[19] And these aren't isolated exceptions. Many other verbs resist being fed into the maw of the locative rule. Here are four other unhappy campers, two of them verbs that like only the content-locative, and two that like only the container-locative. (Following the usual convention in linguistics, I've put an asterisk next to the sentences that sound odd to native speakers.)

Tex nailed posters onto the board.
*Tex nailed the board with posters.

Serena coiled a rope around the pole.
*Serena coiled the pole with a rope.

Ellie covered the bed with an afghan.
*Ellie covered an afghan onto the bed.

Jimmy drenched his jacket with beer.
*Jimmy drenched beer into his jacket.

That's funny. . . . Why should the second sentence in each pair sound so odd? It's not that the iffy sentences are unintelligible. No one could be in doubt as to the meaning of *Amy poured the glass with water* or *Jimmy drenched beer into his jacket*. But language is not just whatever set of ways people can think of to get a message across. Children, in the long run, end up with a fastidious protocol that sometimes rules out perfectly good ways of communicating. But why? How do children succeed in acquiring an infinite language when the rules they are tempted to postulate just get them into trouble by generating constructions that other speakers choke on? How do they figure out that certain stubborn verbs *can't* appear in perfectly good constructions?

An equivalent puzzle arises if you invert the way you think about the problem and make the child the master and the language the slave. How did the English language come down to us with all those exceptional verbs, given that they should have been whipped into conformity by the first generation of children faced with learning them?

There are three ways out of this paradox, but none of them is palatable. The first is that we (and the hypothetical child we have been imagining) have framed the rule too broadly. Maybe the real locative rule is restricted to a subset of verbs sharing an overlooked trait, and children somehow figure out the restriction and append it as a codicil to the rule. But if there is such a trait, it's far from obvious, because the verbs that submit to the rule and the ones that resist it are quite close in meaning. For example, *pour, fill,* and *load* are all ways of moving something somewhere, and they all have the same cast of characters: a mover, some contents that move, and a container that is the goal of the movement. Yet *pour* allows only the content-locative (*pour water*), *fill* allows only the container-locative (*fill the glass*), and *load* goes both ways (*load the hay, load the wagon*).

The second option is that children don't coin these rules at all. Maybe they really do file away in memory just those combinations of verbs and constructions they have heard in the speech of their elders, and conservatively stick to just those combinations. Under this theory, they would be like Marvin in the eponymous comic strip on the following page.

Well, that would certainly solve the problem. Children would never be tempted to say *pour the cup with juice* or *cover an afghan onto the bed* because they would never have heard anyone else say things like that. Verbs

Marvin—NAS. North American Syndicate.

would keep their privileges in perpetuity, because children would learn the constructions on a verb-by-verb basis, just as they learn the words themselves, each a unique combination of a sound and a meaning.

Some linguists have taken this hypothesis seriously, but it doesn't seem to be right.[20] For one thing, it would be surprising if children were *that* conservative, given that they have an infinite language to master and only a finite sample of speech to go on. For another, the English language seems to expand rapidly to accommodate new verbs in new constructions, suggesting that at least by the time they reach adulthood, speakers are not conservative verb-memorizers. Most Americans, on hearing the Britishism *He hoovered ashes from the carpet* (a content-locative), readily generalize to *He hoovered the carpet* (a container-locative). Likewise, when the container-locatives *burn a CD* (put songs onto it) and *rip a CD* (copy songs off of it) came into common parlance, the content-locatives *burn songs onto the CD* and *rip songs from the CD* followed closely on their heels (or perhaps the other way around).[21]

Do only adults make these leaps, or can they be seen in childhood, as children are learning the language? The psychologist Melissa Bowerman, like many psycholinguists, kept meticulous diaries of her children's speech when they were small, recording and analyzing every anomaly. She showed that children really do use verbs in constructions that they could not simply have recorded from the mouths of their parents.[22] Here are three examples of creative content-locatives, and three of creative container-locatives:

Can I fill some salt into the bear?
I'm going to cover a screen over me.
Feel your hand to that.

Look, Mom, I'm gonna pour it with water, my belly.
I don't want it because I spilled it of orange juice.
I hitted this into my neck.

To ensure that these weren't rare errors from unusual children, the psychologist Jess Gropen and I corroborated the finding in two ways. First, we sifted through online corpora of children's speech, where we found similar errors.[23] Second, we used a method for assessing generalizations called the *wug* test, after a classic study by the psychologist Jean Berko Gleason.[24] Gleason showed children a cartoon of a little bird and said, "Here is a wug. Now there are two of them. There are two . . ."—at which point four-year-olds happily filled in the blank with *wugs,* a form they could not have memorized from adults. In our case we told children that *mooping* meant to move a sponge to a purple cloth, turning it green. Sure enough, the kids said we were *mooping the cloth*—a container-locative that they had never heard anyone use before.[25] So much for Moderate Marvin.

There's a third way out. Maybe children do make errors but are corrected by their parents and are thereby chastened into avoiding the offending verb in that construction forever after. This, too, is unlikely. Notwithstanding the widespread belief among psychologists that parents are responsible for everything that develops in their children, attempts to show that parents correct their children's deviant sentences, or even react differently to them, have turned up little.[26] Parents are far more concerned with the meaning of children's speech than its form, and when they do try to correct the children, the children pay little heed. The following exchange is typical:

CHILD: I turned the raining off.
FATHER: You mean you turned the sprinkler off?
CHILD: I turned the raining off of the sprinkler.

And even if parents did occasionally raise an eyebrow at their children's odd usages and the children did take heed, the effect would fall short of what we need to solve the problem. Many of the obstreperous verbs are rare, yet people have strong intuitions about what the verbs can and can't do. People sense that they would never say *They festooned ribbons onto the stage* or *She siphoned the bottle with gasoline,* yet word-frequency counts show that these verbs are literally one in a million.[27] It is unlikely that every English speaker uttered each of the obdurate verbs in each of the offending constructions at some point in childhood (or, for that matter, adulthood), was corrected, and now finds the usage strange on account of that episode.

We have a paradox.[28] From the time they are children, people generalize. They avoid generalizing to certain words (at least as adults). It's not because they were corrected for each overgeneralization. And there is no systematic difference between the words that allow themselves to be generalized and those that don't. These four statements can't all be true.

Why should anyone care about what seems like a small problem in a tiny corner of psycholinguistics? The reason is that the learnability of the locative construction is typical of many paradoxes in explaining language, where partial patterns, too seductive to ignore but too dangerous to apply, are ubiquitous. In *Crazy English,* the language maven Richard Lederer calls some of them to our attention:

> If adults commit adultery, do infants commit infantry? If olive oil is made from olives, what do they make baby oil from? If a vegetarian eats vegetables, what does a humanitarian consume? A writer is someone who writes, and a stinger is something that stings. But fingers don't fing, grocers don't groce, hammers don't ham, humdingers don't humding, ushers don't ush, and haberdashers do not haberdash. . . .
>
> . . . If the plural of *tooth* is *teeth,* shouldn't the plural of *booth* be *beeth*? One goose, two geese—so one moose, two meese? If people ring a bell today and rang a bell yesterday, why don't we say that they flang a ball? If they wrote a letter, perhaps they also bote their tongue.[29]

Each of these oddities defines a scientific problem for linguistics and psychology. The one in the second paragraph—irregular plurals and past-tense forms—is tricky enough that I have written a book and many papers trying to make sense of it.[30] Unfortunately, my favorite solution to that puzzle is of no help here. Irregular forms like *teeth* and *rang* are what linguists call positive exceptions: they exist, even though the usual rule, like "Add -*ed* to form the past tense," fails to generate them. Children can learn them upon hearing them, one at a time. We also have a good idea as to how children use positive exceptions to preempt or block the rule-governed forms, so that they don't say *flang* or *meese*. Conjugations and declensions are neatly organized into paradigms in which each verb ordinarily has a single past-tense form and each noun a single plural form. When the child hears *Boggs flung the ball* or *Vern shot two moose,* those irregular forms stake out their cells in a mental matrix and fend off the rival forms *flang* and *meese* (together with *flinged* and *mooses*).[31]

But verb-construction mismatches are *negative* exceptions: they *fail* to exist despite the fact that a rule *does* generate them. Children have no direct evidence from parents' speech that these forms are ungrammatical. Not hearing them—the proverbial dog that didn't bark—isn't itself evidence, because there are an infinite number of perfectly grammatical forms that they also don't hear, and they can't very well exclude all of them or they would be confined to parrothood. Nor can they use some competing form to preempt them (in the way that *flung* preempts *flinged* and *flang*), because verb constructions, unlike conjugations, aren't arranged into neat cubbyholes. The reason you can't say *pin a board with posters* or *coil the pole with a rope* is not that each of these is repelled by some synonym guarding the grammatical turf in the way that *flung* repels *flang* and *flinged*. There *is* no verb that allows one to talk about covering a board by pinning posters onto it in the container-locative construction.

The verb-learnability paradox has attracted attention for another reason. About halfway down our plunge from a wide view of the human mind to the acquisition of locative constructions, I said that language acquisition is an example of the problem of induction—making valid generalizations about the future from limited data available in the present, whether they involve language acquisition by a child, learning by a computer, or theorizing by the scientist. The pickle we find ourselves in is common to induction of all kinds: how to back off from an overly general hypothesis in the absence of negative data.[32] If you frame a conclusion too broadly, and don't have complete corrective feedback from the world (say, you grow up thinking all swans are white, and never get to New Zealand, where you'd see black swans), you are in danger of never finding out that you are wrong. In this case, a hypothetical child is tempted to generalize that all verbs about moving something somewhere can be expressed in either of two English constructions. Yet somehow children grow into adults who generalize beyond the verbs they have heard while uncannily holding back from some of the verbs they haven't heard. The locative construction (along with similar constructions) presents us with a paradox of a child seeming to learn the unlearnable, and thus became a focus of attention among linguists and computer scientists interested in the logic of learning in general.

Nature does not go out of its way to befuddle us. If some phenomenon seems to make no sense no matter how we look at it, we are probably overlooking some deeper principle about how things work. This is exactly what happened in the paradox of learning locative verbs, and the missing principles are about the kinds of ideas that populate the human mind.

FLIPPING THE FRAME

Of the four apparent facts that can't all be true at the same time—people generalize; they avoid some exceptions; the exceptions are unpredictable; and children don't get corrected for every mistake—the most assailable is the one about the unpredictability of the exceptions, the one that says there is no way to distinguish the verbs that take part in an alternation from those that sit it out. Maybe we just haven't looked hard enough. Often a linguistic pattern that at first seems haphazard turns out to have a stipulation that divides the sheep from the goats. For example, the mystery of why you can't apply *-er* and *-est* to certain adjectives, as in *specialer* and *beautifullest,* was solved when someone noticed that the suffixes apply only to words that are monosyllabic (*redder, nicer, older*) or have at most an insubstantial second syllable (*prettier, simpler, narrower*). Perhaps there is also a subtle criterion that distinguishes the verbs enlisted into the locative construction from the draft dodgers—what the linguist Benjamin Lee Whorf called a cryptotype.[33] If children's rules became sensitive to that criterion, the paradox would vanish. The hidden criterion is unlikely to involve the sounds of the verbs, since in that regard they are pretty much alike; it is more likely to involve their meanings.

The breakthrough, in my mind, came in a paper by a pair of linguists, Malka Rappaport Hovav and Beth Levin, who at the time were working down the hall from me at MIT.[34] Under the influence of Chomsky, linguists had tended to think of rules as operations that cut and pasted phrases, such as moving a prepositional object leftward into the position of the direct object, or moving the direct object rightward into a prepositional phrase.[35] It was this mindset that made it seem so odd that the locative rule would care about the content of the verb, just as it would seem odd if your word-processing program announced that it refused to cut and paste words with some meanings while obediently doing so with others. But what if the rule transformed not the arrangement of the phrases in a construction, but something much more abstract, namely, the framing of events that goes into its meaning?

Imagine that the meaning of the content-locative construction is "A causes B to go to C," but the meaning of the container-locative construction is "A causes C to change state (by means of causing B to go to C)." In other words, loading hay onto the wagon is something you do to hay (namely, cause it to go to the wagon), whereas loading the wagon with hay is something you do to the wagon (namely, cause it to become loaded with hay). These are two different construals of the same event, a bit like the gestalt shift in the classic face-vase illusion in which the figure and ground switch places in one's consciousness:

In the sentences with the hay and the wagon, the flip between figure and ground is not in the mind's eye but in the mind itself—the interpretation of what the event is really about.

Now, at first glance the difference between causing a thing to go to a place and causing a place to change by moving a thing to it may seem as rabbinical as the difference between whether the destruction of the World Trade Center consisted of one event or two events—a question of "mere semantics." But as with the gigabucks at stake in the aftermath of 9/11, mere semantics can matter.

For one thing, this new understanding of the phenomenon is simpler and more elegant—not always a sign that a theory is true, but not something to be ignored either. When reconceived as a conceptual gestalt shift, the locative rule is no longer a matter of cutting and pasting phrases in complicated ways for no particular reason. It can now be factored into two very general and useful rules:

- A rule of semantic reconstrual (the gestalt shift): If a verb means "A causes B to move to C," it can also mean "A causes C to change state by moving B to it."
- A rule for linking meaning to form: Express the affected entity as the direct object.

In the content-locative (*load hay onto the wagon*), we have *hay* as the direct object, because the event is construed as something being done to the hay. In the container-locative (*load the wagon with hay*), we have *the wagon* as the direct object, because the event is now construed as something being done to the wagon. Other linking rules take care of how the other participants are expressed. One rule links the causal agent (the guy pitching the hay) to the subject. The other links miscellaneous participants to oblique objects, each getting a preposition suitable to its meaning. The preposition *into* means "to in," namely, "to an interior portion of"; *onto* stands for "to on"; *with* stands for "a means of changing something."

Though we have replaced one rule with several, the picture as a whole is simpler, because, as we shall see, these rules get reused in different combinations all over the language. And satisfyingly, we can explain *why* the locative rule does what it does. The *wagon* participant *has* to switch from oblique object to direct object (as opposed to being pasted into any old position), because that participant has been reconstrued as "the affected entity," and affected entities, whether they are things that change location or things that change state, are expressed in syntax as direct objects.

I promised you that the reason we are obsessing over locative constructions in a book about human nature is that it tells us things about the way humans think. One of these things was broached in the first chapter: the mind has the power to frame a single situation in very different ways. Here we see that this power is so pervasive that it isn't just recruited in contentious squabbles like invading Iraq versus liberating Iraq or manipulating a ball of cells versus killing a young person, where no one is surprised that different perspectives are possible. Rather, it pervades the way we construe even the simplest, most concrete, and most innocuous events of everyday life, like putting hay in a wagon or breadcrumbs in a turkey.

Scrutinizing the locative construction not only shows that construing and reconstruing is a basic power of cognition, it exposes the elements that make up each construal, and some of their quirks. The gestalt-shift theory implies that the two locative constructions, contrary to first impressions, are not completely synonymous. There must be situations in which one construction truthfully applies and the other does not. That is indeed the case.

When one *loads hay onto a wagon,* it can be any amount, even a couple of pitchforkfuls. But when one *loads the wagon with hay,* the implication is that the wagon is full.[36] This subtle difference, which linguists call the holism effect, can be seen with the other locative verbs: to *spray the roses with water* implies that they all got sprayed (as opposed to merely *spraying water onto the roses*), and to *stuff the turkey with breadcrumbs* implies that it is completely stuffed.

The holism effect is not an arbitrary stipulation tacked onto the rule, like a pork-barrel amendment on a spending bill. It falls out of the nature of what the rule does, namely, construe the container as the thing that is affected. And that, in turn, reveals an interesting feature of the way the mind conceives what things are and how they change. The holism effect turns out not to be restricted to the locative construction; it applies to direct objects in general. For instance, the sentence *Moondog drank from the glass of beer* (where *the glass* is an oblique object of *from*) is consistent with his taking just a couple of sips. But the sentence *Moondog drank the glass of beer* (where *the glass* is a direct object) implies that he chugged down the whole thing. Similarly, you might say *He climbed up the mountain* even if he thought better of it partway and came back down, but if you say *He climbed the mountain* you are suggesting that he reached the summit. Or think about the difference between the sentences in each of these pairs, which at first glance look synonymous:

Peter painted on the door.
Peter painted the door.

Betty put butter on the bun.
Betty buttered the bun.

Polly removed peel from the apple.
Polly peeled the apple.

In each pair, the second sentence, which expresses the affected entity as the direct object, implies that something was done to the whole thing, not just part of it: the door was entirely painted, the bun thoroughly buttered, the apple completely skinned.

But the holism effect is even more sweeping than that. It's not so much a property of the direct object (which is just a position in a sentence) as it is a property of the *concept* that tends to be expressed as a direct object, namely, the entity being affected. In the examples we have been looking at, the affected entity happens to get expressed as the direct object because when a sentence contains a causal agent, the agent generally gets first dibs on the subject slot. But when the agent is not mentioned, the affected entity can be the subject, as in *The ball rolled* or *The butter melted*. And crucially, when the subject does accommodate an affected entity, it is interpreted holistically, just like direct objects. This is best seen in lovely pairs like these:

Bees are swarming in the garden.
The garden is swarming with bees.

Juice dripped from the peach.
The peach was dripping with juice.

Ants crawled over the gingerbread.
The gingerbread was crawling with ants.

The second sentence in each pair presents a sensuous image of an entity so saturated with stuff or bits that the mind blurs the two and apprehends the entire entity as doing what the stuff or bits ordinarily do: the garden swarms, the peach drips, the gingerbread crawls.

But *why* is the content interpreted as a whole in these constructions? The reason is that the English language treats a *changing* entity (a loaded wagon, sprayed roses, a painted door) in the same way that it treats a *moving* entity (pitched hay, sprayed water, slopped paint). A state is conceived as a location in a space of possible states, and change is equated with moving from one location to another in that state-space. In this way, locative constructions illustrate a second discovery in the hidden world down the rabbit hole, the ubiquity of metaphor in everyday language. The linguist Ray Jackendoff has explored the way in which many of the words and constructions used for motion, location, or obstruction of motion in physical space are also used for a kind of metaphorical motion, location, or obstruction of motion in *state*-space:[37]

> Pedro went from first base to second base.
> Pedro went from sick to well.
>
> Pedro was at second base.
> Pedro was sick.
>
> The manager kept Pedro at first base.
> The doctor kept Pedro well.

In the first sentence, Pedro's body actually moved in space, but in the second, he could have been in bed the whole time; only his *health* moved, metaphorically speaking. Concepts of space seem to infect other concepts as well, as we saw in the first chapter when noting the way that people count and measure out events as if they were objects made of time-stuff. People also use space as a model for an abstract continuum when they speak of the *rising* or *falling* of their paycheck, their weight, or their spirits,[38] or when they plot data points, representing anything whatsoever, on graph paper.[39] Whether the ubiquity of metaphor is a revolutionary discovery about the mind, a banal fact about the history of a language, or something in between is the topic of a later chapter. My intent here is to show how the psychology of space sheds light on the holism effect, and tells us something about the psychology of concepts in general.

When the mind conceptualizes an entity in a location or in motion, it tends to ignore the internal geometry of the object and treat it as a dimensionless point or a featureless blob. The linguist Len Talmy notes that a typical preposition or other spatial term specifies a relationship between a figure and a place that is defined by some reference object.[40] Usually the reference object is larger and more prominent, and the figure moves or is located relative to it. (The exception that proves the rule is Beatrice Lillie's quip about the *Queen Mary* when she first saw it, "How long does it take for this place to get to London?") And usually the reference object is specified in more geometric detail. It is conceptualized as having a certain number of dimensions along which it is stretched out: one dimension, like a stick or a string; two dimensions, like a sheet of paper or plywood; or three dimensions, like a couch or a watermelon. And it is conceptualized as having certain axes, parts, cavities, and boundaries that align with those dimensions.

So the figure being positioned and the place where it is said to be located are treated differently in language: the first is reduced to a dimensionless speck, whose internal geometry is ignored; the second is diagrammed, at least schematically. Take the English phrases *on your hand, under your hand,* and *in your hand.* Each picks out an aspect of the geometry of the hand, namely, its top, its bottom, and a cavity it can form. The choice of preposition depends upon that geometry: a marble can be *in one's hand* if the hand is cupped and the marble is on the palm side, but the marble is *on it* if the hand is straightened or if the marble is on the back. The marble can't be *in* one's forearm or shin or trunk at all, since these are conceptualized as one-dimensional dowels. Compare this blueprinting to the stylized treatment of the figure being located. In these examples I have cited a marble, but the figure can have any shape or configuration whatsoever: it can be a marble, a matchstick, a matchbook, or a moth, and these can be upright, sideways, or upside down, and it can still be *in* or *on* or *under* your hand. To be sure, not all prepositions treat the figure as a blob or point: *along* and *across,* for example, require the figure to be elongated. But the most common ones are myopic about the figure being positioned.

This leads us to a deeper explanation of the holism effect. In the locative alternation, when the container (such as *the wagon* in *load hay into the wagon*) gets promoted to direct object, it is also conceptually reanalyzed as something that has been moved in state-space (from the "empty" slot to the "full" slot). And in this reconstrual, it gets compacted into a single point, its internal geometry obliterated. Wagons become loaded, flowerbeds sprayed, turkeys stuffed, not as arrangements of matter in space with niches and hidey-holes that may separately accommodate bits of matter, but as entities that are, taken as a whole, now ready for carting, blooming, or cooking. Indeed, the holism effect is a bit of a misnomer. We're really talking about a *state-change* effect, and ordinarily the most natural way that an object changes state when something is added to it is when the stuff fills the entire cavity or surface designated to receive it. But if an object can be thought of as changing state even when it has stuff in just *one* part, then the container-locative may be used there, too. Thus we can say that a graffiti artist has *sprayed a statue with paint* even if he has colored just one part of it, because a single splotch is enough for people to consider it defaced.

To complete the circle, I have to show how the theory resolves the original paradox. How does the idea that the locative is a gestalt shift explain why some verbs allow the shift while other verbs, seemingly similar to them, do not? The key is the chemistry between the meaning of the construction and the meaning of the verb. To take a simple case, one can *throw a cat into the room*, but one cannot *throw the room with a cat*, because merely throwing something into a room can't ordinarily be construed as a way of changing the room's state. This chemistry applies to more subtle cases as well. Verbs that differ in their syntactic fussiness, like *pour, fill,* and *load,* all pertain to moving something somewhere, giving us the casual impression that they are birds of a feather. But on closer examination each of these verbs turns out to have a distinct kind of *semantic* fussiness—they differ in which aspect of the motion event they care about.

Take the verb *pour,* and think about when you can use it. *To pour* means, more or less, to allow a liquid to move downward in a continuous stream. It specifies a causal relation of "letting" rather than "forcing," and it specifies a manner of motion; these are the bits of meaning that differentiate it from other ways in which liquid moves, such as *spray, splash,* and *spew.* Since *pour* says something about the motion, it can be used in the construction that is about motion; hence we can say *pour water into the glass.* But *pour* doesn't care about how or where the liquid ends up. You can pour water into a glass, all over the floor, or out the window of an airplane, dispersing it into a mist. Nothing predictable happens to the destination of a poured liquid, and so the verb is inconsistent with a construction that specifies how the state of a container has been changed. And thus we can't say *she poured the glass with water.*

Now take the verb *fill. To fill* something means to cause it to become full (it's no coincidence that *fill* and *full* sound alike). It's all about the state of the container: no fullness, no filling. But *fill* is apathetic about *how* the container became full. You can fill a glass by pouring water into it, of course, but you can also fill it by bailing water out of a bathtub, by holding it out the window during a rainstorm, or by letting a leaky faucet drip into it overnight. That's why *fill* is the syntactic mirror image of *pour:* by specifying the change of state of a container, it is compatible with a construction that is about a state-change, and thereby allows us to say *fill the glass with water.* But because it says nothing about a cause or manner of motion of the contents, it isn't compatible with a construction that is all about a motion, and thereby doesn't allow us to say *fill water into the glass.*

Finally, take the verb *load.* What is the common thread connecting what you do to hay and a wagon, bullets and a gun, film and a camera, suitcases and a car, and software and a computer? It's not just that a thing gets put in a place. The thing has to be of the right size, shape, or content to enable the place to do what it is meant to do—fire, take pictures, go on a trip, and so on. You can't *load a camera* by putting hay or bullets in it, nor have you *loaded the car* if you throw Dad's suitcases in the trunk but leave Mom's on the sidewalk. In fact, you haven't even *loaded the camera* if you stuff film behind the lens cap or into the case rather than in the proper place in the film chamber. The verb *load,* then, simultaneously specifies something about how a content has been moved *and* how a container has changed. It thus slides into either the content-locative (*load the film*) or the container-locative (*load the camera*).

How could one test this theory? The most straightforward way is to devise a *wug* test that teaches different kinds of invented locative verbs to children or adults and then see how they use the verbs. When the verb pertains to a manner of motion, people should spontaneously use it in a content-locative construction; when it pertains to a change of state, they should use it in a container-locative. Gropen and I did the *wug* test (a *moop* test, to be exact).[41] In some of the experimental variations, *mooping* referred to moving something in a conspicuous manner, like zigzagging a wet sponge over to a wet purple cloth. In others, the verb also referred to moving the sponge to a cloth, but this time the motion was nondescript and the cloth turned green or pink when the sponge touched it. When the verb described a zig-zagging motion, children and adults were more likely to describe the event as *mooping the sponge,* a content-locative. When it described motion re-sulting in a color change, they were more likely to describe it as *mooping the cloth,* a container-locative. This is just what you would predict if people in-sert a verb into a construction according to which aspect of the event the verb picks out. (For the curious: To make a potion that changes color on demand, boil small pieces of purple cabbage in water, then fish out the cab-bage and let the purple liquid cool. If you add a base to it, such as baking soda, it will turn green; if you add an acid, like lemon juice, it will turn pink. The cabbage juice sold in supermarkets is red because vinegar has already been added.)

So in trying to crack a puzzle in how children infer the syntax of their mother tongue, we were forced to reconceptualize what they had to learn: from an operation for cutting and pasting phrases to a mental gestalt shift in how a situation is construed. This uncovered a number of basic features of our thought processes: that the mind deploys a set of rival frames that can construe even the most plodding everyday event in more than one way; that a frame for thinking about a change of location in real space can be metaphorically extended to conceptualize a change of state as motion in state-space; and that when the mind conceives of an entity as being some-where or going somewhere, it tends to melt it down to a holistic blob.

These conclusions, though, raise new questions. Is the mind so flexible that it can conceptualize any event in any way whatsoever? If so, how could we get anywhere in thinking and talking? And is there more to our basic understanding of motion and change than a blob just being somewhere, going somewhere, or changing somehow?

THOUGHTS ABOUT MOVING AND CHANGING

The flexibility of the human mind—its ability to flip frames, shift gestalts, or reconstrue events—is a wondrous talent. But it makes it difficult to pre-dict how a person will think and talk about a given situation. When I hit a wall with a stick, am I affecting the stick by moving it to the wall, or affecting the wall using the stick as an instrument? When Harold likes Hildy, is he causing himself to think well of her, or is she causing him to approve of her?

If Bill does an impression of John Travolta in *Saturday Night Fever,* is he causing Debbie to laugh in the same way that he causes a balloon to pop by pricking it, or does Debbie have enough free will that "causation" is not an appropriate way to think about it? When Becky shouts across a noisy room to Liz, what is she doing: affecting Liz, creating a message, making noise, sending a message across the room, or just moving her muscles in a certain way? Even the most palpable cognitive distinction—who did something, and who had something done to him—can be mentally flip-flopped, as when a hockey player shouts, "Kiss my elbow!" or when Woody Allen in *Play It Again, Sam* gets roughed up by some bikers and tells his friends, "I snapped my chin down on some guy's fist and hit another one on the knee with my nose."

Cognitive flexibility is in many ways a blessing, but in figuring out how language works, it is something of a curse. Language is supposed to give us a way to communicate who did what to whom. But how can it ever do that if two people can look at the same event and make different assignments of the who, the what, and the whom in the first place? This isn't just a hypothetical worry; it saps the explanatory power from the gestalt-shift theory of how children learn verbs. To regain it, we are forced to drill deeper into the psychology of motion and change.

Here is the problem. If people are mentally agile enough to interpret events in many ways, what's to prevent a child from interpreting the meaning of *to nail* as "to obscure a surface by nailing things to it," or *to coil* as "to cause a long object to have a filament coiled around it"? If the answer is "Nothing," then nothing would prevent a child from saying *Tex nailed the house with shingles* or *Serena coiled a pole with a rope,* and we would be back where we started. With enough cognitive flexibility, *any* verb specifying a movement of some contents could be reconstrued as specifying a change of state of a container. In which case anything goes—any verb could be used in any way.

Of course, we have good reason to believe that English speakers conceive of events in similar ways, because they make similar judgments about how verbs can be used. But how do they arrive at this consensus? There must be some independent criterion that tells a child when a kind of motion can also be construed as a noteworthy state-change, and when the state-change would be too vapid or gerrymandered to deserve being singled out.

And that requires penetrating into the verb to find a deeper layer of meaning that the mind might use to decide which of the cognitively available roles—agent, thing that moves, thing that changes—should be applied in construing *this* kind of event.

To uncover this stratum of meaning, it helps to start small—to lay out microclasses of semantically similar verbs that do or don't undergo an alternation and then try to spot any common threads.[42] Here are some verbs that can take part in the locative alternation—that is, they allow you to say either *smear grease on the axle* or *smear the axle with grease:*

> brush, dab, daub, plaster, rub, slather, smear, smudge, spread,
> streak, swab

And here are some similar verbs that resist it—that is, they allow you to say *pour water into the glass* but not *pour the glass with water:*

> dribble, drip, drop, dump, funnel, ladle, pour, shake, siphon,
> slop, slosh, spill, spoon

What's the difference? They both look like ways of getting some goo into or onto a receptacle. But now think about the physics. In the first list, the agent applies force to the substance and the surface simultaneously, by pushing one against the other. In the second, the agent allows gravity to do the work. It's the difference between causing and letting, between acting on something directly and acting on it via an intermediary force, between expecting something to change as one is doing something in real time and expecting it to change shortly after one has done something. "Moving" and "changing" are not enough grounds for the mind to construe an event in a particular way. It also cares about finer-grained concepts like forcing versus enabling a force, causing versus letting, and before-and-after versus at-the-same-time.

Let's examine another set of microclasses that stand on opposite sides of a linguistic divide. What do these verbs—each of which appears in both constructions—have in common?

> inject, shower, spatter, splash, splatter, spray, sprinkle, spritz,
> squirt

Again, think physics. All involve imparting force to the substance, causing it to zip off into or onto the surface—a form of causation different from both the pressing in the *brush* verbs and the enabling of gravity in the *pour* verbs. And they, in turn, differ from the following list, whose verbs refuse to enter the container-locative construction (that is, one can't say *spit the floor with tobacco juice*):

> emit, excrete, expectorate, expel, exude, secrete, spew, spit,
> vomit

This disgusting list is made up of verbs in which the substance is expelled from inside a volume, though they differ in the kind of volume, the nature of the orifice, what the substance is, and how it is expelled. The geometry of insideness and outsideness is enough to unite this microclass while differentiating it from other microclasses.

There are still other lists. Verbs of sending little particles out in all directions do alternate:

> bestrew, scatter, seed, sow, spread, strew

but verbs of attaching something to something else with a fastener do not:

> attach, fasten, glue, hook, nail, paste, pin, staple, stick, strap,
> tape

Verbs of forcing a substance into a container against the limits of its capacity do alternate:

> cram, crowd, jam, pack, stuff, wad

but verbs of wrapping a flexible one-dimensional object around a rigid one do not:

> coil, spin, twirl, twist, whirl, wind

When we look at verbs that are finicky in the other way—verbs like *fill*, which allow *fill the glass with water* but not *fill water into the glass*—we see that they also fall into microclasses defined by geometry, physics, and human purpose. Here is an inventory, which I include for the verbophiles to read and so that everyone else can get a general impression:

> *To cause a layer to cover a surface. Liquid layer:* deluge, douse, flood, inundate. *Solid layer:* bandage, blanket, coat, cover, encrust, face, inlay, pad, pave, plate, shroud, smother, tile.
>
> *To add something to an object, making it aesthetically better or worse:* adorn, burden, clutter, deck, dirty, embellish, emblazon, endow, enrich, festoon, garnish, imbue, infect, litter, ornament, pollute, replenish, season, soil, stain, taint, trim.
>
> *To cause a mass to be coextensive with a solid or layer. Liquid:* drench, impregnate, infuse, saturate, soak, stain, suffuse. *Solid:* interlace, interlard, interleave, intersperse, interweave, lard, ripple, vein.
>
> *To add an object that impedes the movement of something. Liquid:* block, choke, clog, dam, plug, stop up. *Solid:* bind, chain, entangle, lash, lasso, rope.
>
> *To distribute a set of objects over a surface:* blot, bombard, dapple, riddle, speckle, splotch, spot, stud.

What is going on? Are Anglophones a tribe of overly toilet-trained fusspots? What kind of civilization would care about exactly how things get smeared, sloshed, splattered, spewed, strewn, stuffed, or splotched when deciding how to use a verb? The answer is to be found not in psychosexual stages but in the psychology of construing physical events. Recall that the use of locative verbs depends on what they are thought to pertain to: the way something moves, the way a surface is affected, or both.[43] What these microclasses are telling us is that certain aspects of geometry and physics are salient enough to the minds of English speakers to determine how they construe events.

With the *brush* verbs, the agent applies force simultaneously to the stuff and the surface, so these verbs are naturally construed as affecting both entities, which is why they allow both constructions. A jointly felt force is also present in the *stuff* verbs, where the contents and the container are pressed up against each other, and here too both constructions are welcome. With the *pour* verbs, though, gravity stands between what the agent does and how the surface gets wet, so the agent is less easily construed as acting directly on the container, and these verbs appear only in the content-locative construction. The *attach* verbs also implicate a go-between (the glue, the nail, and so on), thereby separating the agent's act from its effect on the surface by one link, and sure enough, these verbs don't like the container-locative construction either.

Looking at it from the other side, we see verbs that pinpoint how a surface or container changes when something gets barnacled onto it: it gets better or worse (*adorn, pollute*), less tolerant of movement (*block, bind*), saturated (*drench, interlace*), or obscured (*cover, inundate*). As a result of committing themselves to how the surface changes (but not caring about how the infusions and encrustations got there), these verbs accommodate the container-locative (*drench the shirt with wine*) but not the content-locative (*drench wine into the shirt*).

So a deeper look at which verbs participate in the locative alternation has forced us to take a deeper look at what compels the mind to construe physical events in certain ways. And at that depth we have discovered a new layer of concepts that the mind uses to organize mundane experience: concepts about substance, space, time, and force. These concepts encourage the mind to unite events that have nothing in common in terms of what they look like, smell like, or feel like, yet they obviously matter to the mind a great deal. They are so pervasive that some philosophers consider them to be the very scaffolding that organizes mental life, and in chapter 4 I will show how they saturate our science, our storytelling, our morals, our law, even our humor. But we've stumbled upon these great categories of cognition in a more inauspicious way, by trying to make sense of a small phenomenon in language acquisition. This serendipity emboldens me to use similar puzzles as an entrée into two of the other major themes of human thought.

THOUGHTS ABOUT HAVING, KNOWING, AND HELPING

The theory in the last section may seem to have dragged out a lot of paraphernalia just to explain why *pour the glass with water* sounds funny. But the same paraphernalia demystifies other constructions in English, and in doing so it shines a light on some of the other apparatus of thought.

The dative is a pair of constructions, one similar to the content-locative, the other containing two naked objects:

> Give a muffin to a moose.
> Give a moose a muffin.

The first is called the prepositional dative (because it contains a preposition, namely, *to*), the second the ditransitive or double-object dative (because the verb is followed by two objects, not just one). In traditional grammars the two phrases are called the indirect and direct objects; linguists today usually call them simply the "first object" and the "second object." The term *dative*, by the way, has nothing to do with dates; it comes from the Latin word for "give."

The dative has all the ingredients of the learnability paradox we encountered with the locative. First, the two constructions are more or less synonymous. Second, the alternation embraces not just one verb but many:

> Lafleur slid the puck to the goalie.
> Lafleur slid the goalie the puck.

> Danielle brought the cat to her mother.
> Danielle brought her mother the cat.

> Adam told the story to the baby.
> Adam told the baby a story.

This is the kind of pattern that would invite an astute child to pick up the pattern and extract a rule that says, "If a verb can appear in a prepositional dative, then it can appear in a double-object dative, and vice versa."

Third, children do pick up the pattern. Their everyday speech contains many examples of double-object forms that they could not have memorized from their parents:[44]

> Mommy, fix me my tiger.
> Button me the rest.
> How come you're putting me that kind of juice?
> Mummy, open Hadwen the door.

Using *wug* tests, Gropen and I have shown that children who are taught *norp the pig to the giraffe* ("ferry it over in a gondola car") will generalize and say *norp him the horse*.[45] Adults, too, generalize the dative. When *to fax* came into common parlance in the 1980s, it didn't take long for people to say *Can you fax me the menu?* Nor were they conservative when they extended the verb *e-mail* to *I'll e-mail him the directions*.

Fourth—and here's where the paradox arises—the generalization runs up against counterexamples in both directions. There are verbs that appear only with the prepositional dative:

> Goldie drove her minibus to the lake.
> *Goldie drove the lake her minibus.

> Arnie lifted the box to him.
> *Arnie lifted him the box.

> Zach muttered the news to him.
> *Zach muttered him the news.

And there are verbs that appear only with the double-object dative:

> The IRS fined me a thousand dollars.
> *The IRS fined a thousand dollars to me.

> Friends, Romans, countrymen: Lend me your ears!
> *Friends, Romans, countrymen: Lend your ears to me!

And fifth, the promiscuous verbs and monogamous verbs seem to convey the same kinds of meanings. *Slide the puck* and *lift the box* are ways of moving something; *tell a story* and *mutter the news* are both ways of communicating something. The paradox, once again, is how a child can both generalize and uncannily know when to back off from the exceptions, even though the exceptions seem to be arbitrary.

When we faced this paradox with the locative, we solved it by thinking of the alternation as a conceptual gestalt shift between causing to go and causing to change. The dative, too, turns out to involve a gestalt shift, this time between causing to go and causing to *have*. *Give a muffin to a moose* means "cause a muffin to go to a moose," whereas *give a moose a muffin* means "cause a moose to have a muffin."[46]

Once again this may seem to be hair-splitting, because causing-to-go usually results in causing-to-have. With a movable object, you have to cause it to go to someone for that person to have it, and even immovable and intangible possessions can be thought of as moving in a metaphorical sense. In this metaphor, possessions are things, owners are places, and giving is moving. Thus we can say *The condo went to Marv* or *Marv kept the condo* (analogous to *The ball went to Marv* and *Marv kept the ball*) even if the condo was in no danger of physically going anywhere.

Nonetheless, the two construals are cognitively different, because some kinds of causing-to-go do *not* result in causing-to-have. Consider these homonyms:[47]

> Annette sent a package to the boarder.
> Annette sent a package to the border.

With the first sentence, one can apply the dative rule and get *Annette sent the boarder a package.* But with the second, the result is nonsense—*Annette sent the border a package*—because borders, being inanimate entities, can't own packages or anything else. The concept of possession is something that we ordinarily apply only to animate beings. This immediately explains why some verbs refuse to enter the double-object construction. The sentence *Goldie drove the lake her minibus,* for example, would imply that the lake now possesses the bus, which makes no sense.

Not only are some kinds of causing-to-go incompatible with causing-to-have, but some kinds of causing-to-have are incompatible with causing-to-go.[48] When we say *Cherie gave Jim a headache,* we mean that she caused Jim to have it, presumably because she's a nudnik whose antics made his head hurt, not because a headache walked over on little legs from Cherie's head into Jim's. And sure enough, it's less natural (though not impossible) to say *Cherie gave a headache to Jim.*[49]

We can also sense a difference in *meaning* between two dative constructions even when they sound equally natural. When talking about a first baseman, it sounds a bit odd to say *Pedro threw him the ball, but a bird got in the way.* There's no problem, however, in saying *Pedro threw the ball to him, but a bird got in the way.* That's because with many verbs the double-object form implies that the recipient actually possesses the object, not just that it was sent in his direction. For similar reasons, *Señor Jones taught Spanish to the students* is compatible with his fruitlessly lecturing to dullards who don't remember a word. But *Señor Jones taught the students Spanish* carries more of an implication that the students now know Spanish—that they metaphorically possess it.[50]

Speaking of metaphor, the dative construction works with a number of verbs of communication, as in *Ask me no questions, I'll tell you no lies* and *Sing me no song, read me no rhyme.*[51] It's as if we think of ideas as things, knowing as having, communicating as sending, and language as the package.[52] This is sometimes called the conduit metaphor, and it can be seen in dozens of expressions for thinking, saying, and teaching. We *gather* our ideas to *put* them *into* words, and if our verbiage is not *empty* or *hollow,* we might *get* these ideas *across* to a listener, who can *unpack* our words to *extract* their *content.*

Another puzzle from the locative comes back to haunt us here, and in the process of dispatching the problem, we open another window into the machinery of thought. A gestalt shift between causing-to-go and causing-to-have, even with the metaphorical extension to ideas, is not enough to distinguish the verbs that enter the constructions from the verbs that don't. The problem, once again, is the curse of cognitive flexibility: the mind has the potential to construe all kinds of events as changes of possession, and we need to explain why it makes the effort with some kinds of events and not with others. Why can an English speaker *throw someone a box* ("cause him to have it by throwing it to him") but not *lift him the box* ("cause him to have it by lifting it to him")? Why can you *tell him the news* but not *mutter him the news*?

Once again, the problem is that we are standing too far back to make out the cognitive detail that matters. When we get closer, we can make out finer points of meaning that license the mind to construe some kinds of sending and communicating, but not other kinds, as causing-to-have.

Verbs of giving go both ways, logically enough:

> feed, give, hand, lend, loan, pay, sell, serve, trade

So do verbs that indicate imparting force to an object instantaneously, sending it on a trajectory to a recipient, as in *Lafleur slapped him the puck:*

bash, bat, bounce, bunt, chuck, flick, fling, flip, heave, hit, hurl,
    kick, lob, pass, pitch, punt, roll, shoot, shove, slam, slap,
    slide, sling, throw, tip, toss

But as with the locative alternation, physics matters. Verbs that indicate the *continuous* application of force to an object to keep it moving, rather than one quick fillip to send it on its way, don't like the double-object construction nearly as much (questionnaires confirm this difference).[53] That's why it's odd to talk about *lifting him the crate,* and other drawn-out maneuvers:

carry, drag, haul, hoist, lift, lower, lug, pull, push, schlep, tote,
    tow, tug

The distinction between events that are construed as instantaneous, like throwing, and events that are construed as protracted in time, like lugging, matters a lot in language. Linguists call this general realm of meaning—how states and events are distributed in time—"aspect" (not to be confused with the other timekeeper in language, tense). As we shall see when exploring the concept of time in chapter 4, aspectual distinctions matter in many areas of language and reasoning, not just in this construction.[54]

When it comes to communication, the double-object construction is discerning in a different way. It easily accommodates verbs that specify the kind or purpose of a message, as in *ask* (which pertains only to a question) or *read* (which pertains only to something written):

ask, cite, pose, preach, quote, read, show, teach, tell, write

But it puts up more resistance to verbs that specify the manner of speaking:[55]

babble, bark, bawl, bellow, bleat, boom, bray, burble, cackle,
    call, carol, chant, chatter, chirp, cluck, coo, croak, croon,
    crow, cry, drawl, drone, gabble, gibber, groan, growl,
    grumble, grunt, hiss, holler, hoot, howl, jabber, lilt, lisp,
    moan, mumble, murmur, mutter, purr, rage, rasp, roar,
    rumble, scream, screech, shout, shriek, squeal, stammer,

stutter, thunder, trill, trumpet, tsk, twitter, wail, warble,
wheeze, whimper, whine, whisper, whistle, whoop,
yammer, yap, yell, yelp, yodel

It's as if focusing on the manner of producing sound breaks the spell of the metaphor of communication as sending, and forces the mind to construe these acts in physical terms, as merely making noise.

A final world of meaning exposed by the dative construction is the concept of helping or hurting. Many languages have a special marker for a person who benefits from an action, called the benefactive. In English a benefactive is often introduced by the preposition *for*, as in *Gentlemen still open doors for women* or *She bought a house for her fiancé*. Some of these beneficiaries also enjoy the dative alternation: you can *buy a house for your fiancé* or *buy your fiancé a house*, *build a house for your fiancé* or *build your fiancé a house*. But the benefactive relation isn't enough: it's odd to say *Gentlemen open women doors* or *He fixed me my car*. The reason, once again, is that the double-object construction means "cause to have," and in the odd sentences the women don't come to own the door as a result of its being opened and the customer doesn't gain ownership of the car as a result of its being fixed (it was his car in the first place).

A beneficiary can generally fit into a double-object construction only when he or she benefits as a result of receiving something, and even then only with some kinds of verbs. One kind involves doing something that enables someone to have something, as in *Oh Lord, won't you buy me a Mercedes-Benz*. (Other verbs in the family include *earn, find, get, grab, order, steal*, and *win*.) Another has to do with creating something with the intent of giving it to someone, as in *Bake me a cake as fast as you can* (its soul mates include *build, cook, knit, make*, and *sew*).

What's the opposite of a benefactive? A malefactive, of course—the poor sap who is worse off as a result of the action. English sometimes uses the preposition *on* for this purpose, as in *They played a trick on us* and *My horse died on me*. English also has a microclass of malefactives that can enter the double-object construction and that mean "cause or intend someone *not* to have something":

They *fined* her twenty-five cents.
That remark just *cost* you your job.

And *forgive* us our trespasses, as we forgive those who trespass
against us.
You *bet* your life!
They took all the trees and put them in a tree museum, and
charged all the people a dollar and a half just to see 'em.[56]

Other examples include *begrudge, deny, envy, spare,* and *save*. Since these
verbs don't mean "cause to go," most of them, not surprisingly, can't appear as prepositional datives—you can't say *They fined twenty-five cents
from her,* or *on her,* or *of her.*

It isn't invariably true that pure benefactives and malefactives—those
where the change of fortune isn't a side effect of giving or depriving—are
banned from double-object sentences. In a few cases the construction turns
up when someone is helped or harmed but nothing changes hands. One of
them involves idioms with *give* and *do:*

Hymie, give me a hand!
Give me a kiss, just one sweet kiss.
Can you do me a favor?
Someone should give him a good swift kick.

Another involves acts of symbolic dedication:

If you want my hand in marriage, first you'll have to kill me a
dragon.
Cry me a river![57]
God said to Abraham, "Kill me a son."[58]

And yet another is a help-yourself construction common in nonstandard
American English:

Why don't you take yourself a cab and go jump in the lake?
Five more minutes, he'd have chewed himself a hole through
the fence.
Have yourself a merry little Christmas.[59]
I stepped outside to smoke myself a J.[60]
Mercy sakes alive, looks like we got us a convoy.[61]

But my favorite is the Neologizing Imperative Retort, an idiom popular in the English Renaissance that means "Don't think you're doing me a favor by offering or saying such-and-such":[62]

What is this?
"Proud"—and "I thank you"—and "I thank you not"—
And yet "not proud"? Mistress minion you,
Thank me no thankings, nor proud me no prouds.
　　　　　　　—Shakespeare, *Romeo and Juliet*, act III, scene 5

"I heartily wish I could, but—"
"Nay, but me no buts—I have set my heart upon it."
　　　　　　　—Sir Walter Scott, *The Antiquary*

Advance and take thy prize, the diamond; but he answered,
Diamond me no diamonds! For God's love, a little air!
Prize me no prizes, for my prize is death!
　　　　　　　—Tennyson, *Lancelot and Elaine*

It may sound a bit hoity-toity, but the idiom still appears in newspapers and the Internet. My collection includes *UT me no UTs* (the title of an essay protesting the ugly two-letter postal abbreviations for state names, such as UT for Utah), *Comment me no comments, Blog me no blogs,* and even *Jeff Malone me no Jeff Malones,* from a basketball journalist rejecting the suggestion that Malone was all-star material.

Why does the English language—and in fact many languages—use the same construction for giving and benefiting, for denying and harming? Another grammatical metaphor is afoot: TO BE WELL OFF IS TO POSSESS SOMETHING, and TO HELP IS TO GIVE.[63] The blurring of a possessor and a beneficiary in the double-object construction displays the helping-as-giving part of the metaphor. But the more basic part, prospering-as-owning, can be seen in many idioms with the verb *have*. We talk about *having good fortune, having it made, having a good time, having a ball, having it all, having your teeth fixed, having something for dinner, having someone for dinner, having someone* (sexually), and *having someone where you want him.* (Remember Hannibal Lecter, who had an old friend for dinner—with fava beans and a nice Chianti.) The dative construction shows us once again that abstract concepts seem to be represented in the mind (at least in the part of the mind that interfaces with language) in thuddingly concrete ways. Being well off is like having something; knowing something is like having it; having something is like having it near you.

THOUGHTS ABOUT ACTING, INTENDING, AND CAUSING

Cracking open a third construction exposes another major piece of our conceptual infrastructure. Many verbs can appear in both intransitive and transitive forms, though with different participants in the subject slot:

> The egg boiled.
> Bobbie boiled the egg.
>
> The ball bounced.
> Tiny bounced the ball.
>
> The soldiers marched across the field.
> Washington marched the soldiers across the field.

This is called the causative alternation, because in the transitive form the subject causes the object to do what it ordinarily does in the intransitive construction (boil, bounce, march, and so on). And it appears to have the same toxic brew that made the other constructions seem unlearnable:

A tempting pattern. At least two hundred English verbs flip between causative and intransitive forms, including *bend, drop, dry, float, melt,* and *rip.*[64]

Evidence for generalization. Here are some errors that show that children can pick up the causative pattern and apply it to new verbs:[65]

> Go me to the bathroom before you go to bed.
> And the doggie had a head. And somebody fell it off.
> Be a hand up your nose.
> Don't giggle me!
> He's going to die you, David. The tiger will come and eat David
>   and then he will be died and I won't have a little brother
>   anymore.

Children can also be shown to generalize in *wug* tests. Gropen and I taught children that *pilk* meant "do a headstand," and when they saw a toy bear upend a pig, they said the bear *pilked him*.[66] Grown-ups generalize, too: the Macintosh operating system tells its users to *Allow power button to sleep the computer* (cause it to go into "sleep" mode) and to *hover the mouse over the box* (cause the cursor to hover).

Exceptions. Some intransitive verbs resist the intrusion of a causal agent:

> The baby is crying.
> *The thunder is crying the baby.

> The frogs perished.
> *Olga perished the frogs.

> My son came home early.
> *I came my son home early.

And some transitive verbs resist the attempt to strip their causal agent away:

> We've created a monster!
> *A monster has created!

> She thumped the log.
> *The log thumped.

> He wrecked the car.
> *The car wrecked.

Apparent arbitrariness. To take just two examples: You can *march soldiers home* but not *come them home*. And when you boil a lobster, you can say that *the lobster boiled,* but when you make an omelet, you can't say that *the omelet made.*

A frame flip. The gestalt shift behind the causative is less of a mystery than it was for the locative and dative, because here the two constructions obviously are not synonyms. One means that something happens (*The cookie crumbled*); the other means that someone caused something to happen (*She crumbled the cookie*). But the conceptual flip in the causative involves more than splicing a causal actor onto the beginning of a mental movie (that's pretty much all that happens when you express the causal event in a separate verb like *make* or *cause*, as in *She made the cookie crumble* or *She caused the cookie to crumble*). For the causative construction to apply, the causation has to be effected with your bare hands, so to speak, or as directly as one billiard ball clacking into another. It's fine to say *She made the cookie crumble by leaving it outside in the cold*, but it's not so fine to say *She crumbled the cookie by leaving it outside in the cold*. Likewise, you might say that *Darren caused the window to break by startling the carpenter, who was installing it*, but in that scenario it would be perverse to say that *Darren broke the window.*[67] And while one can say that *Fred caused the glass to melt on Sunday by heating it on Saturday*, it seems strange to say that *Fred melted the glass on Sunday by heating it on Saturday.*[68]

The causative construction also prefers the effect to be an outcome that the actor *intended*. Our cookie crumbler can get away with her act being described as *crumbling the cookies* if she wanted the cookies crumbled, was too arthritic to break them herself, and knew that a few minutes in the cold would do the trick. If the effect is not the ultimate goal of the action, the causative can't apply. Though *to butter* means "to cause butter to be on," when the King puts some butter on the royal slice of bread by means of first putting butter on the royal butter knife, we don't say *The King buttered his knife*. That is because the butter is put on the knife as a means to an end, not as the end itself.

On its own, the directness effect would cripple our ability to talk about cause and effect. Direct causation is something of an illusion; under a sufficiently powerful microscope, it vanishes from sight. When I cut an apple, I first decide to do it, then send neural impulses to my arm and hand, which in turn causes the muscles to contract, causing the hand to move, causing the knife to move, causing the knife to contact the surface of the apple, causing the surface to rupture, and so on. Nonetheless, there is a clear sense in which this entire chain, circuitous though it is, is more direct than paying a servant to cut the apple. When we describe an event, we have to choose a grain size below which the subevents are treated as invisible. For a physical event initiated by a person, muscle contractions and every physical event preceding the outcome fall below the grain size, so you can *break a window* with your fist or by hitting a long fly ball. But when the causal chain contains another human actor, such as a butterfingered window-installer, that link is above the grain size, and the main action is no longer seen as causing the outcome in the direct manner required by causative verbs. That's why you aren't said to *break the window* by shouting "Boo!" as it's being installed, even though you're causing it to break.

A series of experiments by the psychologist Phillip Wolff has confirmed that when people use causative verbs, they single out events that are caused directly, intentionally, and without an intervening actor. For example, they judge that a woman *dimmed the lights* only when she slid a dimmer switch, not when she turned on her toaster, that a man *waved the flag* only when he shook a flagpole, not when he raised the flag on a windy day, and that a boy *popped a balloon* only when he pricked it, not when he let it graze against a hot light bulb on the ceiling.[69]

The grain size of the mind's view of the world is adjustable. From a bird's-eye view, we can say that *Henry Ford made cars* or *Bush invaded Iraq*, though the causal chain between anything that Ford did and a Model T rolling off the assembly line had many intervening links. This feature of conceptual semantics inspired Bertolt Brecht's "Questions from a Worker Who Reads":

> Who built Thebes of the seven gates?
> In the books you will find the names of kings.
> Did the kings haul up the lumps of rock? . . .
> The young Alexander conquered India.
> Was he alone?
> Caesar beat the Gauls.
> Did he not have even a cook with him?[70]

Still, the directness effect applies, but at the new grain size. When we agglomerate events through the wide-angle lens of history, which sees only the acts of influential leaders, a causative verb will cut the chain at the link immediately connected to the outcome. Thus we don't say that *Neoconservative intellectuals invaded Iraq*, even if they influenced Bush, nor that Osama bin Laden did, even though Bush would not have been emboldened to order the invasion if it were not for 9/11. Nor do we say that *Florida voters who failed to understand a butterfly ballot invaded Iraq*.

A speaker expects his listeners to share the grain size he has in mind, and when they don't, the result can be a failure to communicate. Piping plovers are adorable little shorebirds that look like they're wearing bow ties and run up and down Cape Cod beaches like windup dolls. They are classified as a threatened species (though they seem to be everywhere), and local authorities take measures to protect their nesting grounds. Still, I did a double take when I saw the following headline in the *Provincetown Banner:* PLOVERS CLOSE PARKING LOT. An image flashed through my mind of little birds dragging a chain across the entrance and waving away traffic. I thought it was the silliest thing I had ever seen until I turned the page and read DOG FECES CLOSES BEACHES.

The causative construction subscribes to a theory of free will. Most verbs of human action cannot take part in the causative construction, even when the actions are, in some sense, compelled by prior events. You can't say *Bill laughed Debbie* (with his Travolta impression), *Judy cried Lesley* (by leaving the party with Johnny and coming back wearing his ring), or *Don Corleone signed the bandleader the contract* (by making him an offer he couldn't refuse). This is true whether the actions are considered voluntary (like signing a contract) or involuntary (like laughing and crying). Human acts are conceptualized as having some hidden cause inside the actor, and hence are not directly causable by an outsider.[71]

The metaphor of human action as coming from an internal essence or impulse is echoed in verbs for physical happenings. Two kinds of verbs readily participate in the causative alternation. One microclass contains verbs of manner of motion or posture, the rock-and-roll verbs:

> bounce, dangle, drift, drop, float, fly, glide, hang, lean, move, perch, rest, revolve, rock, roll, rotate, sit, skid, slide, spin, stand, swing, turn, twist, whirl, wind

The other contains verbs of change of state, like bending and breaking, growing and shrinking, or hardening and softening:

> age, bend, blur, break, burn, char, chill, chip, collapse, condense, contract, corrode, crack, crash, crease, crinkle, crumble, crush, decrease, deflate, defrost, degrade, diminish, dissolve, distend, divide, double, drain, enlarge,

> expand, explode, fade, fill, flood, fold, fracture, fray,
> freeze, fuse, grow, halt, heal, heat, ignite, improve,
> increase, inflate, light, melt, multiply, pop, reproduce, rip,
> rumple, rupture, scorch, shatter, shrink, shrivel, singe,
> sink, smash, snap, soak, splay, splinter, split, sprout, steep,
> stretch, tear, thaw, tilt, topple, warp, wrinkle

But most verbs in the microclass in which an object emits something—a light, a sound, or a substance—resist the causative. You can't *glow a light, whine a saw, bubble a sauce,* or use the other verbs of emission in a causative sentence:

> blaze, flame, flare, glare, gleam, glisten, glitter, glow, shimmer,
>   shine, sparkle, twinkle
> blare, boom, buzz, chatter, chime, creak, fizz, gurgle, hiss,
>   howl, hum, peal, purr, splutter, squawk, swoosh, thrum,
>   vroom, whine, whump, zing
> drip, emanate, erupt, foam, gush, leak, ooze, puff, radiate,
>   shed, spout, sweat

It's as if such outbursts, like human actions, come from within, and thus don't admit some other direct cause at the same grain size.

Also resisting the causative are verbs of going out of existence: you can't say *To die a mockingbird, Decease Bill,* or *Mr. Gorbachev, fall down this wall!*

> decease, depart, die, disappear, disintegrate, expire, fall apart,
>   lapse, pass away, pass on, perish, succumb, vanish

It's not that the *concept* of directly causing something to go out of existence is ineffable. English has a gruesome and well-stocked lexicon of killing and destroying:

> assassinate, butcher, crucify, dispatch, electrocute, eliminate,
>   execute, garrote, hang, immolate, kill, liquidate,
>   massacre, murder, poison, shoot, slaughter, slay

> abolish, annihilate, ban, blitz, crush, decimate, demolish,
>   destroy, devastate, exterminate, extirpate, finish,
>   obliterate, ravage, raze, rescind, ruin, tear down,
>   terminate, waste, wipe out, wreck

Moreover, these verbs of mayhem are as intransigent about losing their causal agents as the going-out-of-existence verbs are about gaining one. You can't say *Bill killed,* meaning he died, or *The building razed,* meaning it collapsed or burned. English allows one to talk both about going out of existence and about causing to go out of existence, but not with the same verb. It's as if the language took the existential stance—perhaps the moral stance—that when something ceases to be on account of old age, peaceful causes, spontaneous combustion, internal rot, or carrying the seeds of its own destruction, that demise is qualitatively different from the outcome of malice aforethought. This isn't a peculiarity of English; many other languages use distinct verbs for dying and killing, even when they allow their other verbs to do double duty for happening and causing to happen.[72]

When I allude to moral sentiments to explain the syntax of causative verbs, I am not trying to enliven a grammar lesson with an eye-catching figure of speech. Morality and causative verbs tap the same mental model of human action. Moral judgments apply most clearly to people who act with the intention of bringing about a foreseen effect. That is also the job description of the subject phrase in a causative construction. The skillful use of transitive and intransitive constructions can thus be used to frame a moral argument.

Though causative constructions ordinarily finger a guilty party, they can jettison their subject when expressed in the passive voice. That makes the passive a convenient way to hide the agent of a transitive verb and thus the identity of a responsible party, as in Ronald Reagan's famous non-confession "Mistakes were made," now a cliché for evasion by a public figure. But the intransitive alternative to a causative goes one step further. It doesn't just *hide* the cause; it refuses to admit that there *was* one. While *The ship was sunk* (passive) entails that there was a perpetrator, possibly unknown, *The ship sank* (intransitive) is consistent with its just happening, perhaps through a lack of preventive maintenance, a stroke of bad luck, or a proverbial "act of God" (though without the god). Media watchdogs sometimes count up headlines with causative verbs in the active voice, the passive voice, and the intransitive form, looking for evidence that a news agency might be attempting to exculpate or incriminate one side or the other in a conflict. For instance, a pro-Israel group noted the prevalence of Reuters headlines like BUS BLOWS UP IN CENTRAL JERUSALEM, which uses an intransitive to downplay the responsible agent. As the linguist Geoffrey Pullum notes in a blog on linguistics and public affairs, " 'Bus Blows Up' is indeed a strange way to describe an incident in which a human being straps explosives to himself, gets on a crowded bus in a city street, and kills 13 people by detonating his payload, clearly intending to murder as many Jews as possible at one go. . . . Reuters describes the event as if the bus had just exploded all on its own."[73]

Incidentally, not all verbs that gain and lose a causal subject are products of the causative rule. The ways in which someone can cause something to happen are so numerous that causative verbs are coined profligately in many spheres of life without necessarily being the outcome of a causative rule. As a result, some causative verbs seem to flout the principles I've been illustrating. But these rebels tend to be specialized: you can *walk someone* if he is a batter, *bleed him* if he is a patient, and *burp him* if he is a baby. And others are not truly causative. To *shine a light* is to aim it, not activate it, and to *drive, sail, walk, waltz,* or otherwise lead a person to locomote somewhere is to accompany him, not coerce him.

Before concluding the chapter with some observations on language as a window into our cognitive nature, let me sum up the linguistic story. Alternations of grammatical constructions reflect cognitive gestalt shifts: cause-to-go and cause-to-change; cause-to-go and cause-to-have; happen and cause-to-happen. These shifts give the two constructions subtly different meanings, reflecting different ways of construing a situation. The pickiness of verbs that seemed to make them impossible to learn may be explained at two levels. At a macroscopic level, the kinds of verbs that refuse to enter into a construction are those whose meanings are flatly incompatible with it (*throwing a cat into the room* is not a way to change a room's state; *driving a bus to the lake* doesn't result in the lake possessing anything; *laughing a person* is inappropriate to an agent with free will). But since people can twist themselves into a cognitive pretzel and construe almost any event in almost any way, to predict usage down to the last verb one has to look at microclasses with similar meanings, which march into a construction or stop at its gates in lockstep. And those microclasses expose a stratum of human cognitive obsessions and metaphors based on them: holistic blobs, which can move or change; force, which can be applied instantaneously or over time; possessions, which include ideas and good fortune; and events, which can just happen or can be caused by an agent in a hands-on, premeditated act.

SMART SPEAKERS OR SMART LANGUAGE?

I have been writing as if the English language were a thoughtful person who always had a good reason for anything she did. We saw a clever rationale for the way different construals of a situation are expressed in different constructions (with the affected entity always as the direct object, whether it is caused to go, caused to change, or caused to have). We saw subtle shadings of meaning across constructions that from a distance looked like synonyms. And we saw implicit theories of physics and psychology in the ways that verb classes were finicky about the constructions they entered into. But languages don't think; people do. Do flesh-and-blood speakers really go through these rationalizations when they learn and use verbs?

The answer has to be both yes and no. On the "yes" side, we know that languages are not designed by a committee but evolve spontaneously in a community. Any predictable connection between form and meaning that isn't a matter of chance—such as the way that the syntax of so many verbs can be predicted from their meanings—must be the brainchild of *some* speakers, at some point in the history of the language. And we know from *wug* tests and other experiments that modern speakers, both children and adults, are sensitive to the major laws connecting the meaning of a construction to its form.[74]

On the "no" side, with most verbs individual speakers don't *need* to work out the rationales if all they want to do is use the verbs in the same way as everyone else. All they need to do is learn the semantics of each microclass (what the verbs have in common) and its syntax (which constructions the verbs are happy in). That's enough to allow them to predict which new verbs can be extended to which constructions, without their having to know why.

The reason that the answer can be both yes and no is that different people may be sensitive to the rationale for different verbs and classes at

different times. In this regard a language may be like other cultural products, which may be used at different times by innovators, early adopters,
early majorities, late majorities, and laggards.[75] The semantic rationale for
a verb frame is most likely to have entered the minds of the innovators and
early adopters: the speakers in a language's history who first extended a
construction to a new kind of verb. This new usage may fall deadborn from
the innovator's lips or be welcomed into a segment of the community with
open arms. The reception is partly capricious (as we shall see in chapter 6),
but when a new combination does catch on, it could involve the later adopters' grasping the rationale with a stroke of insight recapitulating that of the
original coiner, their dumbly memorizing the verb in that construction, or
something in between. The only thing that matters is that when people hear
a verb in a construction, they naturally generalize it to verbs with a similar
meaning.

Can we catch innovators in the act of stretching the language? It happens all the time. Though linguists often theorize about a language as if it
were the fixed protocol of a homogeneous community of idealized speakers, like the physicists' frictionless plane and ideal gas, they also know that
a real language is constantly being pushed and pulled at the margins by different speakers in different ways.

There are times when we all have to push the outside of the grammatical
envelope, because a sentence has to do several things at once and they may
be at cross-purposes. So far I have been showing you how the positions in a
sentence are assigned to particular meanings, such as the first object expressing the person that is caused to have something and the second object expressing the thing that he has. But at the same time, the left-to-right ordering
of words in a sentence must accomplish something else: keep the listener at-
tuned to the information that is "given," merely setting the stage for the
message, and the information that is "new," requiring an update of the listener's understanding of the world. And on top of these two, there is a third
demand on word order: the speaker has to show mercy on the listener's
memory and put lengthier phrases toward the end of the sentence, where the
listener can think them over in peace and quiet. ("Put the new material last"
and "Put the heavy material last" are two of the most important guidelines
for good style in writing and speaking.) In juggling these demands, we sometimes must sacrifice the usual likes and dislikes of the verb.

For example, we've seen that *give a headache to Jim* sounds worse than *give Jim a headache*. But when the monosyllabic "Jim" is replaced by a long, complicated, and unexpected entity, a writer may press the prepositional dative into service to shunt the entity to the end of the sentence. The linguist Joan Bresnan and her collaborators, trawling the Web for liberal usages of the dative, found real-life examples like these:[76]

> The spells that protected her identity also gave a headache to
> anyone trying to determine even her size . . .
> From the heads, offal and the accumulation of fishy, slimy
> matter, a stench or smell is diffused over the ship that
> would give a headache to the most athletic constitution.

The alternatives—*give anyone trying to determine her size a headache; give the most athletic constitution a headache*—would have squeezed a big wad of material into the middle of the sentence and left a little shred of material dangling at the end. Presumably the writers avoided them for that reason.

Our ability to stretch a construction to new verbs in the heat of a conversation or text doesn't mean that the results are accepted as perfectly normal sentences. Speakers differ in how easily they stomach the various generalizations that other speakers make, depending perhaps on their age, birthplace, subculture, or even personality. I can swallow the foregoing sentences with *give a headache to* (barely). But I draw the line at a similar stretching—*kiss it goodbye*—that must have sounded fine to the columnist David Brooks and his copy editors at the *New York Times*, seeing as how he used it no fewer than three times in the course of a single essay on the 2006 Israel-Hezbollah crisis:[77]

> You can kiss goodbye, at least for the time being, to some of the
> features of the recent crises. You can kiss goodbye to the
> fascinating chess match known as the Middle East peace
> process. . . . You can also kiss goodbye to the land-for-peace
> mentality.

I've read these sentences over and over, but they still stick in my craw. So do many other edgy constructions I have heard and jotted down. Apple's *sleep the computer* and *hover the mouse* still sound odd to my ears, though I bet they are unexceptionable to younger Macintosh users who have grown up with them. And here are some other examples of oddball usages from my collection:[78]

Content-locatives:
Women do not invest sexual messages in clothing choice.
She said we just dug up some trash someone littered.

Container-locatives:
He squeezed them [fish fillets] with lemon juice.
We installed twenty-one banks with ISDN lines.

Double-object datives:
Reach me my socks.
When you go I'm going to preach you a great funeral.

Prepositional dative:
[The report] was given a normal and wide distribution, but
we did not brief it to the President.[79]

Causatives:
The year Sidney Poitier won best actor he rose us all up in
the world.
Lectric Shave: Stands up whiskers for a 50% closer shave.

Intransitives:
The bacteria live off the dissolved minerals that exude from
the vent.
Can germs harbor in these things?

All these one-shots stay within the large semantic envelope of the construction types (cause-to-change, cause-to-have, cause-to-happen), more evidence that the semantic rationales are psychologically real. But they push the outsides of the envelopes of the microclasses. And in the future, as some of them are repeatedly spoken, with or without the extenuating circumstances that first prompted their use, they could redraw the envelopes or seed new ones in the minds of receptive speakers—those who are younger, more immersed in the relevant specialty (cooking, politics, computers, business), or less linguistically fussy than people like me. That is how a language changes.

A LANGUAGE OF THOUGHT?

When speakers avoid using a verb in a construction or wince when they hear others use it, they must be sensitive to subtle semantic distinctions, such as between pinpointing a kind of motion and pinpointing a kind of change, or between applying force instantaneously and applying it over time. No one ever teaches these distinctions or includes them in dictionary definitions; it took a long time even for linguists to discover them. And no one has a motive to acquire regulations that prevent them from saying perfectly intelligible things. So where does the sensitivity come from?

All we really need to do to explain this acuity is to suppose that people *represent* the verbs in memory in such a way that verbs with similar syntactic tastes have overlapping definitions. That way, whenever a new verb is learned, it automatically activates its classmates. And for that to happen, verb meanings have to be couched in a language of thought that graphically displays the aspects of meaning that the verbs in a microclass share, while hiding the aspects that distinguish them.

Here is an example. Once children have learned how *pour* is used, they generalize its syntactic tastes to *drip* and *slosh* but not to *spray* and *squirt*. That would happen automatically if their mental definitions boldly displayed the concepts LET and CAUSE, which render *pour, drip,* and *slosh* as similar to one another but different from *spray* and *squirt,* while tucking away the mental movies of liquid in motion that distinguish pouring from dripping or spraying from squirting. This would make *pour* and *drip* "look alike" to the mind's eye (despite the fact that pouring and dripping look different to the real eye) and *pour* and *spray* "look different." And that, in turn, would allow the child to transfer what he has learned about *pour* to *drip* but not to *spray*.

If there is indeed a language of thought, it will have to be quite abstract to make the verbs in a microclass look alike and those in different microclasses look different.[80] It can't just reflect the sights and sounds of the events denoted by the verbs. For instance, as far as sensory experience is concerned, the verbs *hand, carry,* and *bring* seem alike (they could all describe the same event in a movie), whereas the verbs *throw, kick,* and *roll* seem different. Yet the entrance requirements for the dative see it the other way around. The last three are all verbs of instantaneous causation of motion and hence are allowable in the double-object form, whereas among the first three, *hand* is a verb of giving (double object OK), *carry* is a verb of continuous causation of motion (double object not OK), and *bring* is a verb of causation of motion in a direction (double object OK). Similarly, as far as the machinery of language is concerned, *telling* is different from *saying, shouting, talking,* or *speaking* (which belong to different microclasses), but the same as *quoting, leaking, asking, posing,* and *writing.* Likewise, *shouting* is no more similar to *yelling* or *screaming* than it is to *whispering* and *murmuring* (since they are all manner-of-speaking verbs). *Baking a cake* has to be seen as similar to *building a house* and *writing a letter of recommendation* (verbs of creation) but as dissimilar from *warming a cake, burning a cake,* or *reheating a cake* (verbs of state-change). *Betting* has to be thought of as being like *envying, sparing,* and *begrudging* (verbs of future not-having) but unlike *selling, paying,* or *trading* (verbs of giving). In all these cases, the look-and-feel of an event and the things and actions it involves (carrying, talking, cakes, money) must be submerged, and its abstract structure (change, causation, directness, instantaneity) must be highlighted.

What exactly would go into this abstract language of thought? You may have noticed that a few distinctions of space, time, force, substance, and intention kept popping up in the definitions of the microclasses. That is a clue that they make up the structural framework of our conceptual edifices. But it isn't enough to show that these distinctions are important in English. If they truly make up a language of thought—the conceptual infrastructure of *Homo sapiens*—we should expect to see them in languages across the world.

Now, the exact phenomena we have been exploring—the constructions, alternations, and microclasses—are certainly not universal. They are not even constant across dialects of English, across versions of English spoken in different historical periods, or across individual speakers of standard English today. But we shouldn't expect them to be. Even if children are equipped with some kind of universal talent for language, they can't very well burst into language the way Julie Andrews bursts into song at the start of *The Sound of Music.* Children have to listen carefully for the words and constructions that have become entrenched in their community, so they can deploy their abilities in a way that will let them be understood by their compatriots. And the available pool of words and constructions depends on the vicissitudes of local history: the invaders, trading partners, immigrants, snobs, hipsters, and imported brides that shaped the language in the preceding centuries and millennia, and the fads in slurring or exaggerating sounds that swept through the community. If the mind leaves universal fingerprints on language, they will have to be subtler than a fixed list of rules and constructions found in all the world's languages.

Fortunately, the fingerprints are all over the place. Though the constructions we have been examining are not universal, they repeatedly turn up in unrelated languages and language families all over the world, suggesting that people's language-forming abilities, faced with the need to communicate certain kinds of ideas, are channeled to rediscover these constructions. Locative alternations similar to the ones in English, for example, have been documented in German, Spanish, Russian, Greek, Hungarian, Indonesian, Arabic, Berber, Igbo (spoken in Nigeria), Chinese, Japanese, Korean, and Chichewa and Shona (Bantu languages).[81] Dative or dative-like constructions have been documented in non-Indo-European languages from every continent.[82] Languages with a documented causative alternation number in the hundreds, and many surveys have ferreted out their common properties.[83]

Not only do many languages have the constructions we see in English, but when they do, the constructions tend to play out the same conceptual scripts. In most languages, when a verb flips from a content-locative to a container-locative, the holism effect kicks in, just as it does in English.[84] Some languages say it in so many words: when Igbo speakers express a container in a direct object, they add a word meaning "full" to the verb, as if we were to say *pack-full the suitcase with clothes*. Double-object constructions, too, don't express just any change or movement but instead reserve the first object for possessors, recipients, beneficiaries, or maleficiaries, just as in English. And causative constructions in which a single verb alternates between happening and causing-to-happen prefer direct, hands-on, intended causation over more circuitous or impersonal causal chains.[85]

Though the classes and microclasses vary from language to language, the variation is not haphazard. It tends to consist of differences in exactly where a language makes a cut along a continuum that ranges from concepts that clearly mesh with the meaning of the construction to concepts that clearly don't mesh with it. For the dative, one continuum is defined by the ease of conceiving an act as a kind of giving. At one end we find the prototype, verbs like *give* itself, and every language with a double-object construction allows those verbs into it. Some languages stop there, but most also allow verbs of sending. Some go further and allow verbs of instantaneous motion like throwing (this is where English draws the line), and a few more allow verbs of continuous motion like hoisting and pulling.[86] And when we reach the far end of the continuum, where we find pure motion toward an inanimate goal like driving a bus to a lake, few or no languages are so liberal as to allow a double-object form to be used.

Causatives in the world's languages reveal even more about universals of thought. In demarcating the microclasses, languages draw several lines. The most important one runs from events that clearly have a cause tucked inside the changing entity (in which case the pool-cue impetus implied by the causative is inappropriate) to events that clearly need an external prod to cause and shape the change. Thus few languages allow verbs for human actions to be converted into a causative homonym (as in *Bill laughed Debbie,* meaning that Bill made Debbie laugh), presumably because the immediate cause is attributed to something inside the person.[87] Just short of that extreme permissiveness we find languages that allow the causative with verbs of change or motion. Then there are languages like English that draw the line at a few circumscribed *kinds* of motion and change, excluding other kinds like going out of existence or emitting substances. And there are very conservative languages that restrict the causative rule to only the most passive changes of physical state, like breaking, opening, and melting.[88]

We have been looking for universal elements of thought through a narrow linguistic peephole, verb constructions. That means that the variability we have been seeing exaggerates the variability in the scaffolding of thought itself. When it comes to basic concepts, the world's languages are like a game of Whack-a-Mole: if a language whacks a concept out of one of its grammatical devices, the concept tends to pop up in another. Causation is a perfect example. We have been focusing on constructions in which causation is packed concisely into a verb, as in *break the glass* and *slide the puck.* But causation can also be expressed in its own prefix or suffix, as in the English *en-* (*enlarge, enrich, ensure*), *-ify* (*beautify, electrify, falsify*), and *-ize* (*centralize, publicize, revolutionize*). In English these prefixes and suffixes can attach themselves only to adjectives and nouns, but in other languages, such as Hebrew and Turkish, they may be tacked onto verbs in great numbers. In a third mole hole, causation gets a verb of its own, which sisters up with another verb (the one for the event being caused) to form a two-headed verb; the English equivalent would be *Karen made-break the window.* And sometimes causation gets a verb of its own that stands alone in the clause while the effect is demoted to a subordinate clause, as in *Karen made the window break.* Whenever a language contains more than one of these devices, it reserves the more concise one for more direct causation and the more prolix one for less direct causation (as in the English contrast between *dimming the lights* when sliding a switch and *making the lights dim* when turning on the toaster).[89] It's as if the morphemes were laid out like a little diagram depicting the links in the causal chain, and chains with fewer links were conveyed by fewer morphemes.

Causation is just one of several meaning-moles that keep popping up across the world's languages in one grammatical slot or another. The slots in the game include classes and microclasses; prefixes, suffixes, and other grammatical words (such as prepositions, conjunctions, and auxiliaries); and "light verbs" such as *make, do, be, have, take,* and *go* (which in some languages are the only verbs that exist). The concepts that pop up in these slots fall into a fairly short list, more or less along these lines:[90]

- A cast of basic concepts: event, state, thing, path, place, property, manner
- A set of relationships that enmesh these concepts with one another: acting, going, being, having
- A taxonomy of entities: human vs. nonhuman, animate vs. inanimate, object vs. stuff, individual vs. collection, flexible vs. rigid, one-dimensional vs. two-dimensional vs. three-dimensional
- A system of spatial concepts to define places and paths, like the meanings of *on, at, in, to,* and *under*
- A time line that orders events, and that distinguishes instantaneous points, bounded intervals, and indefinite regions
- A family of causal relationships: causing, letting, enabling, preventing, impeding, encouraging
- The concept of a goal, and the distinction between means and ends

These can be said to be the major words in a language of thought. In a later chapter we will see how they shape our understanding of the physical and social worlds.

Of course the full inventory of human thoughts is much, much larger than this. The verb *to butter* has to contain a representation of a butterlike substance, and if someone were to say that Bush has *out-Nixoned Nixon,* he must have in mind some noteworthy trait of the thirty-seventh president. But these and countless other sensory, cognitive, and emotional distinctions are invisible to the part of the mind that sees some verbs as alike and others as different when deciding how to use them in a grammatical construction. Aside from specific people and substances, they include the mood, attitude, and state of mind of the speaker; the rate of a moving object; the symmetry, color, and gender of the participants; and the physical properties of the setting (the temperature, whether it occurs indoors or outdoors, whether it takes place on land, in the air, or at sea).[91] So the concepts behind language are organized in a particular way. The basic conceptual distinctions assemble themselves into a scaffolding of meaning, which has hooks here and there on which to hang images, sounds, emotions, mental movies, and the other contents of consciousness.

Does the brain really distinguish a skeleton of basic concepts relevant to grammar from a larger portfolio of meanings that flesh it out? The neuropsychologist David Kemmerer suggests that it does, based on his studies of different patterns of language loss following damage to the brain. In one study, Kemmerer studied a patient who lost the ability to distinguish *drip* from *pour* from *spill,* three members of a locative microclass that differ in the details of the motion but that share a conceptual skeleton (enabling downward motion of a liquid or aggregate).[92] But the patient was still sensitive to the abstract semantic notions that governed the verbs' behavior in constructions: she knew that *Sam spilled beer on his pants* was grammatical and that *Sam spilled his pants with beer* was not. This wasn't just because the grammatical test was easier. Two other patients, with damage to different parts of their brains, showed the opposite pattern: they could tell the difference between pouring, dripping, and spilling, but couldn't hear anything wrong in sentences with a clash of core concepts like *Sam spilled his pants with beer.* Other studies have shown related dissociations. Some patients, for example, lose an ability to distinguish hot from cold, red from green, or tapping from slapping (distinctions that don't matter to grammar) while retaining the ability to distinguish stuff from shape or contact from cause-and-effect (distinctions that do matter), or vice versa.[93]

Our trip down the rabbit hole has taken us to a semantic wonderland. We encountered a lush profusion of verbs—a dozen verbs of emitting substances, twenty verbs of changing the aesthetics of a surface, and no fewer than sixty-nine verbs of manner of speaking. We have witnessed events that flip in their interpretation the way a pair of faces turns into a vase. We met gingerbread that crawls, basketball players turned into verbs, and slain dragons symbolically possessed as tokens of a lover's dedication. Yet for all the spraying and smearing, crinkling and crumpling, screeching and squealing, and other attributes that distinguish one verb from another, the most memorable inhabitants are the silent and invisible ones we kept coming across as we looked under the verbs: the ethereal notions of space, time, causation, possession, and goals that appear to make up a language of thought.

OUR COGNITIVE QUIRKS

I promised that a good look at verbs would lay bare our ability to flip from one conceptual frame to another, our habit of using some ideas as metaphors for others, and the inventory of fundamental ideas that structure the meanings of sentences and perhaps thought itself. What are we to make of this inventory? We think about them—we think *with* them—all the time, and one might wonder whether they are the inevitable categories with which any intelligent entity, whether human, silicon, or alien, is forced to deal with reality. Let me conclude the trip down the rabbit hole with some hints that this is not so. The basic ideas that govern our thoughts in everyday life can show themselves to be as eccentric as the Mock Turtle or the Queen of Hearts.

The constituents of common sense we have encountered, like causation, force, time, and substance, are not just home editions of the concepts used in logic, science, or our best collective understanding of how to manage our affairs. They worked well enough in the world in which our minds evolved, but they can leave our common sense ill-equipped to deal with some of the conceptual challenges of the modern world. I have in mind not the esoteric paradoxes of quantum mechanics or relativity, but more pedestrian enigmas in which our intuitions seem to be out of kilter with the reality in which we live our lives. Here are some ways in which the core concepts exposed in this chapter can lead us astray in dealing with life's challenges.

Having and benefiting. Let's start with a trite example. Remember the grammatical metaphor for well-being: to be in a good state is to possess something. Overall, few people would disagree with Sophie Tucker when she said, "I've been rich and I've been poor. Rich is better." Yet when it comes to finer gradations of having, sages throughout history have tried to keep us from backsliding into this mindset, reminding us that money can't buy happiness, that only a cynic knows the price of everything and the value of nothing, and that it isn't true that he who dies with the most toys wins. Modern happyologists have confirmed that once people reach a certain baseline of affluence, additional wealth and possessions lead to little if any additional contentment.[94]

Having and knowing. Another misleading conceptual formula is the conduit metaphor, in which to know is to have something and to communicate is to send it in a package. Again, it has a kernel of truth: if information were never transmitted with some fidelity from mind to mind, knowledge could never accumulate in a society, and language itself would be useless. But cognitive science has repeatedly shown ways in which the metaphor falls short. In chapter 1 we saw that language understanding is more than just extracting literal meaning, as George Costanza learned too late when he realized that *coffee* doesn't necessarily mean coffee. And once a meaning is extracted and stored in memory, it does not sit there like a knickknack on a shelf; memory research confirms Twain's observation that people tend to remember things whether they happened or not.[95] Traditional education was dominated by a version of the conduit metaphor sometimes called the savings-and-loan model: the teacher dispenses nuggets of information to the pupils, who try to retain them in their minds long enough to give them back on an exam. Though progressive philosophies of education, which aim to lead children to rediscover knowledge rather than be passive repositories of facts, have had their excesses, it's undeniable that people retain more when they are called on to think about what they are learning than when they are asked to pluck fact after fact out of lectures and file them away in memory.[96]

Having and moving. Languages often treat a possession as a thing at a place, and giving or selling as moving it to a new place, whereupon it is no longer at the original place. For tangible chattels like chickens and cakes this can literally be true, and it is serviceable enough in its metaphorical extension to more abstract goods like money and real estate. But intellectual property is a real conundrum for the metaphor. Though you cannot eat your cake and have it, too, this is not true for information, which can be replicated ad infinitum without loss. Thanks to information technology like file sharing and downloading, a person can gain possession of a song or an image or a piece of software without leaving the original owner bereft of it. The clash of intuitions, between safeguarding an object that can be in only one place at a time and "information wanting to be free," has incited one of the fiercest legal battles raging today: how to extend laws that were originally designed for the ownership of physical goods to the ownership of copyable ideas like words, songs, images, designs, formulas, and even genes.[97]

Time. The model of time that underlies language is not the inexorable ticking of a clock, measuring out the stream of life in constant units. Instead, it coarsely packages stretches of time into instantaneous events (like throwing), protracted processes (like pulling and pushing), and culminations of a process (like breaking a glass). And the part of the mind that interfaces with language keeps track of these gobs of time only with the signposts of before-and-after and at-the-same-time. Missing from this intuitive timekeeper is the idea of time as a continuous, measurable commodity that is coextensive with our existence. One can't help wondering whether this clash of conceptualizations lies behind the frustration felt by the hurried citizens of a postindustrial society as they run up against the more lackadaisical (and possibly more intuitive) notion of time found in the third world, the American South, and the Massachusetts Registry of Motor Vehicles.

Things and locations. When the mind locates one object with respect to another, it is apt to compress the first one into a pinpoint or blob whose shape and parts are no longer discernible, like a thing in a box. We have seen this holistic mentality carried over to *abstract* spaces for qualities or states, like a filled wagon or a garden swarming with bees. I suspect this is one of the reasons people have so much trouble understanding statistical comparisons. One example has been very much in the news. Many researchers have documented that the distributions of talents and temperaments for men and women are not identical. In tests of mental rotation of 3-D objects, for instance, the average score for men is higher; in tests of verbal fluency, the average score for women is higher.[98] Averages are only averages, of course; some women are better spatial thinkers than most men, and some men are more verbally fluent than most women. Yet when people hear about this research, they tend mangle it into the claim that every last man is better than every last woman (or vice versa). People who celebrate the difference write books like *Men Are from Mars, Women Are from Venus* (a clear example of the thing-at-a-place metaphor); people who deplore it accuse the researchers of saying that "a whole group of people is innately wired to fail."[99] It's as if people heard the statistic that women outlive men on average and concluded that every woman outlives every man. The image of one orb floating above another seems to come more naturally to the mind than an image of two overlapping bell curves.

Causality. The prototypical image of cause and effect made plain by language has a person voluntarily acting on an entity and directly bringing about an intended change of position or state. This is not far from the concept of criminal responsibility built into our legal system—the *actus reus* and *mens rea,* or bad act and guilty mind, necessary to establish first-degree murder and other serious crimes. Unfortunately, real life often throws up causal scenarios that don't easily fit this billiard game (many of them entertainingly analyzed in Leo Katz's *Bad Acts and Guilty Minds: Conundrums of the Criminal Law*). A woman wanting to poison her husband puts arsenic in his apple, but he throws it away. A homeless person fishes it out of the garbage can, eats it, and dies. Has she murdered him? What about a homeowner who slams the door in the face of a child fleeing a wild dog, leaving the child to be torn into pieces? Or a man who arrives at the home of a depressed woman with a rope and a box, ties one end of the rope to a rafter and the other into a noose, and convinces her to put her head in the noose and kick the box away?

Conundrums of causality are not just law-school exercises. On July 1, 1881, President James Garfield was waiting to board a train when Charles J. Guiteau took aim at him with a gun and shot him twice.[100] Both bullets missed Garfield's major organs and arteries, but one lodged in the flesh of his back. The wound was minor by today's standards and needn't have been fatal even in Garfield's day. But his doctors subjected him to the harebrained medical practices of the time, like probing his wound with their unwashed hands (decades after antisepsis had been discovered) and feeding him through his rectum instead of his mouth. Garfield lost a hundred pounds as he lingered on his deathbed, succumbing to the effects of starvation and infection eighty days after the shooting. At his trial, Guiteau repeatedly said, "The doctors killed him; I just shot him." The jury was unpersuaded, and in 1882 Guiteau was hanged—another man whose fate hinged on the semantics of a verb.

3

FIFTY THOUSAND INNATE CONCEPTS (AND OTHER RADICAL THEORIES OF LANGUAGE AND THOUGHT)

Anyone who engages in intellectual debate comes to recognize the tactics, ploys, and dirty tricks that debaters use to bamboozle an audience when the facts and logic aren't going their way. There's the appeal to authority ("Spaulding says so, and he has a Nobel Prize"), the ascription of motives ("Firefly is just seeking attention and grant money"), the calling of names ("Driftwood's theory is racist"), and the tainting by association ("Hackenbush is funded by a foundation that once funded Nazis"). Perhaps the best known is the setting up and knocking down of a straw man, a stratagem so versatile that one sometimes wonders whether intellectual life would go on without it.

The beauty of the straw man is that he can be used in so many ways. The most hackneyed is the straw-man boxing match, in which one replaces a formidable opponent with a defeatable simpleton. But there's also the straw-man two-step: first set up the effigy, then acknowledge that he is not so fatuous after all, but frame his reasonableness as a capitulation to one's devastating criticisms.[1] And then there is the sacrificial straw man, useful when one worries about being on the fringe of respectable opinion: set up a fanatical version of one's theory, then distance oneself from it as proof of one's moderation. It is the same strategy that wine dealers use when they stock an exorbitantly priced bottle on every rack. They know that insecure buyers gravitate to the middle of the range, so if there's a hundred-dollar bottle on display, they'll go for the thirty-dollar bottle, whereas if the most expensive bottle had cost thirty dollars, they'd have been content to spend ten.[2]

In a previous book I argued that many intellectuals today subscribe to the extreme view that the human mind is a blank slate with no innate talents or temperaments.[3] As a result, theories that attribute to the human mind faculties that should be unexceptionable, such as sexual jealousy or parental love or an instinct for language, are apt to be viewed as extreme. In pressing this case it would have been a rhetorical godsend to sculpt a sacrificial innatist with views far more extreme than this—say, someone who believed that our standard equipment includes not just a few emotions and thinking skills but tens of thousands of full-blown, concrete concepts like "trombone," "carburetor," and "doorknob." I forbore using this tactic, and not just from intellectual scruples. My own position, I felt sure, was moderate in the first place, and in any case the radical innatist who would have been called in to outflank me was not made of straw but of flesh and blood: my former MIT colleague, the philosopher and psychologist Jerry Fodor.[4]

Fodor is a brilliant, witty, and pugnacious scholar who, among other things, helped to lay the conceptual foundations for cognitive science and to develop the scientific study of sentence comprehension.[5] His notorious theory that we are born with some fifty thousand innate concepts (a conventional estimate of the number of words in a typical English speaker's vocabulary) makes an appearance here not as a player in the nature-nurture debate but as a player in the debate over how the meanings of words are represented in people's minds. In the preceding chapter, I proposed that the human mind contains representations of the meanings of words which are composed of more basic concepts like "cause," "means," "event," and "place." Fodor begs to differ. He believes that the meanings of words are atoms, in the original sense of things that cannot be split. The meaning of *kill* is not something like "cause to die," but rather "kill," full stop. The meaning of *cut* is "cut," the meaning of *load* is "load," the meaning of *trombone* is "trombone," and so on, up to the fifty thousand words that a person knows. And if the concepts beneath word meanings are not assembled out of innate parts during the course of learning, they must themselves be innate. So Fodor's radical innatism does not come from some in-the-marrow conviction that everything is in the genes. It is an implication of his belief that word meanings are indivisible wholes. And both of us can't be right.

It is to Fodor's credit that he pursues his claims to their logical consequences, regardless of how unconventional they might be. As fellow philosopher Dan Dennett put it:

> Most philosophers are like old beds: you jump on them and sink deep into qualifications, revisions, addenda. But Fodor is like a trampoline: you jump on him and he springs back, presenting claims twice as trenchant and outrageous. If some of us can see further, it's from jumping on Jerry.[6]

The last sentence, a play on Newton's famous words, "If I have seen further it is by standing upon the shoulders of giants," explains why this chapter will lavish attention on what would seem to be the crazy idea that the concept of "carburetor" is somehow coded in our DNA. Not only is it a matter of fairness to acknowledge alternatives to the theory I am advancing; it's a matter of clarity and discovery. Much can be gained by contrasting a theory with its alternatives, even ones that look too extreme to be true.

You can only really understand something when you know what it is *not*. In the preceding chapter I presented the theory of conceptual semantics—that word meanings are represented in the mind as assemblies of basic concepts in a language of thought. Your reaction might have been, "So what's the big deal? How *else* could people know how to use the words in their vocabularies?" In this chapter we will see how else. I will try to explain the merits of the theory of conceptual semantics by setting it off from three alternatives.

The first is Fodor's Extreme Nativism (in cognitive science, the term "nativism" refers to an emphasis on innate mental organization; it has nothing to do with the political term for anti-immigrant bigotry). The second is Radical Pragmatics, the idea that the mind does not contain fixed representations of the meanings of words.[7] Words are fluid, and can mean very different things in different circumstances. We give them a meaning only on the fly, in the context of the current conversation or text. And what we draw upon in memory is not a lexicon of definitions but a network of associations among words and the kinds of events and actors they typically convey.[8] The third radical alternative, Linguistic Determinism, upends the view of language and thought I have been assuming. Rather than language being a window into human thought, which is couched in a richer and more abstract format, our native language *is* the language of thought, and so determines the kinds of thoughts we can think.[9]

Aside from clarifying the theory of conceptual semantics, a visit to the arenas in which it competes with alternatives will allow us to explore some additional features of human nature. We will learn how the mind conceives of bodies and persons, how it represents number, and how it deals with the three dimensions of space. And we will see how people use their knowledge of words to tinge their language with attitude and emotion, to impress and amuse their listeners, and to comment on language itself.

A final reason to follow the debates on the mental representation of words is that they are proxy battles in much larger conflicts of ideas. Frequently asked questions like "What is innate and what is learned?" "Is the meaning of our speech and writing determinate or is it relative to a context?" "Does our language constrain what we can think?" and "Are human cultures fundamentally alike or different?" resonate throughout intellectual life, and in *The Blank Slate* I showed that far-ranging moral and political implications are often read into them. We won't resolve those issues here, but a close look at words, which can be tallied and probed more easily than other parts of culture, can illuminate how we might make sense of them.

The opening to this chapter, and my promise to contrast my theory with allegedly radical alternatives, should alert you to the entrance of homunculi made of hay. In explaining the alternatives, I'll do my best to distinguish the Featherbeds, people who broach a radical idea but draw back when jumped upon, from the Trampolines, those who bounce back with full force.

EXTREME NATIVISM

The train of argument that leads Fodor to his extreme conclusion begins innocently enough.[10] Almost everyone in the various nature-nurture debates acknowledges that people have to be born with an ability to represent certain elementary concepts (if only "red," "loud," "round," and so on) and an ability to assemble new ones from this inborn inventory as a result of experience (if only by associating them with one another). For instance, the complex concept "red square" is learned by connecting the simple concepts "red" and "square." The key question is, which concepts are part of the innate inventory, and which are assembled out of them (or at least get their meaning from the way they connect to them)? One way to answer the question is to distinguish the concepts that are patently decomposable into simpler ones (such as the meaning of "the man in the gray flannel suit" and other combinations of words) from the concepts that are patently atomic, without containing anything smaller or more basic (such as "red" or "line," which are triggered directly by the eyes and visual system). On the nurture side, empiricists tend to make do with an abstemious inventory of sensori-motor features, invoking only the process of association to build more complex ones. On the nature side, nativists argue that a larger and more abstract set of concepts, such as "cause," "number," "living thing," "exchange," "kin," and "danger," come to us ready-made, rather than being assembled onsite.

Both sides, if pressed, have to agree that the simple building blocks of cognition—like the keys on a piano, the alphabet in a typewriter, or the crayons in a box—must themselves be innate. Type on a standard type-writer all you want; though you can bang out any number of English words and sentences and paragraphs, you'll never see a single character of Hebrew or Tamil or Japanese. As Leibniz said in amending the slogan of empiri-cism, "There is nothing in the intellect that was not first in the senses . . . except the intellect itself."[11]

What about the concepts that underlie the meanings of words? Both an empiricist and a not-so-extreme nativist would be satisfied with the claim that most of these concepts are built out of more elementary units—perhaps *mother* is mentally represented as "female parent," perhaps *kill* is concep-tualized as "cause to become not alive." These units are innate, or perhaps in turn are decomposable into even more elementary units that are innate. (The buck has to stop with *something* innate, so we can explain why chil-dren, but not chickens or rhubarb or bricks, can learn words and concepts in the first place.) If a unit can't be decomposed into a combination of more basic units, it must be innate, just as the letter *A*, which can't be built out of anything simpler, is innate in a typewriter.

But, Fodor claims, the meanings of most words *can't* be decomposed into simpler units. Definitions always leak. *Kill,* for example, doesn't really mean "cause to become not alive."[12] As we saw in the previous chapter, you can *cause someone to become not alive* on Wednesday by poisoning him on Tuesday, but you can't *kill someone on Wednesday* by poisoning him on Tuesday. Nor can you kill someone by slamming the door in his face as he flees a mad dog, though you can cause him to become not alive that way.

Moreover, Fodor says, philosophers who have tried to reduce complex concepts—such as "know," "science," "good," "explain," and "electron"—into definitions made of more elementary concepts have failed abjectly in their efforts. Finally, he argues, when we turn to the psychology of people using language in real time, we see no sign that they have more difficulty with putatively complex concepts than with putatively simple ones.[13] For example, intuitively we feel it's no more difficult to understand the word *father* than to understand the word *parent,* even though "father" is sometimes held to be a complex concept defined out of the simpler concepts "male" and "parent."[14]

Now, if concepts are undefinable, that means they aren't built out of more elementary concepts, which means they must themselves be elementary concepts, which means they must be innate. That doesn't mean that children come out of the womb brandishing full-blown knowledge of fathers and killing and carburetors. These concepts still have to be triggered by their counterparts in the world, or, as the ethologists say, "released," just as the innate concept of "mother" in a gosling has to be released by the sight of a moving hulk, and the innate concept of "rival" in a male stickleback fish has to be released by the sight of a red spot. Nor does the atomic nature of word meanings mean that people are ignorant of the information traditionally plunked into their definitions. People might know that fathers are parents and are male because they have the rule of inference "If something is a father, then it is male; if something is a father, then it is a parent." These "meaning postulates" would supplement a person's logic system, joining other rules of inference like "If *p or q* is true, and *p* is false, then *q* is true." They just wouldn't be part of the word's meaning.[15]

Fodor allows for a few exceptions. Definitions *can* be supplied for certain jargon words like *ketch* and *sloop,* for mathematically defined terms like *triangle* and *prime number,* and for multipart words like *dishwasher* and *blackness* (since it would be perverse to say that *wash dishes,* a phrase, was cognitively complex while its counterpart as a word, *dishwasher,* was not). But, Fodor concludes, "then there are the other half million or so lexical items that the *Oxford English Dictionary* lists. About these last apparently nothing much can be done."[16] And if nothing can be done to define them, they must be atomic, hence innate, no matter how many they turn out to be—fifty thousand, or perhaps five hundred thousand, or perhaps even more than that, if we consider the words in other languages that are not translatable into a single word in English. *Boiingg!* If that seems to go against evolutionary biology (since one might have thought that natural selection could not have anticipated a need for concepts like "carburetor" and "trombone" before there were carburetors and trombones), so much the worse for evolutionary biology—Fodor, like his archenemies on the empiricist side, blows off Darwinism as a bunch of after-the-fact just-so stories.[17] And if it goes against common sense, so much the worse for common sense. We don't allow common sense to override scientific discoveries when it comes to the ethology of other species, like spiders or fish, so why should we grant it veto power over discoveries about the ethology of human beings?[18] In any case, stranger things have happened in the history of science—just look at the weird stuff that came out of quantum physics. *Boing! Boing! Boing!*

My main brief against Extreme Nativism is that its key premise—that word meanings cannot be decomposed into more basic concepts—is mistaken. But Fodor's dismissal of common sense deserves a comment as well. Fodor correctly notes that history has often vindicated unconventional ideas—after all, they all laughed at Christopher Columbus and Thomas Edison. The problem is that they all laughed at Manny Schwartz, too. What, you've never heard of Manny Schwartz? He was the originator and chief defender of the theory of Continental Drip: that the southern continents are pointy at the bottom because they dribbled downward as they cooled from a molten state. The point is that they were *right* to laugh at Manny Schwartz. Extraordinary claims—and fifty thousand concepts being innate, including "trombone" and "carburetor," is an extraordinary claim— deserve extraordinary evidence. As we shall see, Fodor's evidence is extraordinarily thin.

Why should an innate repertoire of thousands of concepts be considered extraordinary? Well, if you're going to claim that something is innate, it's not unreasonable to check that the claim is consistent with the science of how innate things come into being, namely evolutionary biology. Our best understanding of evolution says that innate things that are expensive, elaborate, and useful (as a fifty-thousand-word vocabulary surely is) have come down to us because they increased the reproductive success of our ancestors.[19] And as I mentioned, it's hard to see how an innate grasp of carburetors and trombones could have been useful hundreds of thousands of years before they were invented.

Fodor's ally, the cognitive scientist Massimo Piatelli-Palmarini, recognizes that this is a problem for the fifty-kiloconcept theory, and has crafted an argument that tries to reconcile it with contemporary biology.[20] (Noam Chomsky, who also believes that many word meanings are innate, has made a similar argument.)[21] Look at the immune system, says Piatelli-Palmarini. Biologists used to think that organisms were "instructed" to have antibodies against the foreign proteins (antigens) borne by pathogens and parasites, perhaps by some process in which plastic antibodies molded themselves against the shape of those proteins. Now we know that the immune system churns out millions of different antibodies, including ones that lock onto proteins our bodies have never encountered and may never encounter (say, the liver tissue of an orangutan, or a parasite found only in central Africa). Our immune response consists of *selecting* the preexisting antibody that best fits an antigen, and then letting it proliferate. (Each antibody was originally produced out of simpler elements, but those elements cannot "see" the foreign antigens, and the initial antibodies are generated from them blindly.) The immune system, then, is adaptive and intelligent, but not because of an ability to be instructed by the environment. Rather, it is innately profligate and wasteful, but contains enough distinct units that the right ones for a given environment can be triggered when they are needed. Perhaps the same is true for the neurobiological system that generates our concepts.

The problem with this argument is that it glosses over a vital difference between the immune system and the brain. Our lavish supply of antibodies is not a sign that the body spends its resources like a drunken sailor. It is an adaptation to the threat posed by the innumerable, rapidly evolving, and malevolent microorganisms that surround us. Organisms keep a vast stock of antibodies around because any gap in coverage would quickly be targeted by an opportunistic germ. It's the same principle that makes airport screeners "wastefully" scrutinize all passengers for weapons rather than just young Arab men. The moment they stopped searching elderly Chinese women, Al Qaeda would find a way to get bombs into the handbags of elderly Chinese women.

The requirements for a conceptual system are completely different. Far from having to cover every conceivable possibility, our concepts have to be constrained to avoid the vast majority of them, so that children can figure out what a word means from a few examples of its use. Word learning is as scandalous an induction problem as the acquisition of syntax or the practice of science, because there are an infinite number of generalizations, most of them wrong, that are logically consistent with any sample of experiences. When a rabbit hops by and an adult says "Gavagai!" *gavagai* could mean "rabbit," "rabbithood," "hopping rabbit," or "undetached rabbit parts."[22] When an emerald is held out and a speaker says "green," it could mean "green," or it could mean "green before the year 2020, blue thereafter" (otherwise known as *grue*). If children were equipped with an inventory of concepts as profligate as their inventory of antibodies, they would possess not just the innate concepts "rabbit" and "green" but also the innate concepts "undetached rabbit parts" and "grue," and would never home in on the correct meanings of words.[23] That undermines one of the main arguments for why *anything* should be innate.

Before we put words into an atom-smasher to see if they break into pieces, let me mention one last conceptual problem with Extreme Nativism. The problem is how we put our concepts to work if, as Fodor says, they have no parts, like so many pebbles. We don't just *have* concepts but *use* them, and for something to be useful in a complex task (a tool, an organ, a bit of software), it must be composed of parts that divide the labor and accomplish more practicable subtasks. If *to melt something* means "to cause something to become molten," then we can solve the problem of how people use the concept "melting" by first examining how they use the concept "cause"—for example, how children recognize examples of causation from trajectories of motion, and how people reason that if X causes Y, Y would not have taken place but for X. Each of these is a more tractable problem than laying out how the entire concept of "melting" works, since it includes those problems and adds new ones on top of them. More important, once the psychology of causation has been worked out, the solution applies automatically to the thousands of *other* verbs that imply causation: *kill, bounce, butter,* and so on (more on this soon). And so on for the other components of meaning. But if *to melt* just means "to melt," it is a mystery how children recognize instances of that concept and how they reason with it—and we are faced with fifty thousand similar mysteries.[24] The avoidance of this problem is all too apparent when we read Fodor, who often seems satisfied to explain a concept by typing the word for it in different cases and fonts:

> The basic idea is that what makes something a doorknob is just: being the kind of thing from experience with which our kind of mind readily acquires the concept DOORKNOB. And conversely, what makes something the concept DOORKNOB is just: expressing the property that our kinds of minds lock to from experience with good examples of instantiated *doorknobhood*. . . . what I want to say is that *doorknobhood* is the property that one gets locked to when experience with typical doorknobs causes the locking and does so *in virtue of the properties they have qua typical doorknobs*.[25]

In fairness, this is not gobbledygook; Fodor is making a coherent, if abstruse, philosophical argument (which I will not try to explicate here). But when it comes to the psychology of concepts, the impression that Fodor is playing a typographical shell game is correct: what the properties of doorknobhood actually are, and how people recognize and reason with them, is left unexplained.

These are some of the theoretical problems with Extreme Nativism. What do the facts of language have to say about it? One immediate problem is well known to linguists: the boundary between the meanings of single-morpheme words (which Fodor says are atomic and innate) and the meanings of multi-morpheme words (which Fodor says are composed out of parts and learned, just like the meanings of phrases and sentences) is often arbitrary. For one thing, the same concept can be expressed by a multi-morpheme word in one language and a single-morpheme word in another, as we saw in the previous chapter when we watched the concept "cause" pop up in different linguistic slots. In English, for example, we have separate morphemes for *see* and *show, come* and *bring, rise* and *raise, write* and *dictate*. In Hebrew, the word for *show* is *cause-to-see*, the word for *bring* is *cause-to-come*, the word for *raise* is *cause-to-rise*, and the word for *dictate* is *cause-to-write*. But no one would want to say that the concept of "bringing" is innate in Americans and Britons but learned in Israelis.

Even within a single language, a concept can switch from multimorphemic to monomorphemic over time. This is how we get most of our irregular forms: speakers slur morphemes together, or hearers fail to discern them, and two morphemes melt into one. *Made* used to be *maked* (*make* + *ed*), and *feet* used to be *foeti* (*fot* [foot] + *-i*).[26] But surely the concepts of "making-in-the-past" and "more-than-one-foot" didn't go from learned to innate among English speakers sometime during the Middle English period. Closer to home, we see transitions from multi- to single-morphemehood whenever an invention becomes commonplace: *refrigerator* → *fridge*, *horseless carriage* → *car, wireless* → *radio, facsimile transmission* → *fax, electronic mail* → *e-mail, personal computer* → *PC*. Did each of these concepts awaken a dormant innate counterpart when people started referring to it with a single word?

All of these facts underscore an important design feature of language. The machinery of syntax allows people to build complex concepts out of simple concepts—say, *remove caffeine from*—whose interpretation depends on the meaning of the words, in this case, "remove," "caffeine," and "from." (Indeed, Fodor makes much of this ability when he is fighting his empiricist foes in other venues, such as in the debate on connectionism.)[27] The machinery of morphology (complex word formation) does the same thing as the machinery of syntax, namely, build complex concepts out of simple ones; in this case, *decaffeinated* can be interpreted from the meanings of *de-, caffeine, -ate,* and *-ed*. But Fodor insists that this machinery must stop at the door of a *single* word—that when people start using *decaf* (without registering its parts) or *Sanka*, a completely different concept must be substituted, indeed, one that is innate. And the ease with which a multi-morpheme word melts and congeals into a single-morpheme word shows that languages do not respect Fodor's boundary.

Admittedly, concepts do shade in meaning depending on whether they are expressed as phrases, complex words, or simple words. For example, components of the meaning of a word are more generic than the same components in a phrase. *To butter* does not literally mean "to apply butter." Any butterlike substance will do: you can *butter your bread with cheap margarine*. And eponymous verbs like *gerrymander, bowdlerize,* and *boycott* long outlive any memory of the person who originally inspired them. But these changes apply across the board as a condition of phrasehood or wordhood, rather than being idiosyncratic to each word (just as you can butter a bread with cheap margarine, you can *paper the walls with vinyl sheeting*). And they consist of tweaks or addenda to the original meanings, rather than complete replacements, like "banana" for "large trolley," which is what could happen if the meanings of simple and complex words were unrelated atoms.

Recall that Fodor also claims that words for complex concepts are no harder to use and learn than words for simple ones. But complex words needn't always be psychologically more taxing, because with practice, the mind assembles packages of elements into chunks and assigns each chunk a single slot in memory and processing.[28] So a speaker may not need more mental resources to use the concept "cause-to-die" than the concept "die." When a concept does exceed the natural chunk size at a stage of development, children indeed have more trouble learning it.[29] The psychologist Dedre Gentner, for example, examined the simple verbs *give* and *take,* the slightly more complex verbs *pay* (give money) and *trade* (give X and receive Y), and the even more complex *spend* (give money and receive X), *buy* (receive X and give money), and *sell* (give X and receive money). The children found the simple verbs easier to act out than the more complex ones, and the most complex ones hardest of all, just as one would expect. And their errors consisted of omitting some of the extra meaning components, like acting out *sell* by giving something but not taking money in return.[30] We can sometimes hear this partial learning in children's speech errors, as when a two-year-old boy accompanying his mother to a cash dispenser asks her, "Are we buying money now?"[31]

The heart of Fodor's argument is his attack on definitions, which, he says, inevitably leave something out of the meaning of the definiendum (the word being defined). The problem with this argument is that a definition (which admittedly is always incomplete) is not the same thing as a semantic representation. A definition is a dictionary's explanation of the meaning of an English word using other English words, intended to be read by a whole person, applying the entirety of his or her intelligence and language skills. A semantic representation is a person's knowledge of the meaning of an English word in conceptual structure (the language of thought), processed by a system of the brain that manipulates chunks of conceptual structure and relates them to the senses. Definitions can afford to be incomplete because they can leave a lot to the imagination of a speaker of the language. Semantic representations have to be more explicit because they *are* the imagination of the speaker of the language. Fodor's attack on complex semantic representations depends on confusing them with dictionary definitions.

Fodor works through only one example: the transitive verb *paint* and its definition "cover a surface with paint":

> To start with a fairly crude point, consider the case where the paint factory explodes and covers the spectator with paint. This may be good fun, but it is not a case of the paint factory (or the explosion) painting the spectators.[32]

As he immediately concedes, this is a *very* crude point, because the semantic representation of *to paint* and other causative verbs usually requires an animate agent. (An explosion at a dairy farm doesn't *butter the cows,* an explosion at the WD-40 plant doesn't *oil the hinges,* and so on.) Dictionary writers don't bother to spell this out because they can count on their readers to fill it in. And that's the point—readers *can* fill in that bit of meaning when they learn that *paint* is a causative verb, because they represent its meaning as including the concept "agent of a causal event."

So Fodor moves on:

> Consider that Michelangelo, though an agent, was not a house-painter. In particular, when he covered the ceiling of the Sistine Chapel with paint, he was not painting the ceiling; he was painting a picture *on* the ceiling. . . . Compare Tom Sawyer and his fence.[33]

Let's pass over the fact that many people do describe what Michelangelo did as *painting the ceiling,* a neat application of the locative rule applied to *paint a picture on the ceiling.* (In a Google search, the phrase "Michelangelo painted the ceiling" gets 335 hits, though admittedly one of them locates the ceiling in the "Cistern Chapel" and another in the "Sixteenth Chapel.") Fodor notes that for someone to paint the ceiling, it has to be the person's primary intention that the ceiling be covered, not just that the ceiling happens to end up covered as a by-product of some other intention, in this case inscribing a picture. It is an acute point, but it has nothing to do with some idiosyncratic meaning of the verb *paint.* As we saw in chapter 2, it is a requirement of *all* verbs in the container-locative construction: they all specify a change of state intentionally effected on a surface or container. (A person who falls into a lake on his way to the well hasn't *filled the bucket,* a person who wraps her shoulders with a bandage to stay warm hasn't *bandaged her shoulders,* and so on.) So when Fodor complains that the definition "is getting sort of hairy," he is ignoring the fact that the same hairiness would apply to thousands of other verbs, and that they can be depilated in a stroke by removing the intentionality proviso from each of them and inserting it *once* in more generic concepts like "act," "cause," and "goal"—the concepts that Fodor denies are recurring parts of the meanings of words.

He gives it one more try:

> Anyhow, this definition doesn't work either. For consider that when Michelangelo dipped his brush into Cerulean Blue, he thereby covered the surface of his brush with paint and did so with the primary intention that his brush should be covered with paint in consequence of his having so dipped it. BUT MICHELANGELO WAS NOT, FOR ALL THAT, PAINTING HIS PAINTBRUSH.[34]

True enough, but that's because the verb *paint*—like thousands of other verbs—distinguishes between means and ends, and Michelangelo's paintbrush-dipping is construed as a means, with the surface-adornment specified as the end.

After these three examples, Fodor hopes his readers are losing patience: "I don't know where we go from here," he sighs.[35] By ending the story there, he keeps readers from discovering that each of his three snags—an animate agent, an intended effect, and a distinction between means and ends—is not, as the atomism theory requires, an irreducible quirk of the verb *paint* alone, but a common feature of many verbs with meanings similar to *paint,* exactly what the atomism theory prohibits.

Let me complete the discussion of Extreme Nativism by splitting the atom—showing how verbs are composed from a smaller number of conceptual particles. I relied on this idea throughout chapter 2, but rather than send you back to those examples, let me present a particularly elegant fissioning, based in part on the work of Beth Levin.[36] It involves the quintessential kind of verb: the simple, transitive action verb, in which X does something to Y. If *any* verb meanings are atomic, these should be.

The atoms are smashed by firing various constructions at the transitive verbs and examining the fragments that fly out of the collision. It begins with a construction called the conative, from the Latin "to try." The conative conveys the idea that an agent is repeatedly trying to impinge on something but is not quite succeeding:

> Mabel cut at the rope. [Compare "Mabel cut the rope."]
> Sal chipped at the rock.
> Vince hit at the dog.
> Claudia kicked at the wall.

The preposition *at* points to the entity that figures in the goal of the person's attempt (note the metaphorical extension of *at* from its more concrete sense of a goal of physical motion, as in *Harry fired an arrow at the tree*). As we have come to expect, not all verbs can enter the construction, even if the combination would make sense. Each of these sounds odd:

> *Nancy touched at the cat.
> *Jeremy kissed at the child.
> *Rhonda broke at the rope.
> *Joseph split at the wood.

So the conative alternation applies to a much smaller class of actions than those that can be attempted. It works for verbs of cutting (*chip, chop, cut, hack, slash*) and verbs of hitting (*beat, bump, hit, kick, knock, slap, strike, tap*) but not verbs of touching (*hug, kiss, pat, stroke, tickle, touch*) or verbs of breaking (*break, crack, rip, smash, split*). Stated more economically, the eligible verbs signify a kind of motion resulting in a kind of contact.

Now consider an alternation called possessor-raising:[37]

> Sam cut Brian's arm.
> Sam cut Brian on the arm.

> Miriam hit the dog's leg.
> Miriam hit the dog on the leg.

Terry touched Mavis's ear.
Terry touched Mavis on the ear.

Like other alternations, possessor-raising involves a conceptual gestalt shift, in this case between construing a person as a kind of immaterial soul who *possesses* his body parts (*cut Brian's arm*), and construing him as an incarnate hunk that *is* his body parts (*cut Brian*).[38] With the first construction, one might well ask who or where "Brian" is, if you can cut any one of his parts—his limbs, his head, his torso—without cutting Brian himself. With the second, one can cut the guy on the head or the chest or even the little toe, and in each case it's Brian, not a body part, that you're cutting. The mind-body dualism here is made plain when you try to use the construction with insensate objects rather than bodies. You can't say *The puppy bit the table on the leg, Sam touched the library on the window,* or *A rock hit the house on the roof,* because unlike bodies, the objects are not thought to be endowed with a unified sentience that permeates every part.[39]

But the reason I have brought up possessor-raising is not the mental model it embodies; it is the construction's choosiness among verbs. There's something not quite right about a sentence like *James broke Thomas on the leg* or *Hagler split Leonard on the lip*. Possessor-raising is available only with verbs of hitting and verbs of cutting, not with verbs of breaking. The common denominator is that the verb must specify a kind of physical contact.

Speaking of physical contact, here is an alternation that is similar to the locative:

I hit the bat against the wall.
I hit the wall with the bat.

She bumped the glass against the table.
She bumped the table with the glass.

Once more, several verbs refuse to belong to a club that would have words like them as members:

I cut the rope with the knife.
*I cut the knife against the rope.

They broke the glass with the hammer.
*They broke the hammer against the glass.

She touched the cat with her hand.
*She touched her hand against the cat.

(Of course we are ignoring the readings in which the knife gets cut, the hammer gets broken, or the hand gets touched.) In this case, the verbs of hitting can enter into the alternation, but not the verbs of breaking or touching. A better way of putting it is that the participating verbs involve motion followed by contact, but not motion followed by contact followed by a specific effect (a cut, a break), and not contact without a prior change of location (as in touching).

Our fissioning requires five shots at the atom; bear with me for the last two. A construction called the middle voice specifies the ease with which an action can be performed on something, as in *This glass breaks easily* and *This rope cuts like a dream.*[40] Once again, not all verbs can take part in it:

*Babies kiss easily.
*That dog slaps easily.
*This wire touches easily.

The middle voice applies to verbs that signify a specific effect of some cause, which is true of breaking and cutting but not kissing, slapping, and touching. No effect, no middle.

Finally, there's the anticausative alternation, which converts a transitive verb to an intransitive by ejecting the causal agent. It's different from the middle voice because it describes an actual event in which an object undergoes a change, rather than the generic property of how easily it submits to the change. For instance, with a transitive verb like *Jemima broke a glass* (a causative), you can use its intransitive counterpart like *At three o'clock, the glass broke.* But not all causative verbs agree to shed their agent:

*Sometime last night, the rope cut.
*Earlier today, Mae hit [meaning "Mae was hit"].
*At three o'clock, Clive touched [meaning "Clive was
         touched"].

The anticausative accepts verbs that specify a particular effect, but only if they signify *nothing but* an effect. In that regard they are more restricted than the verbs that can go into the middle voice, where not just an effect but the means of achieving it (such as cutting) may be specified.

My point—and I do have one—is that these alternations cross-classify simple action verbs in terms of components of meaning that they share, defining a family of crisscrossing microclasses. Here's a summary (the verbs that refuse to enter an alternation are marked with an asterisk):

Alternation	Microclass	Examples
Conative	motion, contact	*hit, cut, *break, *touch*
Possessor-raising	contact	*hit, cut, *break, touch*
Contact locative	motion, contact, no effect	*hit, *cut, *break, *touch*
Middle voice	effect	**hit, cut, break, *touch*
Anticausative	effect, no contact, no motion	**hit, *cut, break, *touch*

The tableau hints at an underlying structure that explains why the members fall neatly into rows and columns. The structure can be exposed when the table is rearranged by verb rather than by alternation:

hit: motion, contact
cut: motion, contact, effect
break: effect
touch: contact

What's particularly elegant about the tableau is that the verbs draw on a common pool of concepts, and none of the concepts was gerrymandered to explain the entrance requirements for one construction alone. Instead, a few concepts kept reappearing in different verbs and different constructions. The same "contact" concept that earmarks the verbs that undergo possessor-raising *also* helps earmark the verbs that enter the conative, the contact-locative, and (by its absence) the anticausative. The same concept of "effect" that distinguishes *cut* from *hit* also distinguishes *touch* from *break;* the same concept of "motion" that distinguishes *hit* from *touch* also helps distinguish *cut* from *break*. No verb needs a meaning component customized for it alone, at least not to distinguish the constructions it can enter. (Of course, many verbs, like *kiss* and *chip* and *snap,* do have a nugget specific to them, but it doesn't exhaust the verb's meaning or affect its syntactic behavior.)

What we see in these exploded fragments is a combinatorial system of meaning inside the verb. The elements that sort the verbs into alternations are not arbitrary markers, like those for gender or declensional class in Latin, because they determine the real-world situations in which speakers use the verbs. For example, it is the "motion" component of *hit* that rules out using the verb to describe causing a bruise by leaning on someone's arm with slowly increasing pressure, and the absence of "motion" from *break* that allows us to say *Sy broke the bicycle* even if he didn't take a sledgehammer to it but was just too heavy for its wheels. Nor are these stipulations just part of our system for reasoning ("meaning postulates"), because they govern how verbs enter syntactic constructions, and therefore are part of the language engine proper; they are not just the common sense that we use in our everyday inferences when dealing with the world.

To sum up: concepts like "motion," "contact," and "cause-and-effect" sort verbs into crisscrossing classes, and thereby must be components of the verbs' meanings. This implies that these words *have* meaning components, which implies that they are not unsplittable atoms, which implies that they need not be innate. And if *hit* and *cut* and *break* aren't innate, then it's all the less likely that *trombone* and *carburetor* are. And this is a happy conclusion. It corroborates our suspicions from common sense and evolutionary biology about the nature of concepts. It boosts our confidence that concepts like "cause" and "motion" really are basic components of our cognitive toolbox. And it shows that saying that some concepts are basic, and possibly innate, is not a slippery slope toward saying that *all* concepts are basic and innate.

RADICAL PRAGMATICS

If you could imagine a theory that is as contrary as possible to Extreme Nativism, it might be Radical Pragmatics. Its disagreement with conceptual semantics is not about whether the mental representations of word meanings are innate, or whether they are atomic, but whether they exist at all. The watchword of Radical Pragmatics might be a quotation from William James: "A permanently existing 'idea' or representation which makes its appearance before the footlights of consciousness at periodical intervals, is as mythological an entity as the Jack of Spades."[41] According to Radical Pragmatics, a permanently existing conceptual structure underlying the meaning of a word is also as mythical as the Jack of Spades, because people can use a word to mean almost anything, depending on the context. The general idea can be illustrated by the Rigelian language in the comic strip *Monty:*

Monty © United Feature Syndicate, Inc.

English, and the other languages of earthlings, just don't take it to that extreme. The nuance in the way people use words, according to Radical Pragmatics, calls for a way of thinking about language and thought that is very different from the image of a dictionary in the head with fixed chunks of conceptual structure packaged with every entry.

"Radical Pragmatics" sounds like an oxymoron, but it alludes to the branch of linguistics called pragmatics, the study of how language is used in context in light of the knowledge and expectations of the conversants. *Radical* pragmatics is the imperialist bloc of the field, which tries to explain as many aspects of language as possible in those terms.[42] The name was coined by Geoffrey Nunberg, the linguist known to many Americans from his newspaper and radio commentaries on language.[43] Other radical pragmaticists include the anthropologist Dan Sperber and the linguist Deirdre Wilson, the psycholinguist Elizabeth Bates, and members of the schools of cognitive science known as Connectionism and Dynamic Systems.[44] The advocates of those schools are the springiest Trampolines, and later in this section we will examine a connectionist model that shows how radical Radical Pragmatics can be.

The touchstone for Radical Pragmatics is the phenomenon shown in the *Monty* cartoon: polysemy, pronounced "poLISSamee," meaning "many meanings."[45] Polysemy refers to a word's having a number of distinct but related senses, and is different from two other ways in which a sound can have multiple meanings.

In homonymy ("same name"), a single word has several unrelated meanings. Homonymy usually arises when an ancestral word budded off new senses in a language's history and current speakers retain no inkling of the original connection. For instance, the word *odd* originally referred to something that stuck out, like the point of a triangle. Then it was extended to refer to something that metaphorically stuck out because it was unusual, and then was extended further to refer to a number that had one unit sticking out from a pair.

Polysemy is also different from homophony ("same sound"), in which distinct words are pronounced the same way, usually because their original pronunciations got merged in the history of the language. For example, *four* and *fore* sound alike today, but *four* originally rhymed with *tour,* and *fore* originally rhymed (more or less) with *flora;* we see fossils of the old pronunciations in the way the words are spelled. Homonymy and homophony are often used in wordplay, as in the nerdy joke proving that a horse has an infinite number of legs:

> Horses have an even number of legs. Behind they have two legs, and in front they have fore-legs. This makes six legs, which is certainly an odd number of legs for a horse. But the only number that is both even and odd is infinity. Therefore, horses have an infinite number of legs.[46]

With polysemy, in contrast, the senses of a word are so tightly linked that it takes a linguist or an artificial intelligence researcher to spot the difference. Here are some examples:

- *Chicken* can refer to a kind of animal (*Why did the chicken cross the road?*) or to a kind of meat (*Try it, it tastes like chicken!*).
- *Newspaper* can refer to an organization (*Jill works for a newspaper*) or to an object (*Here, squish the roach with this newspaper!*).
- *Book* can refer to a body of information (*Abe's book is unconvincing*) or to a physical object (*Abe's book weighs five pounds*).

- *Window* can refer to a pane (*She broke the bathroom window*) or to an opening (*She came in through the bathroom window*).
- *Monkey* can refer to a species (*Monkeys live in trees*) or to an individual (*Monkeys have taken over the island*).
- *France* can refer to a political entity (*France is a republic*), the leaders of the entity (*France defied the United States*), or a patch of land (*France has two mountain ranges*).
- *Construction* can refer to an event (*The construction took nine months*), a process (*The construction was long and noisy*), a result (*The construction is on the next block*), or a manner (*The construction is shoddy*).

Somehow we seem to configure the appropriate meaning for a polysemous word according to its milieu in the sentence and in the conversation or text. People usually don't realize how effortlessly they switch among polysemous senses until they hear them clash in the form of wordplay called zeugma (or syllepsis), where two incompatible senses are juxtaposed. Examples include Benjamin Franklin's "We must all hang together, or assuredly we shall all hang separately," Charles Dickens's "She came home in a flood of tears and a sedan chair," and Groucho Marx's "You can leave in a taxi. If you can't get a taxi, you can leave in a huff. If that's too soon, you can leave in a minute and a huff." And occasionally a speaker will call attention to polysemy to make a point, as when a soldier in *All Quiet on the Western Front* is told that wars begin when one country insults another and he replies, "I don't get it. A German mountain can't insult a French mountain, or a river, or a forest, or a cornfield."[47] But often polysemous senses are mixed without raising an eyebrow:

> Yeats did not enjoy hearing himself read aloud.
> The *Boston Globe* decided to change its size and typeface.
> Don't worry about that review—tomorrow it will be wrapping fish.
> Sally's book, which would make a good doorstop, is full of errors.
> The chair you're sitting in was common in nineteenth-century parlors.

The window was broken so many times that it had to be
boarded up.[48]

Polysemy is everywhere. A *sad movie* makes you sad, but a *sad person*
already is sad. When you *begin a meal,* you eat it (or, if you're a cook, pre-
pare it), but when you *begin a book,* you read it (or, if you're an author,
write it). What makes something a *good* car is different from what makes it
a good steak, a good husband, or a good kiss. A *fast car* moves quickly, but
a *fast book* needn't move at all (it just can be read in a short time), and a *fast
driver,* a *fast highway,* a *fast decision,* a *fast typist,* and a *fast date* are all fast
in still different ways.

Even something as concrete as a color word can change like a chame-
leon. *Red* refers to very different shades when it modifies a grape, a hot
poker, the meat of ungulates, a head of hair, or the face of a man who has
discovered he's been lecturing with his fly open. The polysemy of color
words inspired a poem passed on to me by Saroja Subbiah, which circulated among the Maori staff in a New Zealand government office:

Dear White Fella
When I am born I'm black
When I grow up I'm black
When I am sick I'm black
When I go out ina sun I'm black
When I git cold I'm black
When I git scared I'm black
And when I die I'm still black.

But you white fella
When you're born you're pink
When you grow up you're white
When you git sick you're green
When you go out ina sun you go red
When you git cold you go blue
When you git scared you're yellow
And when you die you're grey
And you got the cheek to call me coloured?

Often polysemy arises when a word is used to refer to something that is merely associated with its usual referent, a device called metonymy. I can say that *Suzie is parked out back* or that *Bradley was rear-ended by a bus,* using people's names to refer to their cars, or *Put Chomsky on the Linguistics shelf* and *You can find Hitchcock at the back of the store,* referring to their works. We can also refer to people by their parts and possessions, as when a nurse says *The gallbladder in 220 needs his dressing changed* or when one waitress tells another *The ham sandwich wants his check.*[49]

What do we make of this apparent semantic chaos? According to Radical Pragmatics, interpretation is a loosey-goosey process, which opportunistically draws on anything and everything the person knows about the world and the current audience. And what we call "word meanings" are not discrete entries in a dictionary, mental or otherwise, but patterns of association among stereotypical events and their typical participants. This allows listeners to bend and shape the words of a sentence to fit their associations together in whatever way yields the most probable message a speaker in that situation would convey.

Radical Pragmatics has an element of truth when it comes to *reference,* the way that words may refer to things in the world. When the term *ham sandwich* can refer to a man sitting at a lunch counter, there seems to be little hope for the logician's dream that the expressions in a language can be mapped onto states of the world according to a fixed set of pointers. But the question we are interested in is whether Radical Pragmatics is right about the human mind. The claim that a word lacks a precise mental representation is certainly at odds with the picture of language that emerged from our look at verb alternations, where we saw three displays of crispness.

First, people abjure what should be perfectly intelligible uses of verbs, like *He clogged hair into the sink, She yelled him her order, We melted at the butter,* and *She broke him on the arm.* If the mind is happy with whatever makes sense in context, why do these sentences have an odor of ungrammaticality?

Second, we saw how the boundaries delineating verbs' use are surgical cuts within classes of similar events; they don't just hover around stereotypes.[50] When I think about a typical scenario in everyday life, like putting water into a glass, my mind has a well-fleshed-out image—a thirsty person walking over to a faucet with a clear tumbler in hand and turning a tap to let water flow into it just short of overflowing. Yet when I *talk* about the scenario, most of this flesh melts away, leaving behind one of several skeletons. If I use the verb *pour*, my field of vision narrows to how the water is caused to move, ignoring its destination; that's the reason we can say *pour the water* but not *pour the glass*. But if I use the verb *fill*, my field of vision narrows to the resulting fullness of the glass, ignoring the trajectory of the water; that's why we say *fill the glass* but not *fill the water*. Even the simple class of transitive action verbs, as we just saw, are crisscrossed with sharp incisions, with *cut, break, touch,* and *hit* falling into different semantic zones. Each verb zeroes in on a necessary aspect or aspects of an event (causation, motion, contact) and disregards the others, no matter how commonly they are associated in our experience. This is quite unlike the stereotypes of typical situations that, according to Radical Pragmatics, govern our use of language.

Third, the incisions between classes don't just envelop clusters of stuck-together features (fast, slow, wet, dry, voluntary, and so on) but are laid down by formulas with a syntactic and algebraic structure.[51] *Cutting*, for example, doesn't just invoke a motion, a contact, and an effect in any old combination. (Moving an egg into contact with a hot pan, causing the egg to split, is not the same as *cutting an egg*.) Rather, the motion has to *include* contact of the cutting implement with the egg *followed by* its motion through the surface of the egg, *resulting in* the cut. Also, word classes are algebraic in the sense that they require variables in specific slots. Consider what it takes to state the common denominator of a microclass, like the manner-of-speaking verbs. What do *whisper, mumble, shout, purr, yammer,* and so on have in common? Not a *particular* manner, because any manner will do. Nor can it be the common denominator among the manners, because there is none: the features of whispering and muttering are nullified by the features of yelling and shrieking. No, the verbs in the class simply have to specify *some* manner, *any* manner. That means the characterization of the class must be something like "All verbs must contain in their definitions the statement 'manner = x' "—a definition of a definition. This raises the bar for how logically sophisticated the mental representation of word meanings must be, and casts doubt on the idea that it is a loose tangle of associations.

■ ■ ■

So if verbs and constructions can be so precise, how do we explain the wild world of polysemy? Actually, it turns out that polysemy is no wilder than other parts of language. It grows out of an interaction between memorized forms and combinatorial operations, the two main ingredients of language (and the topic of my book *Words and Rules*).

Once the extent of polysemy in language became apparent, linguists started to scrutinize the examples, and today they distinguish between irregular and regular polysemy, analogous to the distinction between irregular and regular inflectional forms.[52] Irregular forms, like *come-came* and *mouse-mice,* are idiosyncratic and have to be memorized individually; regular forms, like *walk-walked* and *cat-cats,* are predictable and can be generated by a rule. Some senses of polysemous words are as unpredictable as irregular inflections, and may simply have to be memorized. A person learns that the English word *red* means "the color of fire engines," which gets filed away as one definition in the mental dictionary, and in a separate episode learns that it can mean "the color of Lucille Ball's hair," which gets filed as a second definition. This requires that we credit a person's lexicon with many definitions, but we already know that memory for words is capacious—fifty to a hundred thousand words, and probably at least that many idioms—so a few extra definitions for many of them is not too many to remember. Of course, *someone* had to have been supple enough to extend the word to a new sense for the first time, and others had to have deduced the meaning when the innovator loosed it on the population. But the rest of us may have memorized the fruits of that speaker's creativity, and needn't recapitulate it every time. We saw this division of labor in the last chapter, in the section that compared "smart languages" with "smart speakers."

The reason to believe that many polysemous meanings are memorized, rather than stretched as the need arises, is that they are *conventional*—they are arbitrary practices of a language community, neither deducible nor universal. English may use *red* for a natural hair color, but other languages, like French, reserve a separate word for it (*roux*), just as in English we refer to Marilyn Monroe's hair as *blond* rather than *yellow*. Other words for hair colors—*platinum, ash, strawberry, chestnut, brunette, auburn*—have to be learned one at a time (strawberry blond, for instance, is not the color of a strawberry), so why not this sense of *red*? The same is true for words for skin colors, as the White Fella poem reminds us: it's convention, not color vision, that tells us that a sick Caucasian is *green,* a cold one *blue,* and a scared one *yellow.*

Another kind of evidence that many polysemous senses are acquired by sheer exposure comes from the statistics of language: the more frequent a word is, the more polysemous it is, and vice versa.[53] For example, the common verb *set* (which occurs 372 times in every million words) has more than eighty dictionary definitions; the less common verb *sever* (9 per million) has four, and the rare verb *senesce* (less than 1 in a million) has just one. This is just what you would expect if words by default are precise in meaning, and accumulate additional senses through separate exposures, but the opposite of what you would expect if words by default are diffuse in meaning, and are sharpened with additional exposures through discrimination training.

Laboratory experiments also suggest that the meanings of many polysemous words are listed in the brain as separate senses. The psychologists Devrah Klein and Gregory Murphy used a popular technique in experimental psycholinguistics called priming, in which presenting a word to a person activates it in the person's mind, making it easier for the person to recognize the word (and words related to it) for a few tenths of a second.[54] Klein and Murphy flashed a polysemous noun like *paper* (which can mean either a journal or the pulp it is printed on), accompanied by a modifier that narrowed it to one sense, such as *wrapping paper.* This was the priming word, and the question was what, exactly, it primed in the viewer's mind: that specific sense alone, or some semantic core of the word that embraces all of its polysemous senses. To find out, Klein and Murphy flashed the noun again with a different modifier, either one consistent with the original sense, like *shredded paper,* or one consistent with an incompatible sense, like *liberal paper.* The participants were timed as they figured out the meaning of the second phrase and pressed a button if it made sense. The participants were quicker and more accurate when the phrase had just been primed with the same sense of the word (as in *wrapping paper . . . shredded paper*) than when it had been primed with a different sense of the word (as in *wrapping paper . . . liberal paper*). This suggests that for many polysemous words, each sense is a separate unit stored in the brain, which can rise and fall in activation independently of the other senses. A recent extension of the priming technique, using magnetoencephalography to measure brain activity more directly, suggests the same conclusion.[55]

The opposite of irregular polysemy is regular polysemy, where entire families of words gain new senses in one fell swoop rather than having to be memorized one at a time. Some examples of polysemy are so regular that they don't require a multiplication of word meanings at all, just a more sophisticated analyzer for words in combination. Consider the adjective *good*, which means very different things in *a good knife, a good wife,* and *a good life*. Does this entail that *good* has multiple meanings? Only if the language interpreter in the brain stupidly interprets a phrase by looking for the intersection of each of its components, such as the things in the world that are both "knives" and "good things." A more incisive interpreter could wiggle a probe inside the noun and prise out the component of meaning that is modified by *good*, sparing it from having to saddle the word *good* with dozens of meanings.

What is this meaning component? The computational linguist James Pustejovsky argues that Aristotle got it right when he proposed that the mind understands every entity in terms of four causes: who or what brought it about; what it's made of; what shape it has; and what it's for.[56] The interpreter for adjectives like *good* and *fast* (*a good road, a fast road*) and for verbs like *begin* (*He began his sandwich, She began the book*) digs into the part of the noun's conceptual structure that specifies how the object is intended to be used (roads are for driving on, sandwiches are for eating, books for reading) and concludes that *good* and *begin* refer to that part. When instead a count noun is used as a mass noun (as in *There was sausage all over his shirt*), the interpreter picks out the specification of what the thing is made of; once again, no new noun meaning is necessary. What we have is not polymorphous polysemy, as in the Rigelian phhlëmkes, but a few schemes for probing into a word and selecting one of the components of its meaning.

Somewhere in between the complete irregularity of *red* in *red hair* and the complete predictability of *good* in *good road* we find cases that involve an interaction between the two.[57] The interactions come from lexical alternation rules like the ones that alter the construal of a verb (for example, from causing-to-move to causing-to-change). In the case of noun polysemy, there may be a rule allowing the name of a product to refer to its producer or vice versa (*Honda,* the *New York Times*), a rule that allows a word for an opening to be used for its covering (*door, window*), and a rule that allows a word for an animal to refer to its meat (*lamb, goose, swordfish*). The number of rules needed is large but not onerous.

At this point a defender of Radical Pragmatics could protest that these "rules" are bogus—that they are just snapshots of common sense bending and stretching a word's meaning as the need arises. To show that these rules really are parts of a language engine, one needs to show that they mesh with other mechanisms of language, particularly in ways that would leave common sense and the desire to communicate frustrated.

One way in which regular polysemy interlocks with the other apparatus of language is its sensitivity to the *forms* of words rather than just their meanings. In some cases, polysemy is hemmed in by sound. For example, an adjective for a nationality can be turned into a plural that refers to its people, as in *the Swiss, the Spanish, the Dutch, the French,* and *the Japanese.* But the rule applies only in two circumstances: when the adjective ends with a sibilant (you can refer to *the Swiss* and *the Spanish,* but you can't refer to **the German,* **the Coptic,* or *the Belgian), and when it retains a non-English sound pattern (as in *the Hausa, the Tuareg,* or *the Wolof*). And in some cases polysemy is hemmed in by morphological structure (the composition of a word out of stems and suffixes). For example, nouns referring to forms of government can be extended to actual states, as when we speak of *democracies, tyrannies, oligarchies, monarchies,* and *dictatorships*—but not when the noun ends in -*ism:* you can't refer to a bunch of fascist states as *fascisms,* nor is a map of the world splotched with *communisms, marxisms, maoisms, islamisms,* or *totalitarianisms.*[58] The enmeshing of polysemy with grammar is also visible in one of the ways that Americans and Britons are divided by their common language. When a product gives its name to an employer, the name is singular in the United States (*The Globe is expanding its comics section*) but plural in the United Kingdom (*The Guardian are giving you the chance to win books*).

Regular polysemy is also fenced in by fastidious semantic restrictions. You can use *France* to refer to the land, the state, or the leadership but not to the people: it's odd to say *France eats a lot but stays thin.* You can work for *a newspaper* or *a magazine,* but not for *a book* or *a movie.* Words for edible objects can be used for the gloop that results when they are mashed up—*some carrot, some salmon, some apple, some egg*—but not if the objects come in aggregates rather than individually. That's why Mexican restaurants serve *refried beans,* not *refried bean,* and why the purée served in Indian restaurants is called *lentils,* not *lentil.*[59] So we find that polysemous nouns are like alternating verbs: they don't leap helter-skelter into any intelligible construction but instead enlist or desert in neat microclasses.

There is one more way in which polysemy is enmeshed with actual words, not just plausible meanings. When a language already has a word staking out a meaning slot in a suite of related meanings, the word will preempt any interlopers that may have been sent there by a rule of regular polysemy. (It's similar to what we find with inflection, where irregular *mice* preempts regular *mouses*.) When it comes to verbs for traveling that were stretched from nouns for vehicles, you can *ferry, truck, cycle, canoe,* or *motorcycle* somewhere, but you can't *car* or *plane,* because we already have *drive* and *fly*. And we may eat *chicken,* but we don't eat *cow, calf, sheep, pig,* or *deer,* but rather *beef, veal, mutton, pork,* and *venison* (in New Zealand, *cervena*). Incidentally, many people believe that these doublets—a Germanic word for the animal and a French word for its meat—come from a time when Anglo-Saxon peasants tended animals but only their Norman overlords got to eat them. The theory comes from Sir Walter Scott's *Ivanhoe,* in which Wamba the jester explains to a swineherd:

> Pork, I think, is good Norman French; and so when the brute lives, and is in the charge of a Saxon slave, she goes by her Saxon name [swine]; but becomes a Norman, and is called pork, when she is carried to the castle hall. . . . Old Alderman Ox continues to hold his Saxon epithet while he is under the charge of serfs and bondsmen such as thou, but becomes Beef, a fiery French gallant, when he arrives before the worshipful jaws that are destined to consume him. Mynheer Calf too, becomes Monsieur de Veau in the like manner: he is Saxon when he requires tendance, and takes a Norman name when he becomes matter of enjoyment.

The story is charming, but historical linguists tell us that the theory is not true; the Anglo-Saxon and French words didn't sort themselves out until centuries later.[60] What *is* true is that people aren't licentious in using words with new meanings. They learn the conventional senses of vast numbers of words, and channel most of their generalizations in well-worn grooves that swerve around existing conventional senses.

■ ▦ ▪

But what about the ham sandwich sitting at table 14, drumming his fingers in anticipation of the check? No sane linguist would propose a rule that converts sandwiches into people, to say nothing of those gallbladders who need their bandages changed. Sometimes people really do elasticize words in the manner of Lewis Carroll's Humpty Dumpty, who said, "When *I* use a word, it means just what I choose it to mean." But Dumpty added, crucially, "When I make a word do a lot of work like that, I always pay it extra." When a speaker sculpts a word into a truly unconventional sense, the hearer doesn't effortlessly mold her mental entry around it to recover the literal meaning. Rather, there is friction between the speaker's square peg and the listener's round hole, and that friction *itself* conveys information in a parallel stream. Indeed, a predictable clash between a speaker's new use and a hearer's fixed meaning is what gives language much of its piquancy and fun. It is the source of euphemism and dysphemism (deliberately offensive speech), of subtexts and messages tucked between the lines, of verbal humor and wordplay, and of literary metaphor. Let's pay each a visit.

Euphemism and dysphemism. When a waitress refers to a customer as a sandwich, she is not just saving breath. She is exercising a dry wit, reducing the patron she otherwise fawns over into the undignified commodity that is their only real common concern. Likewise, turning a patient into his diseased organ is part of the black humor of health-care professionals (a counterweight to the pity and squeamishness that could get in the way of their work), as in *CTD* (circling the drain) for a terminal patient, *Code Brown* for bowel incontinence, and *wallet biopsy* for a check on a patient's finances. That's why the waitress or intern would probably not use the ham-sandwich/gallbladder construction within earshot of the patron's or patient's spouse. The same is true for the mental grinder that turns count nouns into mass nouns, as in *After he backed up, there was cat all over the driveway.* Far from being a mere synonym for *flesh,* this use of *cat* verges on sick humor, the kind of thing a sensitive person would avoid saying around the late cat's owner.

In general, referring to a person by a body part, physical trait, or typical accoutrement—that is, by a metonym—is dysphemistic. That is true of many racial epithets (*a slope* or *a slant* for an Asian, *a redskin* for a Native

American, *a wetback* for a Mexican American), of misogynistic terms for women (*a skirt, a broad, a piece of ass*), and of disrespectful terms for job-holders (*a suit* for a corporate bureaucrat, *a jock* for an athlete, *a wrench* for a mechanic). It is an extension of the mentality behind the possessor-raising construction, in which people are conceived to be more than their body parts. By extension, reducing someone to a body part (or other possession) implicitly denies that he or she is a person.

Euphemisms, in contrast, often refer to a person with a hypernym—a category more inclusive than the one that first comes to mind. Though hypernyms are not really examples of polysemy the way metonyms are, their use in emotionally tinged speech is another illustration of how choices among words can make a psychological difference. When Kipling ended a poem with "You'll be a man, my son," and Sojourner Truth repeated "Ain't I a woman?" in her defiant 1851 speech, they were not belaboring the obvious. They were using the terms *man* and *woman* to confer a degree of dignity that might otherwise be denied to a young male or to an African American female. These generic descriptors for each sex have also been proudly asserted in song: in the refrain of "Mannish Boy" by Muddy Waters, "I'm a Man" by Steve Winwood, "I Am Woman, Hear Me Roar!" by Helen Reddy, and "I'm a Woman (W-O-M-A-N)" by Peggy Lee (and later by Miss Piggy). The effect can also be heard in the euphemism *person of color* and in the Yinglish *mensch*, which originally meant "man" but now denotes a person of uncommon maturity and decency.

The use of hypernyms to exalt can be seen in names for inanimate objects. The suits in marketing often try to dress up their products with pretentious monikers like *driving machine* (car), *photographic instrument* (camera), *beauty bar* (soap), and *dental cleaning system* (that would be a toothbrush). The reason that hypernyms for people and products sound more dignified is not completely clear. Perhaps it is a mirror image of the indignity that arises from referring to someone by a physical part or trait. An abstract essence or archetype is seen as somehow purer and nobler than the familiar concrete particulars that earmark an entity as belonging to a specific class. But whatever the reason, a general rule is that metonyms derogate, hypernyms elevate.[61]

Subtexts. Many humanities scholars have written books called *The Invention of X* or *The Construction of X*, where *X* is not the kind of thing that can literally be invented or constructed; examples include tradition,

romantic love, the human, America, and reality. These scholars are not trying to dilute the meaning of *invention* or *construction* to mere "origin,"
which is what would happen if listeners applied the most expected interpretation in context. They are trying to awaken readers into realizing that what
they might have thought was a natural entity is in fact a historical creation,
and hence can be re-created—a subtext that would vanish if the meanings
of *invention* and *construction* didn't stand their ground in readers' minds.
The same whiplash was exploited by the writer Pamela McCorduck in her
book on artificial intelligence called *Machines Who Think.*

Wordplay. When humorists force their listeners to switch from an
anomalous sense to an interpretable one, the contrast in dignity is a trigger
for mirth, as in this dialogue from W. C. Fields:

> DON: Oh Bill, it must be hard to lose a relative.
> w.c.: It's almost impossible.[62]

In *The Importance of Being Earnest,* Lady Bracknell exploited the same
ambiguity: "To lose one parent, Mr. Worthing, may be regarded as a misfortune; to lose both looks like carelessness." The drollery is set off by the
clash among the senses of the polysemous word *lose:* "misplace a possession," "suffer the death of a loved one," and "shake off a pursuer." If comprehension automatically molded a plastic meaning into the most suitable
sense, there would be no clash between the decorous and the undignified
senses, a clash that is necessary for the jokes to work. (And no one would
understand why Mae West said, "Marriage is a great institution, but I'm
not ready for an institution yet.")

Literary metaphor. When a writer uses a striking metaphor, as in Nabokov's "I was the shadow of the waxwing slain / By the false azure of the windowpane" or Tom Lehrer's "Soon we'll be sliding down the razorblade of
life," he or she is not just pressing new words into service to communicate
a proposition (like "I was dejected" or "Life is hard"). The writer is exploiting the literal meaning of those words to shock listeners into apprehending
the topic in a more emotionally intense way than they would in the normal
course of events, and to recognize that the artist is deliberately shaking
them out of that complacency. We will return to this point in chapter 5
when we visit literary metaphors and explore how they differ from dead
metaphors. For now it is enough to note that in literary metaphor, as in

dysphemism, subtext, and wordplay, the friction that arises between a conventional sense and an unexpected one shows that the mind *has* a conventional sense, not a globule that morphs into whatever interpretation is most sensible at the moment.

Does Radical Pragmatics really entail that word meanings are will-o'-the-wisps that can be blown about by the context? When its proponents repudiate word meanings to explain polysemy, do they really pay the price of being unable to explain how people register the literal meanings of unexpected word uses? One way to answer these questions is to look at a computer simulation of Radical Pragmatics devised by the connectionist modelers James McClelland and Alan Kawamoto.[63] The merit of a computer simulation is that one can see the implications of a set of assumptions uncensored. Their model proves to be a genuine Trampoline—a full-strength implementation of Radical Pragmatics, presented without apologies for its surprising behavior.

McClelland and Kawamoto wanted to model the resolution of polysemous words in context, such as the different senses of *with* in *Luke ate his pasta with a fork* and *Luke ate his pasta with clam sauce,* or the different roles of the subject in *A ball broke the window* and *A boy broke the window.* In accord with Radical Pragmatics, they proposed that fixed representations of meaning are too rigid and clumsy for the task, and that artificial neural networks, which associate features with features rather than manipulating structured representations, are suitably flexible to do the job.

The modelers set up a network that was designed to take the words of a sentence as input and produce an interpretation of who did what to whom as output. The input consisted of a bank of a thousand neuronlike units, each representing a feature of the verb's meaning (such as "intense action" or "causes a chemical change") or a feature of one of its sentence-mates (such as "subject is soft," "subject is medium-sized," "object is hard," or "object is female"). The output consisted of 2,500 units, each representing a feature of the role played by a participant in one interpretation of the sentence, such as "causal agent is round" (which would be suitable to a ball-breaks-window event) or "instrument of a shredding event is hard" (which would be suitable to a knife-cuts-paper event). The generic nature of the features was intentional, allowing each feature to figure in many related interpretations. The model had no representations of individual word meanings, just a dense array of connections between input and output, which were strengthened in a training regimen. A teacher presented the model with thousands of sentences together with their correct interpretations, and the model came to learn what kinds of events tended to be done by what kinds of participants. As a result, it correctly forced the word *with* in *eat pasta with a fork* to mean "instrument" (since it had learned that hard things tend to be used as instruments), and it correctly forced the *with* in *eat pasta with clam sauce* to mean "accompaniment" (since it had learned that soft things tend to be used as food).

Unfortunately, the model played a price for this nimbleness. Its knowledge of word meaning was so malleable that when it encountered a sentence that was in any way out of the ordinary, it pounded its meaning into conformity with the nearest stereotype it had soaked up from its training. When given *The wolf ate a chicken,* it interpreted the meal as "cooked chicken meat," because that's what *chicken* usually refers to when it comes after the verb *eat.* When given *The plate broke,* it interpreted *plate* as "a vase or window," since those tend to break. When told that *John touched Mary,* it interpreted *touch* as "hit," since most instances of touching result from a motion. And when given *The bat broke the window,* which is ambiguous between an animal flying into it and an object being swung against it, it bred a chimera that meant "A bat [the animal] broke the window using a baseball bat"—the one interpretation that people *don't* make. That's what you can get when meanings are molded by expectations and context rather than assembled from rules and entries: an affectionate man falsely accused of beating his wife, and a club-wielding *Vespertilio pipistrellus.*

Radical Pragmatics, I think, is at odds with a fundamental design spec of language. Language is a lever with which we can convey surprising facts, weird new ideas, unwelcome news, and other thoughts that a listener may be unprepared for. This leverage requires a rigid stick and a solid fulcrum, and that's what the meaning of a sentence and the words and rules supporting them must be. If meanings could be freely reinterpreted in context, language would be a wet noodle and not up to the job of forcing new ideas into the minds of listeners. Even when language is used nonliterally in euphemism, wordplay, subtext, and metaphor—*especially* when it is used in those ways—it relies on the sparks that fly in a listener's mind as the literal meaning of a speaker's words collides with a plausible guess about the speaker's intent. In chapter 8 we will see that much of our social life is tacitly negotiated by means of these clashes.

LINGUISTIC DETERMINISM

In Isaac Bashevis Singer's children's story "The Elders of Chelm and Genendel's Key," adapted from a body of Yiddish folklore about a mythical town of fools, the elders grapple with a shortage of sour cream that imperils the celebration of an upcoming holiday on which blintzes are eaten. The elders "pulled at their beards and rubbed their foreheads, signs that their brains were hard at work," until one of them came up with an idea: "Let us make a law that water is to be called sour cream and sour cream is to be called water. Since there is plenty of water in the wells of Chelm, each housewife will have a full barrel of sour cream." The narrator notes that as a result of this law "there was no lack of 'sour cream' in Chelm, but some housewives complained that there was a lack of 'water.' But this was an entirely new problem, to be solved after the holiday."[64]

The whimsy of the story comes from our knowledge that people understand reality independently of the words used to describe it. But while we all know when drawing on common sense that thoughts can't be pushed around by words, many people hold the opposite belief when they intellectualize. The idea that the language people speak controls how they think—linguistic determinism—is a recurring theme in intellectual life. It was popular among twentieth-century behaviorists, who wanted to replace airy-fairy notions like "beliefs" with concrete responses like words, whether spoken in public or muttered silently. In the form of the Whorfian or Sapir-Whorf hypothesis (named after the linguist Edward Sapir and his student Benjamin Lee Whorf), it was a staple of courses on language through the early 1970s, by which time it had penetrated the popular consciousness as well. (While writing this book, I had to stop telling people that it was about "language and thought" because they all assumed it was about how language *shaped* thought—the only relation between the two that occurred to them.) The cognitive revolution in psychology, which made the study of pure thought possible, and a number of studies showing meager effects of language on concepts, appeared to kill the hypothesis by the 1990s, and I gave it an obituary in my book *The Language Instinct*. But recently it has been resurrected, and "neo-Whorfianism" is now an active research topic in psycholinguistics. Several recent studies purporting to show that language determines thought have been widely reported in the media.[65]

Linguistic Determinism is the third radical theory that I will contrast with conceptual semantics. According to conceptual semantics, the meanings of words and sentences are formulas in an abstract language of thought. According to Linguistic Determinism, the language we speak *is* the language of thought, or at least structures it in major ways. Let me say at the outset that language surely *affects* thought—at the very least, if one person's words didn't affect another person's thoughts, language as a whole would be useless. The question is whether language *determines* thought—whether the language we speak makes it difficult or impossible to think certain thoughts, or alters the way we think in surprising or consequential ways. The nagging problem with Linguistic Determinism is that the many ways in which language might be related to thought tend to get blurred together, and banal observations are often sexed up as radical discoveries. When *Newsweek* reported a study on whether there are more words for snow in Eskimo languages than in English, the writer, Jerry Adler, noted:

> It is not hard to see why this mundane observation should have emerged as one of the handful of facts most liberal-arts majors retained from their educations. Simple to grasp, it had implications so profound that anyone who stayed awake through his introductory-psychology course could feel like another Descartes. For if the Eskimos use many different words for things that English speakers lump into one category, does it not follow that they actually perceive the world differently? That Eskimos do not grasp the unity among all forms of frozen precipitation, while non-Eskimos do not see the differences, at least until they try to lift a shovelful of slush? Like, is that heavy—the idea, not the shovel—or what?[66]

As it happens, words for snow in languages like Yupik and Inuit are probably no more numerous than in English (it depends on how you count), but that hardly matters.[67] The idea that Eskimos pay more attention to varieties of snow *because they have more words for it* is so topsy-turvy (can you think of *any other reason* why Eskimos might pay attention to snow?) that it's hard to believe it would be taken seriously were it not for the feeling of cleverness it affords at having transcended common sense. Not only does a Whorfian explanation of Eskimo words for snow reverse cause and effect, but it exaggerates the depth of the cognitive difference between the peoples involved in the first place. As *Newsweek* noted, even if an Eskimo typically does pay more attention to varieties of snow, all it would take is a shovelful of slush to get a non-Eskimo to notice the differences.

There are many ways in which language connects to thought, some banal, some radical. Because the radical versions are the ones that excite people but the boring versions are the ones consistent with the discoveries, it's essential to sort them out. Let me begin with five banal versions of the Whorfian hypothesis:

1. Language affects thought because we get much of our knowledge through reading and conversation. It's unlikely, for example, that people could know the concept of "Tuesday," the fact that Caesar conquered Gaul, or the doctrine of original sin unless they learned them from other people via language. This version of the Whorfian hypothesis is utterly trite because it's pretty much a restatement of what language is—a means of communication—together with the observation that language is important in human affairs, which no one can doubt.

2. Slightly more interesting is the fact that a sentence can *frame* an event, affecting the way that people construe it, in addition to simply conveying who did what to whom. The preceding chapter teemed with examples: the choice of construction can determine whether listeners think of an event as causing water to move or causing a glass to become full, whether they think of it as merely having happened or as having been caused to happen, and so on. The ability of words to frame an event has long been used in rhetoric and persuasion (*pro-choice* and *pro-life, redistribution* versus *confiscation, invading* versus *liberating*), and its effects are easy to document. The psychologists Amos Tversky and Daniel Kahneman, for example, showed that doctors will opt for a cautious public-health program (as opposed to a risky one) when it is framed as saving the lives of 200 people out of 600 who are vulnerable, but will eschew the same program when it is framed as resulting in the deaths of 400 people out of the 600.[68]

Naturally it is fascinating to see how languages provide the means to frame events, and that is a major goal of this book. But the fact that they can

do so is just an extension of the observation that we use language to communicate (version 1). No one is *forced* to construe a situation the way a
speaker frames it (that's why we laugh when Woody Allen says that he
snapped his chin down on some guy's fist) any more than they are forced to
believe anything else a speaker tells them (which is why we laugh when
Chico Marx says, "Who are you going to believe, me or your own eyes?").
When we have nothing to go on but a speaker's words, we might be persuaded by the way he or she frames things, just as we may be deceived by a
false eyewitness report. But as we will see in chapter 5, people have the cognitive means to *evaluate* whether a framing is faithful to reality; the framing
does not lock their minds into one way of construing the world.

3. The stock of words in a language reflects the kinds of things its
speakers deal with in their lives and hence think about. This, of course, is
the obvious non-Whorfian interpretation of the Eskimo-snow factoid. The
Whorfian interpretation is a classic example of the fallacy of confusing
correlation with causation. In the case of varieties of snow and words for
snow, not only did the snow come first, but when people change their
attention to snow, they change their words as a result. That's how meteo-
rologists, skiers, and New Englanders coin new expressions for the stuff,
whether in circumlocutions (*wet snow, sticky snow*) or in neologisms (*hard-
pack, powder, dusting, flurries*). Presumably it didn't happen the other way
around—that vocabulary show-offs coined new words for snow, then took
up skiing or weather forecasting because they were intrigued by their own
coinages.

But could the causal arrow point in the other direction? It's certainly
plausible. When linguistically curious people encounter an unfamiliar
word or construction, they may make an effort to learn it, and in doing so
may pay attention to an aspect of the world that would otherwise pass them
by. But even in this scenario, it's people's interests and knowledge and reasoning that guide the process more than the word itself. When people hear
a name, they need to be interested in the family of concepts it belongs to in
order to bother to learn its meaning. People can hear hundreds of names
for kinds of birds (*vireo, kinglet, murre,* and so on), but unless they are bird-
watchers, the words will go in one ear and out the other. And in order to
learn the words, people have to develop the underlying concepts. When I
encounter the word *neutrino* and learn what it means, I'm really learning

some physics, not just some English. The same point holds for more pedestrian concepts.[69]

4. Since language works by evoking meanings, and since meanings are continuous with thoughts arrived at by other means (such as seeing and inferring), then if one uses the word *language* in a loose way to refer to meanings (as opposed to the actual words, phrases, and constructions that make up a language), then language affects thought—language *is* thought—by definition. This is uninteresting because it is just a slack way of using the word *language*, one that makes it impossible even to state the idea that we think in a medium other than language.[70]

5. When people think about an entity, among the many attributes they can think about is its name. That means that if people are asked a nebulous question with no correct answer and no real-life consequences, they can base their response on the name of the thing they are asked about. For example, if I give you three chips evenly spaced along the color spectrum (say, blue, greenish blue, and green) and ask you to choose the two that belong together, then with nothing else to go on you may pick the two chips that can be described by a single word in your language.[71] This is the most common kind of experiment done to test the Whorfian hypothesis. Technically, it is an example of language affecting thought, since language affects whatever thinking goes into interpreting the experimenter's ambiguous question. But it says little about reasoning in problems that do have a correct answer, or in problems that are ambiguous but not so perfectly balanced that the word can be given the deciding vote.

Now here are two versions of the Whorfian hypothesis that are more interesting. But as we shall see, they too fall well short of Linguistic Determinism.

6. Any computational system must have the means to store the intermediate products of its computations.[72] Computers, for example, swap information in and out of the central processing unit, storing it temporarily in RAM or on a hard disk (hence the flickering icon and annoying delays when you use your PC). A familiar example in everyday human computation consists of jotting down a carried digit at the top of a column of figures

while adding them up. When an intermediate product is stored in a human mind rather than on a disk or on paper, psychologists call it working memory. The two most vivid forms of working memory are mental images, also called a visuospatial sketchpad, and snatches of inner speech, also called a phonological loop.[73] People often use their phonological loop to keep a phone number in mind, to do mental arithmetic, and to keep track of left and right when following directions or remembering locations. The fact that language has a physical side—sound and pronunciation—makes it useful as a medium of working memory, because it allows information to be temporarily offloaded into the auditory and motor parts of the brain, freeing up capacity in the central systems that traffic in more abstract information. If a language provides a label for a complex concept, that could make it easier to think about the concept, because the mind can handle it as a single package when juggling a set of ideas, rather than having to keep each of its components in the air separately. It can also give a concept an additional label in long-term memory, making it more easily retrievable than ineffable concepts or those with more roundabout verbal descriptions.

I suspect that the dollop of additional retrievability and manipulability that a concept gains when it has its own word is a version of the Whorfian hypothesis that has a grain of truth to it and is not completely boring. But it is a far cry from Linguistic Determinism. For one thing, the stock of words in a language is not a closed inventory, permanently cramping its speakers' thoughts, but is constantly being expanded as people respond to cognitive needs by coining jargon, slang, and specialized words (as we will see in chapter 6). For another, many of the verbal strings that we use to offload cognitive demand are not part of the language itself. This will be an important point when we examine a recent claim for heavy-duty linguistic determinism, so let me elaborate on it here.

One of the ways in which the mind can use a snatch of language as a handle for a concept is in a verbal mnemonic. People often concoct mnemonics for arbitrary lists or easily confusable concepts, such as "Every Good Boy Deserves Favor" for the notes EGBDF of a musical staff, and "Red sky at morning, sailors take warning; red sky at night, sailors delight" for amateur weather forecasting. A particularly useful set of mnemonics is put to use in mental arithmetic. Many cognitive scientists believe that the human mind inherited two systems for keeping track of quantities from our mammalian ancestors.[74] One is an analogue estimation system, in which quantities are gauged in an approximate manner by relating them to some continuous magnitude in the head, such as a vague sense of "amount of stuff," or the extent of an imaginary line. The second system keeps track of exact quantities, but only up to a small limit, around three or four. Neither of these is adequate to thinking about quantities that are both exact and large, like 9 or 37 or 186,272. For that one needs to learn a number system in childhood and arithmetic operations in school. And when those operations are performed mentally, rather than on paper, they use snatches of language as lookup tables and scratchpads. For example, a mental echo of one stanza stored in memory ("eight times seven is fifty-six") can trigger another ("six plus nine is fifteen").

In a beautiful experiment on how stretches of language are and are not put to use in mental arithmetic, the cognitive scientists Stanislas Dehaene, Elizabeth Spelke, and their colleagues trained bilingual speakers of Russian and English to add pairs of two-digit numbers, presented to them in one of their two languages.[75] Then they tested them either in the language in which they had been trained or in their other language. Dehaene and Spelke predicted that when people did approximate arithmetic, such as estimating whether 53 plus 68 is closer to 120 or to 150, they would tap into their analogue estimation system (the mental number line), whereas when they did exact arithmetic, such as determining whether the sum of 53 and 68 is 121 or 127, they would mutter words to themselves—in this case, words in the language in which they had been trained. So any reliance on words would have been evident in a slowdown when the participants were tested in the language they had not been trained in, and also in a slowdown when they were given new problems as opposed to being retested on the old ones. And sure enough, the language of testing made no difference when the participants had to *estimate* sums, but it did make a difference when they had to calculate them *exactly*. In a nice follow-up, Dehaene and Spelke scanned the brains of monolingual people doing the same two tasks, and found that areas dedicated to spatial cognition in both hemispheres of the brain were active when the participants were estimating, but areas associated with language in the left hemisphere were active when they were calculating exact sums.

Is this an example of how one's thinking depends on one's language? Not really. For one thing, the abilities deployed in the experiments have nothing to do with the English and Russian languages themselves. They depend on mnemonics that exploit the resources of language (namely, strings of silently murmured speech), but those mnemonics aren't a natural part of any language; they had to be invented separately in the history of the culture and learned in school well after the spoken vernacular was mastered. Also, a functioning language system is not *necessary* for mathematical reasoning. In a paper called "Agrammatic but Numerate," the neuropsychologist Rosemary Varley and her colleagues tested three men with severe damage to the areas of the left hemisphere of the brain responsible for language.[76] The men could not speak or understand sentences, and even had trouble with spoken and written number words. Nonetheless, they had little trouble adding, subtracting, multiplying, and dividing multidigit numbers, including negative numbers, fractions, and bracketed expressions such as $50 - [(4 + 7) \times 4]$ (which resemble embedded syntactic structures such as *The man whom the woman likes is bald*).

The use of a mental echo-box in arithmetic and other forms of conscious reasoning is, I think, the main reason that many people report that they "think in" their native language. But these echoes are not the main event in thinking; most information processing in the brain is unconscious. Ray Jackendoff has noted that the conscious portion of thought seems to fall at an intermediate level in the hierarchy from raw sensation to abstract knowledge.[77] In vision, for example, we are vividly aware of the surfaces in front of us and of their colors and textures, but not of the wobbly, distorted shapes projected onto our retinas, or of the abstract categories to which the objects belong ("tool," "vegetable," and so on). And in language, he notes, people are most aware of the phonological level of language—the sequences of syllables making up words and phrases—not of the raw hisses and hums in the speech wave, or of the abstract syntactic and conceptual structures that give them meaning.[78] The result is that the sounds of language are the manifestations of thought that are most pungently present in our waking awareness, even if they are the tip of an iceberg of mental computation.

Here is another interesting version of the Whorfian hypothesis:

7. Every language forces speakers to pay attention to certain aspects of the world when they are composing or interpreting sentences. In English, for example, you have to worry about tense—the relative times of the event

you're speaking about and the moment you're speaking—whenever you open your mouth to utter a sentence. Other languages, such as Turkish, force speakers to indicate whether they witnessed the event firsthand or learned about it through hearsay. Another example: English spatial terms like *in* and *on* distinguish support from containment (more or less); Korean spatial verbs ignore this distinction but care about whether the fit between content and container is loose (like fruit in a bowl or flowers in a vase) or tight (like a Lego piece snapped onto another one, a cassette in its case, or a ring on a finger). One more: English verbs (such as *float*) can combine the fact of motion with a manner of motion, relegating the direction of motion to a prepositional phrase, as in *The bottle floated into the cave.* Spanish and Greek verbs tend to combine the fact of motion with a direction of motion, relegating the *manner* of motion to an afterthought, as in *The bottle entered the cave, floating.*

So one way in which language *has* to affect thought is that speakers attend to different things as they select words and assemble them into a sentence—an effect called "thinking for speaking."[79] The question is whether a lifelong habit of attending to certain distinctions and ignoring others spills over into thinking for *thinking*—that is, reasoning about objects and events for purposes other than just describing them. Do English speakers, compared with the speakers of other languages, have trouble grasping the distinction between witnessed and nonwitnessed events, or loosely versus tightly fitting containers, or motion in a direction? To ask this question is to answer it; clearly we command these distinctions as we negotiate the social and physical world. So while "thinking for speaking" is probably the most active topic in neo-Whorfian studies, the researchers have shied away from these tests of Linguistic Determinism and aim at far weaker possibilities. For example, they ask whether English speakers, when given hazy tasks like picking an odd man out in a series of actions, are less likely than Korean speakers to pick the action that differs in the tightness of a container. Some experiments find effects of language in these judgments; others don't.[80]

It shouldn't be surprising that the effects of thinking-for-speaking on thinking itself are small at best. Concepts like how things fit together and whether you saw an event with your own eyes or learned about it through hearsay are so important to human life that it seems unlikely that the historical accidents that shaped a given language would outweigh the cultural and cognitive resources we use to keep track of these things. And it's not even obvious that a lifetime of coding a distinction in a language should make the distinction more available in reasoning. It's just as likely that the opposite could happen. When a thought process becomes automatic, it gets deeply embedded in the language system as a cognitive reflex, and its internal workings are no longer consciously available, any more than we have conscious access to the finger motions involved in tying our shoes.

A couple of real-life examples can clarify why thinking-for-speaking may have few effects on thinking itself. Take the semantics of tense. Tense is a prominent feature in the grammar of English, and by Whorfian reasoning a lifetime of practice should make English speakers acutely sensitive to the relative order of when something happened and when one is speaking. Yet detectives and prosecutors know otherwise: automatic computation of the order of events is encapsulated in the language system, leading suspects to betray themselves with their words. In 1994 Susan Smith, who had drowned her two sons and claimed that they had been kidnapped, incriminated herself by telling reporters, "My children wanted me. They needed me. And now I can't help them." Her reflexive use of the past tense betrayed her knowledge that they were already dead.[81] The same part of English grammar may have earned Scott Peterson a lethal injection: prosecutors pointed to the guilty mind that led him to refer to his wife and unborn son in the past tense before their bodies had been found.[82] So even with a strong motive—literally life or death—to think clearly about the order of events, an English speaker's habit of computing tense was of no help in on-the-spot reasoning about what they should and should not say.

Finally we get to the radical versions of the Whorfian hypothesis—to genuine Linguistic Determinism:

8. The words and grammatical structures of a language have a profound effect on how its speakers reason, even when they aren't actually speaking or listening.

9. The medium of thought consists of actual words and sentences in the language the person speaks. Therefore, people cannot conceive of a concept that lacks a name in their language, and the direction of causation runs from language to thought: the ineffability of a concept in one's mother tongue creates a permanent blind spot in one's ability to entertain it.

10. If two cultures speak languages that differ in the concepts they can express, their beliefs are incommensurable, and communication between them is impossible.

These are, of course, arresting claims, which would have far-reaching implications if they were true. So at this point your straw-detectors should be on high alert. But Linguistic Determinism is not in want of real spokespeople, indeed, Trampolines. Whorf himself famously said:

> We dissect nature along lines laid down by our native languages. . . . We cut nature up, organize it into concepts, and ascribe significances as we do, largely because we are parties to an agreement . . . that holds throughout our speech community and is codified in the patterns of our language. The agreement is, of course, an implicit and unstated one, *but its terms are absolutely obligatory.*[83]

Similar pronouncements have been made by many oracles in philosophy and literary criticism:

> We have to cease to think if we refuse to do it in the prisonhouse of language.
>
> —Friedrich Nietzsche

> The limits of my language mean the limits of my world.
>
> —Ludwig Wittgenstein

> Man acts as though he were the shaper and master of language, while in fact language remains the master of man.
>
> —Martin Heidegger

> Man does not exist prior to language, either as a species or as an individual.
>
> —Roland Barthes[84]

These sentiments are not confined to the humanities. After *Science* magazine carried a story about how English has become the lingua franca of science, it published a letter that said:

Language often leads thought. What will we be losing when all scientists write and think in a language that hems the descriptions of facts and theories into a single Subject-Verb-Object (SVO) order? I do not think that one universal SVO language in science, to the exclusion of others, should be underestimated in its potential for severely skewing how scientists look at the world, time, space, and causality, perhaps unconsciously closing off areas of investigation in a way that even the most pervasive Kuhnian paradigm does not.[85]

And this flagship of American science recently published a study on numerical abilities in a South American tribe, whose author, the psycholinguist Peter Gordon, wrote:

> Is it possible that there are some concepts that we cannot entertain because of the language we speak? At issue here is the strongest version of Benjamin Lee Whorf's hypothesis that language can determine the nature and content of thought. . . . The present study represents a rare and perhaps unique case for strong linguistic determinism.[86]

Most other neo-Whorfians are more featherbeddish. Their titles and summaries are teases for Linguistic Determinism, such as "language can affect the way you think," "language can restructure cognition," and (in academese) "language is thought of as potentially catalytic and transformative of cognition."[87] But the teases don't distinguish the mundane ways in which language affects thought (which is what the experiments show) from the sexy ones (the versions of Linguistic Determinism that people find intriguing).

A genuine demonstration of linguistic determinism would have to show three things. The first would be that the speakers of one language find it impossible, or at least extremely difficult, to think in a particular way that comes naturally to the speakers of another language (as opposed to merely being less in the *habit* of thinking that way). Another would be that the difference in thinking involves genuine reasoning, leaving speakers incapable of solving a problem or befuddled in paradox, rather than merely tilting their subjective impressions in inkblot-style judgments. And most important, the difference in thinking must be *caused by* the language, rather than arising from other reasons and simply being reflected in the language, and rather than both the language and the thought pattern being an effect of the surrounding culture or environment.

Let me conclude the chapter by examining three recent dramatic claims for linguistic determinism. They show how diverse kinds of evidence can be brought to bear on the hoary question of language and thought, and they are an excuse for us to shine a light on three fundamental categories of thought—objects, number, and three-dimensional space—each of which has been implicated in a recent study on language and thought.

In an ingenious study on the mental life of infants, the psychologists Fei Xu and Susan Carey showed that babies as old as ten months don't sort objects into kinds based on their shapes when they try to keep track of them. For a baby, it seems, an object is an object is an object.[88] Xu and Carey rigged up a display in which a duck emerged from the right edge of a screen, then went back behind the screen. Then another toy, such as a truck, emerged from the left edge and returned. This drama was repeated several times until the infants grew bored, whereupon the screen fell to the floor, revealing the toys. In one version of the experiment, the fallen screen revealed a truck next to a duck, just as you or I would expect. But in another version, a bit of stage magic left only a truck (or only a duck) standing. This is physically impossible, yet the ten-month-olds were unsurprised, looking at the truck or duck exactly as long as they would have if it had been only that toy that they had previously seen. As far as the babies were concerned, the same entity must have been emerging from both sides of the screen, and the fact that it sometimes looked like a truck and sometimes looked like a duck didn't bother them. When the experiment was done with twelve-month-old babies, though, the results were different. When the screen fell to reveal a single toy, the older babies stared at it puzzled, just as you or I would, with our knowledge that a truck is one thing and a duck is another.

What happened in those two months? Xu and Carey noticed that twelve months is the age at which most babies first respond to words. Perhaps, they suggested, it was learning the words that *caused* the older babies to distinguish the toys from each other in keeping track of how many there were. Xu and Carey supported this version of Linguistic Determinism by showing that during the transitional months, the babies who were capable of understanding some words tended to be the babies who were surprised when the falling screen magically revealed a single toy, whereas the babies without any words tended to be among those who were unsurprised. They also showed that a voice-over that named the objects ("Look, a truck! Look, a duck!") nudged nine-month-olds into noticing that the toys were different.

But on the face of it, a strong Whorfian theory of how we learn to distinguish kinds of objects is unlikely. Deaf people who grow up without a signed or spoken language certainly don't act as if they fail to distinguish bicycles, bananas, and beer cans when keeping track of the things around them.[89] And all of us can distinguish various whozits, thingamabobs, and whatchamacallits in our closets and drawers whose names we never learned. So another way of interpreting the baby experiment is to flip the causal arrow around. Babies learn words for things once they are old enough to distinguish the things in their minds. Indeed, it's hard to think of how babies *could* learn the name for a thing *unless* they could think about that kind of thing as differing from other kinds of things.

In recent years, one of the most powerful sources of evidence on language and thought has come from the exploding new field of animal cognition.[90] The coup de grâce for the idea that language is necessary to sort objects into kinds came from a series of experiments by the psychologists Laurie Santos, Marc Hauser, and their collaborators, performed on a colony of rhesus macaques in Cayo Santiago, a beautiful island off the coast of Puerto Rico that is swarming with monkeys.[91] An experimenter would get the attention of one of the monkeys and administer a version of the truck-duck experiment, but with objects that capture the attention of a typical monkey, such as a carrot and a squash. The monkeys were surprised when, after a carrot and a squash took turns emerging from behind the screen, only one or the other was there when the screen was removed—just as twelve-month-olds, and you and I, would be surprised. But monkeys, of course, don't know a word of English. Other studies show that baby monkeys (four months old) act like baby humans, unsurprised at the magic, suggesting that the primate brain has to develop to a certain point of maturity before it can individuate objects by their kind. It must be this accomplishment that allows babies to learn words, not the other way around.

The most unabashed claim for Linguistic Determinism in recent years is Peter Gordon's study of the number sense in an Amazonian native people.[92] As we have read, Gordon argued for "the strongest version" of the Whorfian hypothesis, and that's how the study was reported in the press in 2004. The Pirahã tribe of Brazil, like many other hunting-and-gathering peoples, count with only three number words, meaning "one," "two," and "many." Even these are used imprecisely, a bit like our expression *a couple*, which technically refers to two but is often used for other small numbers. The physicist George Gamow opened his delightful 1947 book *One, Two, Three . . . Infinity* with a joke about two Hungarian aristocrats who hold a contest on which of them can call out the largest number. The first one concentrates for a few minutes and then says, "Three." The second ponders the challenge for fifteen minutes and then says, "You win." As Gamow notes, the story is probably a malicious slander of aristocrats, but the conversation could well have taken place among many pre-state peoples. They would have been bested by an American child of kindergarten age, a testament to the magnificent accomplishment of the Western system of numbers that most of us take for granted.

I used to be baffled by the prevalence of "one, two, many" counting systems among illiterate peoples until I asked the anthropologist Napoleon Chagnon (who had studied another Amazonian tribe, the Yanomamö) how they get by. He said that in their everyday lives the Yanomamö don't need exact numbers because they keep track of things as individuals, one by one. A hunter, for example, recognizes each of his arrows, and thereby knows whether one is missing without having to count them. It's the same habit of mind that would make most of us pause if someone asked us how many first cousins we have, or how many appliances in our kitchen, or how many orifices in our head.

Recall that in addition to the universal human ability to represent sets of individuals, people can keep track of small exact numbers (up to three or four), and can also estimate much larger quantities, though only approximately (this was the analogue number system documented by Dehaene and Spelke in their study of bilinguals and brain scans). These two components of the number sense are present in babies and monkeys, and of course in all human societies.[93] More sophisticated systems capable of tallying exact large numbers emerge later, both in history and in child development. They tend to be invented when a society develops agriculture, generates large quantities of indistinguishable objects, and needs to keep track of their exact magnitudes, particularly when they are traded or taxed.

Gordon showed that the Pirahã are shockingly inept at any task that requires keeping track of exact numbers from three to nine. They can't look at a few nuts on a table and line up the same number of batteries beneath them, or draw a line on a piece of paper for every battery they see, or watch a few nuts being put in a can and guess, as they are being removed one at a time, when the can is empty. The responses of the Pirahã were not random: the more items they saw, the more items they indicated, on average. But the responses were highly inexact, and they got worse as the numbers grew larger. (All of these are signatures of the analogue estimation system—which reinforces the notion that *this* component of the number sense exists independently of number words.) Gordon concluded that the lack of precise number thoughts among the Pirahã is caused by their lack of precise number words—the "rare and perhaps unique case for strong linguistic determinism."

But as the cognitive scientist Daniel Casasanto put it, this is a case of "crying Whorf": it depends on a dubious leap from correlation to causation.[94] It can't be a coincidence that the Pirahã language just happens to lack big number words (unlike the English language) and the Pirahã speakers just happen to hunt and gather in remote stone-age villages (unlike English speakers). A more plausible interpretation is that the lifestyle, history, and culture of a technologically undeveloped hunter-gatherer people will cause it to lack both number words and numerical reasoning. (Indeed, Daniel Everett, the linguist who studied the Pirahã for twenty-three years, rejected Gordon's conclusion and attributed their limitations in numerical reasoning to general patterns in their culture.)[95] The reason the non-Whorfian interpretation is plausible is that we don't find modern urbanized societies that lack an elaborate system of number words, nor do we find hunter-gatherer societies that have them. Granted, a people could hardly have developed into an urban civilization without number words and number concepts, so we wouldn't expect a modern society to lack number words and still be modern. But that's just the point—when the need arises, both number words and numerical reasoning are soon developed from existing cognitive resources.

It's not that it's *impossible* for a kind of language to part company with a kind of society, a circumstance that would make Whorfian hypotheses forever untestable in principle. Languages evolve and diverge in numerous ways because of the internal dynamics of pronunciation and grammar and the vagaries of history. For these reasons, similar societies can have different kinds of languages, such as Hungarian and Czech or Hebrew and English. For Linguistic Determinism to be true, these typological differences alone—and not any correlated differences in the type of society—would have to channel the thoughts of the respective societies and speakers in different directions. In the example at hand, there would have to be peoples that were prevented from developing the suite of cultural practices that included counting *because* of the historical accident that their language happened to lack number words, while similar peoples, who were lucky enough to speak a language with number words, took off into mathematical sophistication. In reality, history shows that when societies become more settled and complex, whether on their own or under pressure from their neighbors, they quickly develop or borrow a counting system, regardless of their language type.[96]

Could there be a control group for the Pirahã—a people whose culture was similar to theirs but whose language differed in its inventory of number words? Such a people would afford a true test of Linguistic Determinism, unconfounded by culture. Amazingly, not only does a control group exist, but it was described in a different paper in the very same issue of *Science*.[97] The Mundurukú are *also* an illiterate hunter-gatherer people in Brazilian Amazonia, but their language has number words up to five. This is not, however, enough to grant them exact number *concepts* up to five. Dehaene, together with the linguist Pierre Pica and their collaborators, showed that the Mundurukú, like the Pirahã, used number words (other than "one" and "two") approximately: the words for three, four, and five were not invariably used when those quantities had to be named, and they were sometimes used when nearby quantities had to be named. And as with their fellow Amazonians, the Mundurukú's ability to visualize the result of a subtraction (say, when they saw a computer animation of five dots going into a can and four coming out, and had to guess how many were left) was imperfect for numbers greater than three, and got worse and worse for larger and larger numbers. So the presence of additional number words in their language did little or nothing for their exact number sense.

If the Mundurukú had numbers for three, four, and five, why didn't they use them exactly? The investigators put their finger on the problem: the Mundurukú don't have a counting routine. It's tempting to equate a use of the number *five* with the ability to count five things, but they are very different accomplishments. Counting is an algorithm, like long division or the use of logarithmic tables—in this case an algorithm for assessing the exact numerosity of a set of objects. It consists of reciting a memorized stretch of blank verse ("one, two, three, four, five, . . .") while uniquely pairing each foot in the poem with an object in the spotlight of attention, without skipping an object or landing on one twice. Then, when no object remains unnoticed, you announce the last foot you arrived at in the poem as the numerosity of the set.[98] This is just one of many possible algorithms for ascertaining numerosity. In some societies, people pair up the objects with parts of their body, and I know several computer programmers who count like this: "Zero, one, two, three, four. There are five." Now, the counting algorithm we teach preschoolers, like the more complex mental arithmetic we teach school-age children, co-opts words in the language. But it is not part of the language, like subject-verb agreement, nor does it come for free with the language. So in the case of the number sense, the proper comparison—similar cultures, different languages— *refutes* Linguistic Determinism rather than supporting it. The prerequisite for exact number concepts beyond "two" is a counting algorithm, not a language with number words.

Turning from objects and numbers to space, we find the centerpiece of the neo-Whorfian movement, a set of studies by the anthropologist Stephen Levinson and his colleagues aiming to show that a language's spatial terms determine how its speakers use the three dimensions of space to remember the locations of objects.[99] Levinson's group examined Tzeltal, a language spoken in the Chiapas region of Mexico by a Native American people, descendants of the Mayans whose civilization flourished from A.D. 250 to 900. Tzeltal has no general words for "left" or "right." The closest it has are terms for the left or right arm or leg, but the terms are rarely used to refer to the left side of an object, table, or room.[100] Instead, the Tzeltal speakers describe spatial arrangements relative to the mountain slope that dominates their villages. The spatial vocabulary of Tzeltal includes words that mean "up-the-slope" (which is roughly southward), "down-the-slope" (roughly northward), and "across-the-slope." These coordinates are used not just when traipsing up and down the mountain but also when on flat terrain or indoors, and even when describing the arrangements of small objects. According to Levinson, Tzeltal speakers say "The spoon is downslope of the teacup," not "The spoon is to the right of the teacup."[101]

Levinson and his colleagues write that "a speaker of such a language cannot remember arrays of objects in the same way as you and I."[102] They note that Tzeltal speakers confuse mirror images, but are uncanny in knowing which way is north or south—even indoors, and even when they are blindfolded and spun dizzy, as if they had a compass embedded in their heads (as some species of navigating birds do). For example, in one anecdote, a Tzeltal speaker arrives one night at a hotel in an unfamiliar city far from home and asks her husband whether the hot water comes out of the "upslope" or the "downslope" tap.

In several experiments, Levinson's group sat people at a table looking at three toys—say, a fly, a fish, and a frog—arranged in a row from left to right. Then they swiveled the participants around 180 degrees to face a table that had been behind them, handed them a set of toys, and asked them to lay the toys out to be "the same" as they were on the first table:

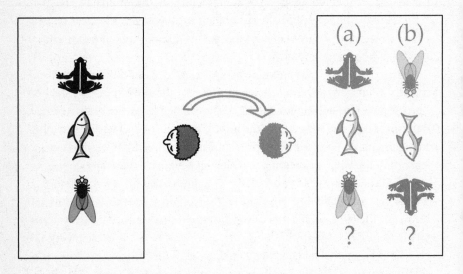

The task is ambiguous. "The same" could mean "the same arrangement relative to the environment," in which case the fly would now be on the person's right, but from a bird's-eye view at the same end of the table, shown as (a) in the diagram. Or it could mean "the same arrangement relative to the person," in which case the fly would still be on the person's left, though from a bird's-eye view it would be at the *opposite* end of the table, as in (b). The Tzeltal speakers tended to arrange the row of toys on the second table in the same way with respect to the environment, as in (a). But speakers of Dutch (which, like English, has general terms for "left" and "right") preserved the left-to-right arrangement from their vantage point, flipping them with respect to the world, as in (b). Therefore, Levinson concluded, "use of the linguistic system . . . actually forces the speaker to make computations he or she might otherwise not make."[103]

As we saw, a true demonstration of Linguistic Determinism would have to show three things: that the speakers of a language find it impossible, or at least extremely difficult, to think in the way that the speakers of another language can; that the difference affects actual reasoning to a conclusion rather than a subjective inclination in hazy circumstances; and that the difference in thought is caused by the difference in language, rather than merely being correlated with it for some other reason such as the physical or cultural milieu. Despite its status as the current poster child for neo-Whorfianism in psycholinguistics, the demonstration fails all three tests.

To understand what is really going on here, we need to look first at how space *can* be thought and spoken about.[104] People don't have GPS receivers in their heads to receive signals from satellites in geosynchronous orbit. Instead, they have to pick a reference frame that can be reliably identified by different people (or by the same person at a different time) and then specify an object's direction and distance relative to it. For the up-down dimension, gravity is a ubiquitous and ever-present reference frame. But the other two directions are problematic, because there are no compasses or "you are here" displays strewn through the world to orient us in the north-south and east-west directions.

One option is to look for a geocentric reference frame: a north-south or east-west axis aligned with landmarks, a ridge, or some other feature of the terrain that is anchored to the planet. A geocentric frame has the advantage of staying put, so something "pointing east" will always point east regardless of where you are standing. But it has the disadvantage of being unavailable when you're indoors or away from home, and often of being stable in just the wrong way. Any object or part that can move will maintain a constant location relative to some *other object* it is attached to, not relative to the world. The handlebars are always at the front *of the bicycle* (not north or south or east or west), and the cold-water tap is always on the right side *of the sink,* regardless of which direction the bicycle and the sink are facing.

Describing the shape or parts of a movable object calls for an object-centered reference frame: a coordinate system that is skewered through a salient object, allowing its parts or other objects to be located relative to its top, bottom, front, and sides. This, too, has advantages and disadvantages. Though it is useful for recognizing shapes and for keeping track of the arrangements of objects, an object-centered frame has trouble distinguishing the two horizontal dimensions consistently. Some objects have a natural front and back (a bicycle, a television set, a refrigerator), but many do not, like a tree or a floor lamp. Worse, apart from a few man-made shapes like cars and letters, almost everything in the world lacks a consistently distinguishable left and right side. Photographs are often published in mirror-reversed form without anyone noticing, and in 2000 the U.S. Postal Service inadvertently printed a stamp showing the Grand Canyon as it would appear in the looking glass. Unlike a previous snafu involving the same stamp design (whose caption located the natural wonder in Colorado rather than Arizona, a geocentric error that led to the recall of one hundred million stamps), this new error, they figured, would make no difference in how the stamp looked to people, so they left it uncorrected. [105]

That sets the stage for a third kind of coordinate system, the egocentric frame, in which people mentally crucify themselves on axes that define top and bottom, front and back, and left and right relative to their own bodies, which, conveniently, they always carry around with them. One problem with an egocentric frame is that people do move around, so the frame is unhelpful in locating things that aren't attached to us, unless we agree to stand in a certain place facing a certain way. Another is that our bodies and brains are mostly symmetrical, and as a consequence we have the devil of a time distinguishing left from right. Children often write letters backwards, and have trouble remembering which shoe goes on which foot. Adults are liable to misremember which way Lincoln is looking on the penny and whether Whistler's mother is facing left or right. [106]

Researchers in spatial cognition believe that people (and many other animals) are born with an ability to use all three reference frames, depending on the task and the circumstances. A simple case in which we can feel ourselves flipping between an egocentric and an object-centered frame is when we see a particular shape as an upright diamond and then see it as a tilted square. [107] A dramatic illustration is this display by the psychologist Fred Attneave, in which the shape in the top right can be perceived as either a diamond or a square, depending on whether it is mentally grouped with the horizontal or the diagonal line of shapes and hence spitted with the horizontal or the diagonal axis:[108]

Our swordsmanship in parsing the visual world has a counterpart in the multiple frames made available by our language. Many English spatial terms, such as *front* and *right,* can be used either in an egocentric way (*to the right of the bicycle*) or in an object-centered way (*the bicycle's right pedal*). The language also has a lexicon of geocentric terms. Not only are there words for the compass directions, but we have words such as *uphill, downhill, seaward,* and *shoreward,* and phrases like *toward the lake* and *away from the hills.* The psycholinguist Lila Gleitman tells her audiences about an island whose inhabitants, like the Mayans, locate many places and directions with the help of geocentric terms. The island is Manhattan, and the terms are *uptown, downtown,* and *crosstown.* Similarly, in Boston the subway system reckons directions with the terms *inbound* and *outbound.*

Given the usefulness of the three kinds of frame, each making up for deficiencies in the others, it would be surprising if a particular people lacked the ability to use one simply because of accidents in the history of their language. And indeed, the very fact that English, like Tzeltal, has geocentric terms, which its speakers understand perfectly well, pulls the rug out from under a strong Whorfian interpretation of the differences between Mayans and Americans in arranging toys on a table. English speakers are certainly capable of using geocentric coordinates, and many do so effortlessly. I know several people who can point to north in a windowless room through dead reckoning, and I once had a house renovated by a contractor who spoke of every fixture, even the small ones, in terms of north, south, east, and west (which were prominent because of a north-south coastline visible through the windows). Similarly, when I visited Southern Utah University, located between parallel mountain ranges in the basin-and-range topography of the American West, I learned that the faculty and upperclassmen referred to locations in buildings by their compass directions (*the northeast lounge, the south lecture hall*), much to the confusion of the freshmen. Even among English speakers, then, geocentric terms may be used when the terrain offers a conspicuous visual frame of reference. So the availability of *left* and *right* in English, and the absence of terms like "up-the-slope" and "across-the-slope," does not seem to "restructure" the cognition of English speakers; they still have the ability to locate things geocentrically when the circumstances make it useful.

What about the speakers of Tzeltal? The psychologists Peggy Li, Linda Abarbanell, and Anna Papafragou conducted several experiments in Chiapas to see if the absence of "left" and "right" and the presence of "up-the-slope" and similar terms really restructured *their* cognition, leaving them unable to locate objects egocentrically.[109] They blindfolded the Mayans and spun them in a chair, asking them to retrieve a coin that had been hidden in one of two boxes. In one experimental condition the boxes were placed on the floor, so their locations were constant in a geocentric frame. In another condition each box was at the end of a beam that was attached to the side of the chair and spun around with it, so the boxes' locations were constant in an egocentric frame. The Mayans could do either task—indeed, they were a bit better at finding the coin that followed them around egocentrically, despite their language. So Tzeltal speakers *can* use an egocentric frame when the situation calls for it, just as English speakers can use a geocentric frame.

Of course, in Levinson's original experiments, the English and Tzeltal speakers did behave differently when swiveling around and lining up toys on a second table. But as we have seen, this is an inkblot test with no correct answer: "make it the same" could mean the same with respect to the person's body, *or* it could mean the same with respect to the world, and the experimenters refused to tell the participants which they meant (since of course they didn't mean either). If the Whorfian effect is simply a matter of tilting people one way or the other in a toss-up, it should be easy to make the difference go away, or at least shrink considerably, by providing hints as to which option makes more sense in the circumstances. Li, Abarbanell, and Papafragou showed that when the task called for it, the Tzeltal speakers could indeed go either way. They could be trained to look at objects on one table and duplicate their arrangement on a second table *either* preserving the north-south orientation *or* preserving the left-right orientation. And, in collaboration with the psychologists Lila Gleitman and Randy Gallistel, the threesome turned Americans into Mayans by simple expedients like testing them outdoors (making the terrain more salient), gluing a landmark like a duck pond to one end of the table, or walking the participants around to the other side of the same table rather than swiveling them around to a different table.[110]

Strike three for Linguistic Determinism comes, as it did for Eskimo snow and Pirahã number, from a test capable of distinguishing causation from correlation. Even when we acknowledge a difference between Tzeltal and English speakers—namely, that Tzeltal speakers are more likely to extend a geocentric reference frame to a tabletop scale—we can state the difference in two ways:

1. Tzeltal speakers habitually reckon directions with respect to the terrain, and this is reflected in their language (non-Whorfian interpretation).
2. The Tzeltal language has terms for directions relative to the terrain, and this causes Tzeltal speakers to reckon directions that way (Whorfian interpretation).

How do we decide which is correct? We can see if there is some feature of the habitat or lifestyle of the Tzeltal speakers that would lead them to pay more attention to north and south than to left and right, independent of their language. As it happens, there isn't just one such feature but many. Unlike college students in the United States and the Netherlands, Tzeltal speakers live their lives in the shadow of a large mountain slope. They are farmers, and spend much of their time outdoors, many of them trudging up the slope semiannually between their farming plots in the highland and lowland regions.[111] And they seldom travel outside the territory. All of these lifestyle choices make the local terrain far more present in their consciousness than it is for peripatetic, indoor-dwelling Americans and Netherlanders. On the other side, Westerners spend a much greater portion of their lives reading, which plunges them into a land of text in which the left-to-right direction is unignorable. (The U.S. Postal Service would never have tolerated a stamp with a mirror-reversed photograph of a sample of text—say, the HOLLYWOOD sign.) So we have plenty of reasons to expect Mayans to rely more on the terrain and Americans and Dutch to rely more on left and right relative to their bodies, without even glancing at their languages.[112]

The knot was cut by Li, Abarbanell, Gleitman, and their collaborators, who noticed that another Mayan people living in Chiapas speak a language, Tzotzil, that does use terms for left and right to refer to relative directions. In other ways, though, their culture is similar to that of the Tzeltal speakers.[113] And sure enough, the Tzotzil speakers arrange toys just as the Tzeltal speakers do. Once again we find that it is the culture and environment, not the language, that leads to differences in how readily one or another mental ability is put to use.

So the new studies recruited to support Linguistic Determinism are consistent only with a mundane version of the Whorfian hypothesis in which speakers of different languages tilt in different directions in a woolly task, rather than having differently structured minds. And even those differences may have been caused not by their language but by features of their culture and environment that are reflected in their language.

The reason I have gone through the prominent new Whorfian claims is not as an exercise in debunking. I did it partly to show how people's perennial curiosity about language and thought can be addressed scientifically, and partly for the opportunity it offers to peer at the mind's ability to reason about objects, numbers, and the three dimensions of space. But I also wished to reinforce a major theme of this book: that language is a window into human nature, exposing deep and universal features of our thoughts and feelings; the thoughts and feelings cannot be equated with the words themselves. So let me conclude with some positive arguments for a language of thought as part of a bigger picture of how the mind works, and a view of where we stand with regard to the three radical theories.

One reason that the language we speak can't be *too* central in our mental functioning is that we had to learn it in the first place. It's not hard to imagine how language acquisition might work if children could figure out some of the events and intentions around them and tried to map them onto the sounds coming out of their parents' mouths.[114] But how a raw stream of noise could conjure up concepts in the child's mind out of nothing is a mystery. It's not surprising that studies of the minds of prelinguistic infants have shown them to be sensitive to cause and effect, human agency, spatial relations, and other ideas that form the core of conceptual structure.[115]

We also know that human thoughts are stored in memory in a form that is far more abstract than sentences. One of the major discoveries in memory research is that people have poor memories for the exact sentences that gave them their knowledge. This amnesia for *form,* however, does not prevent them from retaining the *gist* of what they have heard or read.[116] In a classic experiment, people were presented with sets of related sentences such as *The tree was in the front yard, The ants ate the jelly, The tree shaded the man, The jelly was sweet, The jelly was on the table,* and so on. Soon afterward they were given a list of sentences and asked to tick off the ones they had seen. When faced with sentences that were consistent with a composite of the meanings of the original ones, such as *The ants ate the sweet jelly* or *The tree in the front yard shaded the man,* they swore that they had seen them before, even more confidently than with the sentences they *had* seen.[117] This suggests that stretches of language are ordinarily discarded before they reach memory, and that it is their meanings which are stored, merged into a large database of conceptual structure.

Another reason we know that language could not determine thought is that when a language isn't up to the conceptual demands of its speakers, they don't scratch their heads dumbfounded (at least not for long); they simply change the language. They stretch it with metaphors and metonyms, borrow words and phrases from other languages, or coin new slang and jargon. (When you think about it, how else could it be? If people had trouble thinking without language, where would their language have come from—a committee of Martians?) Unstoppable change is the great given in linguistics, which is not what you would expect from "a prisonhouse of thought." That is why linguists roll their eyes at common claims such as that German is the optimal language of science, that only French allows for truly logical expression, and that indigenous languages are not appropriate for the modern world. As Ray Harlow put it, it's like saying, "Computers were not discussed in Old English; therefore computers cannot be discussed in Modern English."[118]

Perhaps the deepest reason that the effect of language on thought must be limited is that language itself is so badly suited as a medium of reasoning. Language is only usable with the support of a huge infrastructure of abstract mental computation. Not only are sentences cluttered with information that is tailored for auditory communication—such as the sounds of speech, the ordering of words in time, and many devices for engaging the attention of a listener—but it fails to contain information that is essential for lucid inference.[119] The most obvious shortcoming is polysemy. No thinker who is compos mentis could fail to distinguish an opening in a wall from a pane of glass, a folio of wood pulp from a newsgathering corporation, an assembly process lasting ten months from an edifice spanning ten stories, or a species of animal from a beast in the flesh. (Experimental psychologists will test anything, and indeed they have shown that people don't blur the senses of a polysemous verb.)[120] Yet that is what the common words of English, if used as an internal medium of thought, would force a thinker to do. It won't do to appeal to the mind's ability to resolve polysemy in context, because we're talking about the very part of the mind that would *do* the resolving, and that part has to distinguish the categories that are shmooshed together in a word.

In that regard, each of the radical theories about language and thought refutes one of the others in a game of rock-paper-scissors. Differences among languages, the point of pride for Linguistic Determinism, is a headache for Extreme Nativism, which assumes that concepts are innate, hence universal. The precision of word senses, which Extreme Nativism uses to discredit definitions, casts doubt on Radical Pragmatics, which assumes that one's knowledge of a word is highly malleable. And polysemy, which motivates Radical Pragmatics, spells trouble for Linguistic Determinism, because it shows that thoughts must be much finer-grained than words.

The theory of conceptual semantics, which proposes that word senses are mentally represented as expressions in a richer and more abstract language of thought, stands at the center of this circle, compatible with all of the complications. Word meanings can vary across languages because children assemble and fine-tune them from more elementary concepts. They can be precise because the concepts zero in on some aspects of reality and slough off the rest. And they can support our reasoning because they represent lawful aspects of reality—space, time, causality, objects, intentions, and logic—rather than the system of noises that developed in a community to allow them to communicate. Conceptual semantics fits, too, with our commonsense notion that words are not the same as thoughts, and indeed, that much of human wisdom consists of not mistaking one for the other. "Words are wise men's counters," wrote Hobbes; "they do but reckon by them; but they are the money of fools." Centuries later, Siegfried Sassoon invoked a similar association when he wrote:

> Words are fools
> Who follow blindly, once they get a lead.
> But thoughts are kingfishers that haunt the pools
> Of quiet; seldom-seen . . .[121]

4

CLEAVING THE AIR

Ever since I nearly missed a trip some years ago when an alarm clock failed to go off, I've set *two* alarms the night before a flight, one on my personal digital assistant (PDA) and the other on the clock radio in my bedroom or hotel. Since I find the chime of the PDA to be less rude an awakening than the honking of an alarm clock, I set the PDA to go off a minute before the clock. For many mornings, then, over a span of years, I have experienced the chime of the PDA followed a number of seconds later by the blare of the clock. And according to a well-known theory of the perception of causation associated with the philosopher David Hume, I should think that the chime *causes* the blare.[1]

Of course I think nothing of the sort. The cause of the clock's noise-making, I firmly believe, was my fiddling with the buttons before I went to sleep. I think this despite the fact that the interval between cause and effect can vary between eight hours and three, despite the fact that the alarm doesn't always go off (since there are so many things that can go wrong in setting a digital alarm clock), and despite the fact that I have only the vaguest idea of how a digital clock works (I think it has something to do with charges in silicon chips).

Yet despite the tenuous connection between twiddled buttons and blaring alarms (and the more immediate connection between the PDA's chime and the clock's alarm), my conviction of the true cause remains unwavering. That's why, when the alarm clock fails to go off, I don't shake my PDA or hold it up to the light but instead think back to my interaction with the clock the night before. Maybe I'm not smart enough to set a digital alarm clock (I failed to notice the P.M. light, or confused the A and B alarms, or set the alarm to MUSIC but left the radio dial between stations). Maybe the designers of the clock are not smart enough to make an appliance that a typical person can set. Maybe a part of the clock—a wire or chip inside it— is burned out. Maybe the clock's workings were addled by cosmic rays, or gremlins, or the moon rising in Sagittarius. But somehow, I feel sure, the happenings in the clock have *some* intelligible cause, which is to be found not in whatever happens to precede them but in some force or mechanism with causal powers.

People assume that the world has a causal texture—that its events can be explained by the world's very nature, rather than being just one damn thing after another. They also assume that things are laid out in space and time. "Time is nature's way to keep everything from happening at once," according to a graffito, and "Space is nature's way to keep everything from happening to *me*."[2] But in people's minds, time and space are much more than that. They seem to have an existence even when there are no events to keep apart; they are *media* in which the objects and events of our experience must be situated—and not just real objects and events, but imagined ones too.

The human imagination is a wondrous concocter. We can visualize unicorns and centaurs, people who are faster than a speeding bullet, and a brotherhood of man sharing all the world. But there are many things we *can't* imagine, at least not in the form of a mental image.[3] It's impossible to visualize an apple next to a lemon with neither one to the right, just noncommittally "next to" each other (though of course we can *talk about* that arrangement, as I just did). And as with Alice's comment on the Cheshire Cat (that she had often seen a cat without a grin, but never a grin without a cat), we can't imagine an object that is symmetrical or triangular but that does not otherwise have a particular shape (in the case of a triangle, equilateral or isosceles or scalene).[4] We know that elephants are big and gray, take up space, and are at a particular location at any given time. But while I can imagine an elephant that isn't big and isn't gray, I cannot imagine an elephant that doesn't take up space or isn't located somewhere (even if I have it floating around in my mind's eye, it is *somewhere* at every moment).[5] In the old joke, a tourist seeking directions is told by a local, "You can't get there from here." We laugh because we know that it's in the nature of space that all its locations are connected. And as the cognitive psychologist Roger Shepard has noted, people often wish that they had an office with additional space, so they would have more places to put their books. But they never wish they had an office with additional *dimensions,* so they would have more ways to *arrange* their books. Continuous three-dimensional space is an ever-present matrix in which the objects of our imagination must be located.

Our mind's eye is also sentenced to live in a world of time. Just as we can imagine an empty space devoid of objects but cannot imagine a set of objects that aren't located in space, we can imagine a stretch of time in which nothing happens but cannot imagine an event that doesn't unfold in time or take place at a given time. We can imagine time slowing down, speeding up, going backwards, or stopping altogether, but we can't imagine time having two or three dimensions. In fact, it's not even clear that we do imagine time slowing down or stopping so much as we *simulate* those possibilities by imagining things moving at half throttle, or halting in freeze-frame, while time marches on as usual.

You might wonder whether these features of our experience come from the design of the mind or from the nature of the perceptible universe. After all, the world exists in three dimensions, unfolds in time, and obeys causal laws (at least on the scales detectable by our sense organs), and perhaps the mind simply reflects its observable surroundings. But there is a crucial difference between space, time, and causality as they are represented in our minds and as they exist in reality. Our intuitions of these entities are riddled with paradoxes and inconsistencies. But *reality* can't be riddled with paradoxes and inconsistencies; reality just *is.*

Take space. It has to be either finite or infinite, yet neither possibility sits well with our intuitions. When I try to imagine a finite universe, I get Marcel Marceau miming an invisible wall with his hands. Or, after reading about manifolds in books on physics, I see ants creeping over a sphere, or people trapped in a huge inner tube unaware of the expanse around them. But in all these cases the volume is stubbornly suspended in a larger space, which shouldn't be there at all, but which my mind's eye can't help but peek at.

An infinite universe might seem more congenial, since the mind's eye can fly through space indefinitely, with new expanses always materializing in the nick of time. But an infinite space, too, has disturbing implications.

Would an infinite amount of space have an infinite amount of matter in it? It's not just possible but likely: physicists have recently discovered that at large scales matter is distributed evenly throughout observable space.[6] That raises the possibility that an infinite space would be studded with an infinite number of universes. Since a given set of elementary particles can be in only a finite number of states and positions, there are only a finite number of possible arrangements of matter in a given volume. Combined with an even distribution of matter through space, this would imply that there are only so many possible universes, which would in turn mean that universes would repeat themselves in an infinite multiverse. If so, then about 10 to the 10^{28} meters away there is an exact replica of you reading an exact replica of this book, and somewhere else a replica of you that decided to put it down, and in still another universe one that is named Murray, and in yet another a replica with a hair sticking out—indeed, an infinite number of doppelgängers in their doppelgänger universes. This seems too much to stomach, yet it is an implication of the apparently innocuous intuition that space and matter go on forever.

Time, too, doesn't want to be either finite or infinite. It's hard to conceive of time coming into existence with the Big Bang, since we are apt to cheat and imagine a primeval empty space in which a little cosmic time bomb sits waiting to explode. Nor can we fathom an empty time stretching indefinitely in the past before it. At best we can rewind a blank and silent videotape, let the tape play for a moment, then rewind it some more, and so on, never really encompassing an infinity of pastness. Nor can we make sense of what time would mean in the absence of matter and energy. Nothing in that nothingness could distinguish one moment from the next, so we would have no way of understanding why the Big Bang went off at the moment it did go off, as opposed to a few trillion years earlier or later or never. Not to mention the disturbing possibility that if time goes on forever, a rerun of every possible event that has happened will happen again an infinite number of times, a cosmic version of *Groundhog Day*.

As with space and time, the causal grid that we imagine connecting all events cannot stand up to too much scrutiny. I set the alarm, causing it to go off later. But who set *me,* causing me to set the alarm? On the one hand, I can consider myself a heap of clockwork, the neurons in my brain impinging on one another like tiny gears and springs. Yet when I make an uncoerced decision it certainly *feels* like I'm choosing whichever option I want, rather than being the helpless housing of a chain of machinery. Nor can a bystander predict any but the most banal of my choices. On the other hand, I can make no sense of a free will that mysteriously ups and does things without a prior trigger or spark. How does it work? If it's truly random, how can it make choices that are sensible in context? And how can we hold it responsible for its choices if they occurred by chance? But if its choices do respond to the context, including our contingencies of moral credit and blame, in what sense is it truly free?

Space, time, causality. We can't think without them, yet we can't make sense of them. These ruminations on the infrastructure of our experience are not, of course, original; I have taken them (with some twists and embellishments) from the German philosopher Immanuel Kant (1724–1804).[7] Kant said he was awakened from his "dogmatic slumber" by reading Hume, particularly his skeptical probing of causality. Hume wrote that we have no justification for our belief that one event must follow another in the world. All we have is an *expectation* that one will follow the other, based on similar experiences in the past. In accord with the rest of his associationist psychology, Hume suggested that a causal intuition is just a habit stamped into the mind when we repeatedly observe one event and notice that another often follows it. One problem for Hume's account is why observers don't think that one alarm causes the other after hearing the sequence repeatedly. But the problem that roused Kant was that it can't explain our conviction that causes and effects are explicable by lawful forces that govern our universe. As William James put it in a later century, Hume's observer lived in "a world of mere WITHNESS, of which the parts were only strung together by the conjunction 'and.' "[8]

Real observers, Kant concluded, must live in a world of whatness, whereness, whenness, and becauseness, imposed by the way that a mind such as ours can grasp reality. Our experiences unfold in a medium of space and time, which isn't *abstracted* from our sensory experiences (the way a pigeon can abstract the concept of redness when it is trained to peck at a red figure regardless of its shape or size) but rather organizes our sensory experiences in the first place. We are not just a passive audience to these experiences but interpret them as instances of general laws couched in logical and scientific concepts like "and," "or," "not," "all," "some," "necessary," "possible," "cause," "effect," "substance," and "attribute" (the last two pertaining to our concept of matter, such as the ability to conceive of a melting ice cube and the puddle it turns into as the same stuff). These concepts must arise from our innate constitution, because nothing in our sensory experience compels us to think them. Observe as many falling apples as you want; nothing forces you to posit that they are objects tugged by universal gravitation, rather than your just sitting back and enjoying the spectacle like the patterns in a kaleidoscope. You can stare at a cow till the proverbial cows come home; nothing you observe will ever compel you to think "It's not a giraffe" or "All cows are mammals" or "At least one kind of animal eats grass" or "It must have had a mother" or "It can't be the cow that died last week."

Though space, time, and causality (together with logic and substance) organize our world, the paradoxes that infect these concepts—space and time being neither finite nor infinite, choices being neither caused nor uncaused—prove they are not part of the self-consistent world but part of our not-necessarily-consistent minds. There *is* a world, to be sure; it impinges on our sense organs, filling our minds with sensory content and thereby preventing our thoughts from being hallucinations. But since we grasp the world only through the structures of our minds, we can't, wrote Kant, truly know the world in itself. All in all, it's not a bad bargain. Though we can never directly know the world, it's not as if one could know the world without *some* kind of mind, and the minds we are stuck with harmonize with the world well enough for science to be possible. Newton, for example, wrote that in his theory "absolute, true and mathematical time, of itself, and from its own nature flows equally without relation to anything," and that "absolute space, in its own nature, without relation to anything external, remains always similar and immovable."[9] For Kant these are the mind's supports for negotiating reality, and it is futile to try to think without them or around them. He chides us with an analogy: "The light dove, cleaving the air in her free flight, and feeling its resistance, might imagine that its flight would be still easier in empty space."

This chapter is about space, time, causality, and substance as they are represented in language, in the mind, and in reality. I have framed the chapter with ideas based on Kant because the conceptual scaffolding that he said organizes our experience is also conspicuous in the organization of language. One could imagine a hypothetical language whose constructions were dedicated to kinds of sensory experience, like sights and sounds, to the major players in human ecology, like plants, animals, tools, and kin, or to human obsessions, like food, exchange, or sex. But real languages appear to be organized by Kantian abstract categories. We see them in the basic parts of speech: substance in nouns, space in prepositions, causality in verbs, time in verbs and in markers for tense. We saw them (in chapter 2) in the way that verbs enter constructions, which are selective about how something moves, whether it is a substance or an object, whether an event is instantaneous or protracted, and who or what caused it. And we see them in the everyday metaphors that pervade our language and reasoning, as when we say the price of gas can *rise and fall* like a balloon, when we try to count the events of 9/11 like sticks of butter, when we say that two cities can be *an hour apart* as if they were alarm clocks, and when we talk of Sonia *forcing Adam to be nice* or even *forcing herself to be nice* as if she were closing a jammed drawer. So even when our thoughts seem to be engaged in pure levitation, we find them cleaving the air, getting their traction from the invisible yet omnipresent conceptions of space, substance, time, and causality. To understand human nature, we need to take a good look at those conceptions.

This is not to say that Kant himself is a reliable guide to our current understanding of the nature of thought and its relation to the world. Many philosophers today believe that Kant's rejection of the possibility of knowing the world in itself is obscure, and most physicists dispute his blurring of the mind's experience of time and space with our scientific understanding of time and space.[10] Contrary to everyday experience, our best physics holds that space is not a rigid Euclidean framework, but is warped by objects, may be curved and bounded, is riddled with black holes and possibly wormholes, has eleven or more dimensions, and measures out differently depending on one's reference frame.[11] Time is not the steady dynamic flow of our experience but the fourth dimension of a static space-time, or perhaps the solution to a connect-the-dots game in a multiverse of all possible universes, each linked to the one that "succeeds" it like the next frame in a movie.[12] In all of these cases our best scientific understanding of time and space is wildly out of line with the mind's inclinations. Many physicists say that space and time, in the sense of empty media into which objects and events are slotted, don't exist at all, any more than something called "the alphabet" exists above and beyond the twenty-six letters that make up the alphabet.[13]

Also, Kant was a famously murky writer, and even today the experts disagree on whether he was making claims about the mind of *Homo sapiens* or giving specifications for a generic rational knower. I can't see how he could not be making claims about our minds, at least implicitly, and at least one Kant scholar, Patricia Kitcher, has argued that Kant was not just a great philosopher but an ambitious and prescient cognitive psychologist.[14] But whether Kant actually meant the ideas that today are often associated with his name, or only inspired them, at least two of those ideas are invaluable in making sense of the mind.

Kant tried to forge a synthesis of empiricism and rationalism which, in rough outline, works well in today's nature-nurture debate. The mind is not a mere associator of sensory impressions (as in the empiricism of his day and the connectionism of ours), nor does it come equipped with actual knowledge about the contents of the world (as in some versions of the rationalism of his day and in the Extreme Nativism of ours). What the innate apparatus of the mind contributes is a set of abstract conceptual frameworks that organize our experience—space, time, substance, causation, number, and logic (today we might add other domains like living things, other minds, and language). But each of these is an empty form that must be filled in by actual instances provided by the senses or the imagination. As Kant put it, his treatise "admits absolutely no divinely implanted or innate *representations*. . . . There must, however, be a ground in the subject which makes it possible for these representations to originate in this and no other manner. . . . This ground is at least innate."[15] Kant's version of nativism, with abstract organizing frameworks but not actual knowledge built in to the mind, is the version that is most viable today, and can be found, for example, in Chomskyan linguistics, evolutionary psychology, and the approach to cognitive development called domain specificity.[16] One could go so far as to say that Kant foresaw the shape of a solution to the nature-nurture debate: characterize the organization of experience, whatever it is, that makes useful learning possible.[17]

Also strikingly modern is Kant's characterization of space and time as media in which sensations are arrayed. Logically speaking, the visual field can be characterized as a large database of specks and lines, with each entry specifying a color, a brightness, a position, an orientation, and a depth. But psychologically speaking, position in space turns out to be very different.[18] Space is an ever-present medium into which visual content is placed, not just one of several entries in an object's database record. Recall the thought experiments on what we can visualize, like a horse's body with a man's trunk, and what we cannot, like a man and a horse standing next to each other with neither on the left.[19] Location is not just a mandatory feature of an object in the mind's eye but also the main attribute the mind uses to individuate and count objects. For example, we see this array as three objects—one that is striped in the leftmost position, one that is gray in the rightmost position, and one that is both striped and gray in the middle position:

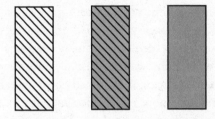

Theoretically, we *could* have seen the array as two objects: one that is striped and in the left and center positions, and one that is gray and in the center and right positions. But we don't see it in that manner, because the mind does not use color or surface marking as pegs to differentiate objects the way it uses location in space. Similarly, we can focus our spotlight of attention on a region in space, even an empty one, as when a basketball player, staring into his opponent's eyes, cocks an internal spotlight of attention to an empty place on the floor where he expects a teammate to appear. But experiments show that we have much more difficulty tuning our attention to all the patches of a given color or surface marking wherever they may be found.[20] Even the primary visual areas in the brain show the special organizing function of space. Each patch of cortical real estate is dedicated to a fixed spot in the visual field, and contours in the world are represented as contours across the surface of the brain, at least on a large scale.[21] Time also has a presence in the mind that is more than just any old attribute of an experience. Neuroscientists have found biological clocks ticking in the brains of organisms as simple as fruit flies. And just as we see stuff that is connected in space as an object, we see stuff that is connected in time as a motion, such as a trajectory or gesture, or, in the case of sound, as a melody or stretch of speech.[22]

So there is a model of space and time in human perception and imagination, and various models of space and time in reality (as it is characterized by the best physics of the day). But in the course of this chapter we shall see that the model of space and time expressed in *language* is unlike any of them. For starters, language is not an analogue medium but a digital one. Though we experience space as continuous and three-dimensional, and time as continuous and inexorably flowing, there is nothing three-dimensional or flowing about expressions for space and time in language, which are staccato strings of sounds. To preview the simplest of the examples we will meet, objects are located as *near* or *far,* events in the past tense or the present tense, with nothing to indicate the precise measurements of a yardstick or stopwatch. Also, the semantics of language pick out disembodied *aspects* of reality and combine and comment on them. I was able to describe in words the arrangements that you couldn't visualize, using expressions like *next to* and *symmetrical* which are agnostic about exactly how matter fills space. I can also describe an event without committing myself to a moment in time using a tenseless phrase like *for Bill to leave.* The selectivity of semantics allows our minds to swing through a universe of abstract concepts that is unanchored in the perceptual media for space and time that organize our immediate experience. Presumably this is the mental currency allowing modern scientists and mathematicians to describe space and time in utterly unintuitive ways.

As we shall see, the models of space and time (and substance and causality) embedded in language are foreign to physics and logic, the benchmarks commonly used by philosophers and psychologists to assess our cognitive performance. Nor are our cognitive models simply readouts of our sense organs or our neural clocks. They *are,* though, readouts of major aspects of human nature. Each of these modes of understanding has been shaped for distinctively human purposes, and they allow us to carve substance, space, time, and causality at the joints that matter most to our physical and social goals. Though Kant did not anticipate that our fundamental categories of understanding might be warped by their origin in what he called "the crooked timber of humanity," the peculiarly human versions of these categories organize our lives in far-reaching ways. They determine the kinds of entities we count and keep track of, the compartments into which we sort people and things, the way we manipulate the physical environment to our advantage, and the way we ascribe moral responsibility to people for their actions. For these reasons, the eccentric conceptions of substance, space, time, and causality with which we cleave the air propel not just our abstract cogitation but the daily course of our lives—our commerce, our politics, our legal disputes, even our humor.

GRINDING, PACKAGING, AND PIGEONHOLING: THOUGHTS ABOUT SUBSTANCE

Space, time, and causality, as important as they are in relating our thoughts to one another, are abstract frameworks, seldom consciously pondered except by philosophers and physicists. What we consciously think about are the actual entities that live in space and time and impinge on one another. And the most basic entities in our thoughts are the ones named by nouns— our concepts of people, things, and stuff. Nouns are the easiest words to identify across languages, are usually the first words learned by babies, and are the labels of the most stable and best-understood human concepts.[23] But for all that apparent simplicity, a look at the meanings of nouns leads us down another rabbit hole. Nouns are not mere pointers to hunks of matter in the world. When a human mind apprehends a person, an object, or a substance, it can construe it in very different ways, and this suppleness carries over to our thoughts about more vaporous entities.

The best way to appreciate what nouns mean is to begin with some examples that seem to mean nothing at all. Think about these sentences (many collected by the linguist Anna Wierzbicka):[24]

Boys will be boys.
A deal is a deal.
What difference does it make what kind you get? Coffee is
    coffee.
A man is a man, tho' he have but a hose upon his head.
Let bygones be bygones.
A woman is only a woman, but a good cigar is a smoke.
Que será, será; whatever will be, will be.
East is East and West is West, and never the twain shall meet.
You must remember this: a kiss is just a kiss, a smile is just a
    smile.

Let Poland be Poland.

A horse is a horse, of course, of course.

There is a joke about a woman who goes to see a divorce lawyer. He asks her, "How old are you?" "Eighty-two," she says. He continues, "And how old is your husband?" "Eighty-five," she answers. "And how long have you been married?" "Fifty-seven years." The lawyer can hardly believe it: "But why do you want a divorce now?" The woman says, "Because enough is enough!"

In their literal forms, these sentences look like empty tautologies, but of course they are not. Any speaker knows what they mean: a reminder that some entity has the essential qualities of its kind, despite one's hopes or forgetfulness to the contrary. "Boys will be boys" means that it's in the nature of young men to do things that are pointless, reckless, or tasteless. I last heard it when students on the men's rowing team sculpted a giant penis out of snow in the middle of Harvard Yard.

Since sayings composed from the X-is-X formula are not circular, the first X and the second X must mean different things. Sometimes a noun refers to something, serving as a pointer to an entity in the world that the hearer must identify. At other times a noun indicates a class or kind, characterized by a definition or stereotype. This distinction, between referring and predicating, is basic to language. A name, like *Canada* or *Luciano Pavarotti,* quintessentially refers to something, though names can be converted into category labels in expressions like *Every producer is searching for another Pavarotti.* Isolated nouns like *boy* and *coffee* are, by default, categories or kinds (boys in general, coffee in general), though they can be turned into referring expressions when they are plugged into phrases, like *that boy* or *the coffee grown in Brazil.* A basic sentence—perhaps a basic thought—refers to something in the subject and says something about its properties in the predicate.

In this book I have been insisting that the meanings distinguished by grammar single out major kinds of human thoughts and thus have real consequences in our lives, consequences that people care about, fight over, and pay for. Names for things are a prime example. We have already seen that the semantics of proper nouns animates the literary question of what we would mean by *William Shakespeare* if someone else turned out to have written his plays, and the practical question of how you would get your identity back if someone stole all your identifying information. Here are three other slices of life in which nouns matter.

The distinction between predicating and referring can be given a price. The most successful new corporation in this century so far is Google, which made its fortune by actually *selling noun phrases*. The problem with earlier Internet portals was that no one knew how to make money from them: users hated the banner ads and seldom clicked through to the advertisers. There is a saying in advertising that half of every ad budget is wasted, but no one knows which half—most people who see an ad have no interest in the product or service. The Google guys, Larry Page and Sergey Brin, had the brainstorm that the words people type into a search engine are an excellent clue to the kinds of things they might buy, making a search engine a good matchmaker for buyers and sellers. So together with the results of an untainted Web search, Google displays a few commercially sponsored sites relevant to the search term at the edges of the screen. Companies pay for this privilege by bidding in a continuous auction for the terms most likely to send eyeballs to their site. As a connoisseur of plurals, I was intrigued to learn that they cost more than singulars. *Digital camera* can be bought for seventy-five cents a click, whereas *digital cameras* fetches a dollar and eight cents. The advertisers know that the plural is more likely to be typed by people who are planning to *buy* a digital camera, though they don't know why.[25] The reason is that a bare noun like *digital camera* is generic, and is likely to be typed by someone who wants to know how they work. A plural like *digital cameras* is more likely to be referential, and typed by someone who wants to know about the kinds that are out there and how to get one.

A more aggressive use of corporate linguistics can be found in companies that are victims of their own success and need to reclaim the names of products that have come to be used generically as common nouns (the nouns are sometimes called "generonyms," and their transition from proper noun to common noun "genericide"). Few people realize that *zipper, aspirin, escalator, granola, yo-yo,* and *linoleum* used to be trademarked names for the products of particular companies. Today, the fear of genericide haunts the proprietors of *Kleenex, Baggies, Xerox, Walkman, Plexiglas,* and *Rollerblade,* who worry about competitors being able to steal the names (and the reputation they have earned) for their own products. Writers who use the names as verbs, as common nouns, or in lowercase type may find

themselves at the receiving end of a stern cease-and-desist letter. I suggest
that they reply in the manner of Dave Barry:

> I want to apologize in a sincerely legal manner to Jockey Inter-
> national Inc., which manufactures Jockey brand wearing ap-
> parel. Recently, I received a certified letter from Charlotte
> Shapiro, a Jockey brand corporation attorney, noting that, in a
> column concerning the issue of whether or not you can eat your
> underwear, I had incorrectly used the official Jockey brand name
> in the following sentence: "Waiter, are these Jockeys fresh?"
>
> Ms. Shapiro points out that the word "Jockey" is an official
> trademark, not a generic word for underwear, and it must be
> used "as an adjective followed by the common name for the
> product." Thus, my sentence should, legally, have read: "Waiter,
> there's a fly in these Jockeys!" . . .
>
> I have nothing but the deepest respect for the Jockey corpo-
> ration and its huge legal department. So just in case I may have
> misused or maligned any brand names in this column, let me
> conclude with this formal statement of apology to Nike, Crafts-
> man, Kellogg's, Styrofoam, Baggies, Michael Jordan, and any
> other giant corporate entity I may have offended: I'm really
> sorry, OK? So don't get your Jockeys in a knot.[26]

It's not just the owners of trademarks that get their Jockeys in a knot
when they hear a cherished referent identified with a common noun. People take even greater umbrage when they hear *themselves* labeled with a
common noun. The reason is that a noun predicate appears to pigeonhole
them with the stereotype of a category rather than referring to them as an
individual who happens to possess a trait. Logicians would be hard-pressed
to specify the difference, but psychologically it matters a great deal. You can
innocuously describe someone's hair as *blond, brunette,* or *red* (adjectives),
but it's a trickier business to refer to the whole person, particularly a woman,
as *a blonde, a brunette,* or *a redhead* (nouns). The terms seem to reduce the
woman to a sexually attractive physical feature, and to typecast her, according to old stereotypes, as flighty, sophisticated, or hot-tempered.[27] Since
metonyms derogate and hypernyms elevate (see chapter 2), nowadays we
refer to someone as *a woman with blond hair* rather than as *a blonde,* unless

the conversation was specifically about hair. An increased regard to the dignity of the individual has also led to the retirement of nouns for people with infirmities such as *cripple, hunchback, deaf-mute, mongoloid, leper,* and even *diabetic*. And today there is a movement in psychiatry to avoid calling someone *a schizophrenic* or *an alcoholic* and instead to refer to him or her as *a person with schizophrenia* or *a person with alcoholism*. A sensitivity to the typecasting power of nouns led the director and medical scholar Jonathan Miller to speak for many people of his ethnicity when he said, "I'm not a Jew. I'm Jew-*ish*. I don't go the whole hog."

It's still safe to refer to ordinary objects and substances with common nouns, and as we do so we display another kind of mental agility. At first glance, the conceptual distinction between an object and a substance seems to be captured in the linguistic distinction between a count noun and a mass noun.[28] Count nouns like *apple* and *pebble* tend to be used for bounded hunks of matter; mass nouns like *applesauce* and *gravel* tend to be used for substances without their own boundaries. The two kinds of noun are sharply distinguished by the grammar of English. We can enumerate and pluralize count nouns (*two pebbles*) but not mass nouns (*two gravels). When we refer to quantities, we have to use different quantifying words: *a pebble* is fine, but *a gravel is not; we talk about *many pebbles* but not **many gravel,* and we talk about *much gravel* but not *much pebble or *much pebbles. And mass nouns can appear in public naked—*Gravel is expensive; I like gravel*—whereas count nouns generally cannot—**Pebble is expensive;* *I like pebble.

An important clue to the mental model of matter behind mass nouns is that in some ways they act like plurals of count nouns. They share some of their quantifiers (*more applesauce, more pebbles*), their ability to appear naked in a sentence (*I like applesauce; I like pebbles*), and their ability to appear with spatial words like *all over,* as in *Applesauce was all over the floor* and *Pebbles were all over the floor* (compare **A rock was all over the floor*).[29] The overlap in grammar reflects a similarity in the way we conceive of substances (the things typically labeled with mass nouns) and multitudes (the things typically labeled with plurals), which together may be called aggregates.[30] Substances and multitudes both lack intrinsic boundaries, and can spill out into any shape. They can coalesce: put some pebbles together with some pebbles and you still get pebbles; put some applesauce together with some applesauce and you stil get applesauce. And they can be divided: half a load of pebbles is still pebbles; half a bowl of applesauce is still applesauce. None of this is true of the typical referent of a count noun, like a horse. No one is in doubt as to the boundary where a horse leaves off and the air around it begins, and when you put two horses together, or cut a horse in half, the result is not a horse—a point that, when applied to babies, is essential to the story about the wisdom of Solomon.

Where a plural differs from a mass noun is that it is conceived as a set of *individuals,* which can be identified and counted. This gives us a taxonomy of everything there is in the physical world.[31] A singular count noun like *pebble* stands for something that is bounded (delineated by a fixed shape) and not made up of individuals. A plural like *pebbles* stands for something that is unbounded and made up of individuals. A mass noun like *applesauce* stands for something that is neither bounded nor made up of individuals. All this suggests that our basic ideas about matter are not the concepts "count" and "mass" but the mini-concepts "bounded" and "made up of individuals." If so, we should see a fourth possibility: things that are both bounded *and* composed of individuals. We do indeed. These are collective nouns such as *committee, bouquet, rock band,* and those twee words for groups of animals that schoolchildren are forced to memorize but that no one ever uses, like *a gaggle of geese* and *an exaltation of larks.*

It may seem as if count and mass nouns are simply labels for hunks and goo, but that underestimates both our language and our minds. Within a language, it's often unpredictable whether a kind of matter is referred to with a count or a mass noun. We have *noodles* (count) but *macaroni* (mass), *beans* (count) but *rice* (mass), and both *hairs* and *hair,* leading Richard Lederer to ask in *Crazy English* why a man with hair on his head has more hair than a man with hairs on his head.[32] The choices differ somewhat from language to language—*spaghetti* is mass in English and count in Italian—and across historical periods of the same language. English speakers used to eat a substance called *pease,* as in the nursery rhyme *Pease porridge hot, Pease porridge cold.* But some grammatically zealous listener in the mists of history misanalyzed it as the plural form *peas,* from which it was a short step to *pea,* the count noun we use today. (The mathematical linguist Jim Lambek once speculated that a grain of rice will someday be called a *rouse.*) People who have learned English as adults have terrible trouble with all this.

My grandfather used to say that he combed his hairs, which is what one does in Yiddish, French, and many other languages.

Presumably the reason that languages often make arbitrary choices about the counthood or masshood of a kind of matter is that the mind can construe an aggregate either as a multitude of individuals or as a continuous substance. After all, when you grind rock into smaller and smaller pieces, from boulders to rocks to pebbles to gravel to sand to dust, there is a gray area in which people can construe the aggregate either as a collection of small things or as a continuous medium, depending on how close they stand, how recently they renewed the prescription for their eyeglasses, and perhaps even their personality (as in the person who can't see the forest for the trees). In that gray area, a language (or, more precisely, past speakers of the language) decides on a word-by-word basis which construal it forces upon current speakers when they use a word.

It's not just nouns that care about boundedness and individuals; verbs do, too. As we saw in chapter 2, verbs like *pour* require aggregates, like water or pebbles; verbs like *smear* and *streak* apply to substances; and verbs like *scatter* and *collect* apply to multitudes. This is because the concept of an action depends on the number and kind of things it affects, as in the difference between *eat* and *drink, throw* and *scatter, murder* and *massacre*. (The biologist Jean Rostand once remarked, "Kill one man, and you are a murderer. Kill millions of men, and you are a conqueror. Kill them all, and you are a god.")[33] The choices can differ among languages and even dialects, such as American and British English. I am always momentarily startled when my British editor offers to *collect* me at the hotel, as if he thinks of me as a bunch of smithereens.

The mind's power to construe matter as countable units or amorphous stuff is not just exercised at the intermediate settings of a rock grinder. *Anything* can be construed in these two different ways. We can always look at a cup (count) but think about the plastic composing it (mass), or look at some ice cream (mass) and think of the shape it assumes, such as a scoop or a bar (count). With many kinds of matter, previous speakers of the language have been considerate enough to bequeath us a distinct word for each construal. We have *butter* (mass) and *pat* (count), *gold* (mass) and *ingot* (count), even *shit* (mass) and *turd* (count)—a case in which taboo words adhere to the grammar of the rest of the language. As we shall see in chapter 7, not all of our profanity is so fastidious.

With all these examples of a language forcing speakers to construe an item as an individual thing or as continuous stuff when they use a word in a sentence, one might wonder whether our ability to *think about* matter in those ways depends on having first mastered the count-mass distinction—a version of Linguistic Determinism advanced by the logician W. V. O. Quine. The psychologists Nancy Soja, Susan Carey, and Elizabeth Spelke devised an experiment to find out. They presented two-year-olds (an age at which children show no signs of distinguishing count and mass nouns in their speech) with either an unfamiliar object, such as a copper plumbing tee, or a curved glob of an unfamiliar substance, such as pink hair gel.[34] They taught each child a word for the item by saying, "This is my tulver"—a sentence frame that is noncommittal about whether the noun is count or mass. Then they showed the toddlers two items—one of the same shape but a different substance, the other of the same substance but a different shape—and asked them to "point to the tulver." The question was whether the children treated what we construe as the object differently from what we construe as the substance without the benefit of clues from the English language.

Here is what happened. When the children had originally been shown what we think of as an object, like the copper tee, they pointed to an object of the same shape but a different substance, such as a plastic plumbing tee, not to the same substance with a different shape, namely a pile of copper bits. But when they had originally been shown what we think of as a substance, like the hair gel, they pointed to the same substance regardless of its shape, such as three smears of hair gel, and not to the same shape of a different substance, such an identically curved glob of hand cream. So well before children know how the English language distinguishes individual objects from portions of a substance, they distinguish them on their own, and generalize words for them accordingly. Names for solids with a noteworthy shape are taken to apply to objects of that kind; names for nonsolids with an arbitrary shape are taken to apply to substances of that kind.

Not only is a language unnecessary for inculcating in children the distinction between objects and substances, but it doesn't have a stranglehold on how its speakers construe matter when they are adults. Speakers can defy a language's stipulations by mentally packaging the referents of mass nouns (*I'll have two beers*) or by grinding the referents of count nouns (*There was cat all over the driveway*).[35] People also package mass nouns into kinds, as when they refer to different *woods* (like oak, pine, and mahogany) or *creams* (like Pond's, Nivea, and Vaseline—whoops, Pond's™ Cold Cream, Nivea™ Creme, and Vaseline™ Intensive Care™ Lotion). As we saw in chapter 3, this packaging and grinding is not without consequences: *We labeled the bloods,* for example, while common among medical workers, sounds odd to everyone else, and using *cat* as a mass noun for cat flesh is insensitive to the dignity of animals. But the fact that it can be done at all shows that the language doesn't dictate the construals available to the minds of speakers.

The intuitive materials-science behind the count-mass distinction assumes a Play-Doh world in which objects are molded out of a substance: rocks are made of rock, glasses are made of glass, beers are made of beer, cats are made of cat. The model breaks down when an object can't be construed as having been formed from a scoop of material. A television isn't made out of something called *television,* so we can't say that a steamroller left television all over the road. The distinction also breaks down when we put a substance under a powerful enough microscope. We use the word *rice* to refer to a cup of it, a grain of it, or even a fragment of a grain of it, but as we zoom closer and closer we reach a point at which we aren't seeing *rice* anymore (presumably there are no rice molecules, or rice atoms, or rice quarks). Perhaps if humans could see the crystals, fibers, cells, and atoms making up matter, we would never have developed a count-mass distinction in the first place. Practitioners of homeopathy, in which a substance is diluted so many times that (according to chemists) not a molecule remains, can be accused of taking the mental model of matter behind mass nouns far too seriously.

The count-mass distinction in our minds is not just unfettered by the object-substance distinction in the world; it is unfettered by the physical world altogether. It is best thought of as a cognitive lens or attitude by which the mind can construe almost anything as a bounded, countable item or as a boundariless, continuous medium. We see this in a distinctive kind of mass noun that does what count nouns usually do, namely, refer to bounded lumps of matter like chairs and apples. These are mass hypernyms (superordinates) such as *furniture, fruit, clothing, mail, toast,* and *cutlery.* Though they don't refer to a substance—chairs and tables aren't made of out of some ingredient called "furniture," nor are postcards and letters stamped out of a substance called "mail"—the words can't refer directly to the individual objects they stand for, either. They require a special classifier noun, as in a *stick* of furniture, an *article* of clothing, or the general-purpose classifier *piece:*

Dennis—NAS. North America Syndicate.

"WOULD YOU LIKE A PIECE OF TOAST FOR BREAKFAST?"   "I'D RATHER HAVE A **WHOLE** ONE, THANKS."

As Dennis will discover, a *piece of toast* (or mail or clothing or fruit or furniture) is not a piece at all. But we need to use *piece* as a classifier to bite off a chunk of fruithood or furniturehood or toasthood for us to identify and count (just as we use classifiers to bite off chunks of substances, as in *a sheet of paper, a blade of grass,* or *a stick of wood*). In English, mass nouns for objects tend to apply to classes of things that are heterogeneous in size and shape but are often acted upon collectively, like furniture in a van, fruit in a basket, clothes in a suitcase, or mail in a sack. But in some languages, like Chinese, *all* nouns behave like mass nouns, standing for the concept itself rather than for separate incarnations of it, and speakers may not count or pluralize them without the use of a classifier, as in "two tools of hammer" or "three rods of pen."

If count nouns and mass nouns can be applied to just about anything, why do languages bother with them? One reason is that they allow us to agree on how to isolate, count, and measure things. Imagine that someone asked you to "count everything in this room." What exactly would you count? The chairs? The chair legs? The colors? The walls? Should you add "1" for the room itself? The task is meaningless until some kind of unit is specified, and that's what count nouns do (it's no coincidence that they are called "count" nouns). Nor can you compare amounts without committing yourself to a count or a mass term. If Sally has one big stone and Jenny three much smaller stones, who has more? Again, the question by itself is unanswerable; it depends on whether you mean "more stone" or "more stones." Even four-year-olds know that these questions call for different answers (according to an experiment by the psychologists David Barner and Jesse Snedeker),[36] and an understanding of the difference in how to quantify matter is essential to our getting the joke in this cartoon:

Monty © United Feature Syndicate, Inc.

For the same reason, the simple judgment of whether two things are "the same" depends on our agreeing on the same *what*—a cup and a pile of cup shards can be the same *ceramic,* though they are not the same *cup.* The count-mass distinction, then, helps us agree on which individuals we treat as mental entities to be tallied and tracked, and which we treat as mere incarnations of a category.

If the count and mass nouns are cognitive attitudes rather than reflexes to kinds of matter, we should see them applied to entities that are not made of matter at all. Indeed we do. The count-mass distinction pops up in many ghostly realms of thought populated by things that don't have mass or take up space. We distinguish discrete *opinions* (count) from continuous *advice* (mass), *stories* from *fiction, facts* from *knowledge, holes* from *space, songs* from *music, naps* from *sleep, falsehoods* from *bullshit.*

Is the ability to construe abstract entities the way we construe things and stuff a late achievement of the mature mind, the result of extensive exposure to abstract count and mass nouns? The psychologist Paul Bloom has shown that the answer seems to be no: it comes naturally to children as young as three.[37] When kids hear a rapid string of chimes and are told, "These are feps—there are really a lot of feps here" (count noun), and are then asked to "make a fep" with a stick and a bell, they are likely to ring it once. When they are told, "This is fep—there is really a lot of fep here" (mass noun), and are then asked to "make fep," they are more likely to ring it multiple times. This corresponds exactly to what they did when the words referred to a physical aggregate, like lentils—they respond to *give a fep* with one lentil, and *give fep* with a handful. So children distinguish count nouns from mass nouns in the same way whether they refer to evanescent events or to physical objects (a feat of mental agility that, we shall see, underlies the semantics of time). Other experiments have shown that children can count other entities that aren't discrete objects, including collections, lobes, actions, holes, and puddles.[38]

So while our ability to think about things and stuff is surely rooted in our perception of lumps and gunk in the physical world, we easily extend it to the world of ideas. As a result we can publicly identify, track, and tally the contents of our consciousness, no matter how airy. Indeed, the ability to quantify the incorporeal is a signature of mental life. How do I love thee? Let me count the *ways*. Ten Jews, eleven *opinions*. There must be fifty *ways* to leave your lover. How many *times* must a man look up, before he can see the sky? Four be the *things* I'd been better without: love, curiosity, freckles, and doubt.[39] And of course, how many *events* took place in New York on the morning of September 11, 2001?

A GAME OF INCHES:
THOUGHTS ABOUT SPACE

"A game of inches" is an expression that has been applied to baseball, football, golf, and sex. But it really applies to any activity that involves moving in space, where a missed step or turn can be a matter of life and death. Assessing the layout of the world and guiding a body through it are staggeringly complex engineering tasks, as we see by the absence of dishwashers that can empty themselves or vacuum cleaners that can climb stairs. But our sensorimotor systems accomplish these feats with ease, together with riding bicycles, threading needles, sinking basketballs, and playing hopscotch. "In form, in moving, how express and admirable!" said Hamlet about man.

Yet when it comes to the *language* of space, we don't seem quite so express and admirable. It is said that a picture is worth a thousand words because a verbal description can leave people stymied in their attempts to form a mental image of the scene. Here are some examples I collected from a few days of reading the newspaper:

- "The first step in construction will be to protect the sheared-off column remnants from the twin towers with a pool-type liner and 12 inches of fill." [Is the liner above or below the fill? Are the column remnants only 12 inches high? Do they support the liner like tent poles, or pierce it like tent stakes?]
- "Designed to protect beachfront land by controlling beach erosion, rock walls are already in place on much of the town's bayside shores." [Are they parallel to the shore, or perpendicular to it?]
- "A 40-foot opening was made in the center of the steel wall that is at the end of the canal." [Does the wall block the end of the canal, like locks, or flank it, like a bus door?]
- "The maintenance crew apparently left a pressurization controller rotary knob out of place, according to the official connected to the investigation." [Did they remove the knob, or just set it wrong?]
- "To form an 'I' floodwall, sheet pilings, a sort of steel fence, are driven into the compacted dirt of the levee. Then reinforcing steel rods are threaded through the top of the piling, and concrete is poured to encapsulate the top of the piling and form the wall." [Huh?]

This imprecision can have serious consequences. Several terrible airplane crashes have been caused by miscommunication between pilots and air traffic controllers over the location and direction of aircraft.[40]

The discrepancy between our smooth and precise movement in space and our coarse and ambiguous language for space is rooted in the design of the brain, which has several systems for keeping track of the 3-D world. There is a complex network for sensorimotor coordination that includes the cerebellum ("little brain"), the basal ganglia, and several circuits straddling the central fissure of the brain. The system is mostly analogue and codes locations precisely, but it is largely invisible to conscious thought.[41] Within the visual brain, there is the "what" system, which runs along the bottom of the brain from back to front. It registers the shapes of letters, faces, and objects, and damage to it can cause dyslexia or agnosia, as in the man who mistook his wife for a hat (the title of the famous book by the neurologist Oliver Sacks). And there is the "where" system, which runs from the back of the brain up toward the top. It allows us to keep track of objects' locations, and damage to it can cause the syndrome called neglect, in which a person may fail to notice the furniture on one side of a room, fail to eat off one side of the plate, and fail to shave one side of his face. As with the rest of the brain, the "what" and "where" systems are duplicated in the left and right hemispheres, though the two sides are not identical in what they do. The "where" system in the right hemisphere is better at assessing analogue spatial relations, like whether two objects are exactly an inch apart. The "where" system in the left hemisphere is better at assessing digital spatial relations, like whether two objects are touching or whether one is to the right or the left of the other.[42]

It's no coincidence that the two major divisions of the primate visual brain are named for English interrogative pronouns. Of course it's the brain that came before the pronouns: we ask about *what* and *where* because our brain is designed to keep track of things and places. The distinction is mirrored in the vocabularies of most languages, where one finds a large class of nouns naming objects of different shapes (keyed to the "what" system in the left hemisphere) and a smaller class of words or morphemes specifying paths and places (keyed to the "where" system in the left hemisphere).[43] In English the distinction is extreme. A look at any visual dictionary shows that English has an enormous number of words for shapes, perhaps as many as ten thousand:[44]

Zippy—Bill Griffith. King Features Syndicate.

In comparison, English has only eighty or so spatial prepositions:

> about, above, across, after, against, along, alongside, amid(st), among(st), apart, around, at, atop, away, back, backward, behind, below, beneath, beside, between, beyond, by, down, downstairs, downward, east, far from, forward, from, here, in, in back of, in between, in front of, in line with, inside, into, inward, left, near, nearby, north, off, on, on top of, onto, opposite, out, outside, outward, over, past, right, sideways, south, there, through, throughout, to, to the left of, to the right of, to the side of, together, toward, under, underneath, up, upon, upstairs, upward, via, west, with, within, without

Space is also encoded in nouns like *edge* and *vicinity,* in verbs like *enter, spread,* and *cover,* and in suffixes like the ones in *homeward* and *Chicago-bound.*[45] Many languages rely on these forms more than on prepositions or their equivalent. But with all these devices, locations are generally carved up far more coarsely than shapes.[46]

The disparity comes in part from an inherent difference between the geometry of shape and the geometry of location. Specifying a shape can require as many pieces of information as the shape has facets, nooks, and crannies. But specifying the disposition of one object relative to another requires only six pieces of information. In theory, a language could locate any object exactly by building its prepositions out of six syllables: one each for the distance from a reference frame in the up-down, left-right, and near-far directions (perhaps using logarithmically scaled units anchored by a common object or body part), and one each for the angles of pitch, roll, and yaw (perhaps using angular increments of a sixteenth of a turn). In fact no language works that way. Language describes space in a way that is unlike anything known to geometry, and it can sometimes leave listeners up in the air, at sea, or in the dark as to where things are.

The first quirk is that spatial terms are highly polysemous.[47] It never occurs to most English speakers that our preposition *on* picks out not one spatial relationship (say, a thing resting on the top of another thing, like a book on a table) but several. Just think of the different meanings of *on* in a picture *on* a wall, a ring *on* a finger, or an apple *on* a branch. Even a language as close to English as Dutch treats these with different prepositions: *op* for a book on a table, *aan* for a picture on a wall, and *om* for a ring on a finger. Still worse is a preposition like *over,* which has more than a hundred distinct uses, including *Bridge over troubled water, The bear went over the mountain, The plane flew over the mountain, Amy lives over the hill, Barney spread the cloth over the table,* and *The book fell over.*[48] If you've ever wondered, as Richard Lederer does in *Crazy English,* why someone in love is *head over heels* (since our head is always over our heels—why isn't it *heels over head?*), the answer is that *over* can refer to a path of motion (as in *The cow jumped over the moon*), not just a location, so the smitten one is being depicted in mid-handspring.

It's not that English always lumps where other languages split. Many languages fail to distinguish *on* from *over,* using a single term for superadjacency, or fail to distinguish *in* from *under.*[49] And it's not that anything goes: languages tend to have terms for contact, vertical alignment, attachment, containment, and proximity, as if there were a cognitive alphabet of spatial relationships more basic than the prepositions of a given language. Then, when the languages lump various spatial relations together in one preposition, they adhere to a universal sense of which relations are most similar. For instance, English *on* collapses a book on a table (vertical alignment plus contact) with a picture on a wall (attachment), presumably because both involve a force keeping one thing in contact with another. For similar reasons, Berber *di* collapses attachment (a picture on a wall) with containment (a toy in a box), again seeing both as ways in which one object impedes another's movement; the two languages differ only in whether to divide the impediments in terms of verticality or containment. Spanish collapses all three in *en.* But no language collapses vertical alignment with containment while excluding attachment, or overness with aroundness while excluding on-ness, because those collections make no cognitive sense.[50]

Another way that spatial language comes up short is that its distinctions are digital—indeed, usually binary. In many languages the most basic spatial distinction is between something that is near the speaker and something that is some distance away, as in *here* and *there.* The distinction is relative, not absolute; as Stephen Levinson points out, *Put it there* has a very different meaning to a crane operator and to a brain surgeon. Most of the world's languages divide the space around the speaker into just these two regions, though about a quarter of them (including Spanish) make a three-way distinction among "near me," "far from me," and "in between," and a very few (such as Tlingit, a language spoken in Yukon) go to four, adding "very far from me." No language has spatial terms that measure out distance in actual units (though of course a culture with a counting system can spell it out with nouns and adjectives, as in *five thousand two hundred and eighty feet*).

The most common spatial distinctions in the world's languages are either-or: you're in or you're out, quite literally. It's not just that languages hack space into zones with fuzzy boundaries. Many of the spatial relationships they care about are *inherently* qualitative, involving distinctions that may loosely be called topological.[51] A topologist is said to be a mathematician who can't tell a doughnut from a coffee cup, because topology deals with qualitative traits like contact, containment, connectedness, and holes—properties that would not change if the world were made of Silly Putty and could be stretched without breaking. Among the more or less topological concepts coded in language are contact, containment, and attachment. The difference between topology and continuous spatial relationships is captured in Groucho Marx's sweet nothing, "If I held you any closer I'd be on the other side of you."

The spatial terms that *are* continuous are still oblivious to the smoothly varying quantities that go into distance, size, and shape.[52] The linguist Len Talmy notes that we use the same preposition *across* in describing an ant crawling across a hand and in recalling a bus trip across the country. Yet the ant displays a stepping motion and completes its journey within the span of attention of the viewer, who observes the motion, godlike, from above. And the bus glides along the highway in a trip that is endured over many days and places and can be pieced together only in memory. Despite the dissimilarity of the experiences, we apply a single preposition to both. This talent for geometric abstraction is visible in the language of children. The writer Lloyd Brown noted that his young daughter once said of two dogs trotting in tandem, "Look at those dogs running like a hook-and-ladder" (the long fire truck with a steerable cab at the back). Another time she asked him for a box of crayons "that looks like an audience"—not the flat box of eight, but the larger one with the crayons stepped in rows like a pitched balcony.[53]

These feats are possible because the part of the mind that interfaces with language treats objects *schematically,* in terms of the way they stretch out along each of the three dimensions of space.[54] In reality every morsel of matter has a length, a width, and a thickness, but when we speak of these morsels we pretend that some of the dimensions aren't there. In the simplest case we can think like geometers and conceive of a point as having zero dimensions, a line or curve as having one, a surface as having two, and a volume as having three. But we can also conceive of more complex shapes by combining and ranking the dimensions. An object is thought of as having one or more primary dimensions, which are what really count in reasoning about it, together with one or more secondary dimensions. A *road*, a *river*, or a *ribbon* is conceptualized as an unbounded line (its length, which serves as its single primary dimension) fattened out by a bounded line (its width, which serves as a secondary dimension), resulting in a surface. A *layer* or a *slab* has two primary dimensions, defining a surface, and a bounded secondary dimension, its thickness. A *tube* or a *beam* has a single primary dimension, its length, and two secondary dimensions, plumping out its cross-section.

Our minds can also focus on the boundaries of an object as if they were objects themselves. A geometer would say that a 3-D volume must be bounded by a 2-D surface, a surface by a 1-D edge, and a line by a 0-D point. But the mind sees more than that. We can also think of a *stripe*, which is 2-D, as bounding a 2-D surface, like the rim of a plate or the border of a rug; the fact that the stripe's *primary* dimension is 1-D is enough to make it the boundary of a surface. Likewise, an *end* is thought of as having zero primary dimensions and as bounding an object with one primary dimension. The word *end* thus embraces a set of entities that are completely different in Euclidean geometry: the 0-D point that bounds a line, the 1-D edge that bounds a ribbon, or the 2-D surface that bounds a beam. An *edge* works the same way, except that it has one primary dimension rather than zero. Often an end or an edge is thought of as including a teeny bit of the adjacent line or surface. That's why we can snip the *end* off a ribbon, or plane the *edge* off a board, feats that are, strictly speaking, geometrically impossible. When the 2-D boundary of a 3-D solid is granted a small bit of the adjacent stuff, we call it a *crust*. Though the word is most familiar from our encounters with bread, we also apply it to scabs and planets, despite their unignorable differences in size, composition, and edibility. The common thread is the dimensional geometry we apply to them.

The apparatus of primary and secondary dimensions, when applied to the negative space left behind when a portion of matter is scooped out of something, gives us our lexicon of nothings: the many words for nicks, grooves, dents, dimples, cuts, slots, holes, tunnels, cavities, hollows, craters, cracks, clefts, chambers, openings, and orifices.[55] The cognitive similarity of things and holes gives rise to many enigmas and paradoxes. We have already encountered the polysemy of *window* and *door* (which can be either openings or coverings). Philosophers worry about how to fit holes into an ontological taxonomy of all the kinds of things there are in the universe, because a hole can be tall, which makes it a kind of object, but that implies that it should be able to be heavy, too, like matter in general, of which objects are a subtype. But a "heavy hole" is as nonsensical as a "happy table" or a "green idea."[56] Head-scratching about holes—real things in the mind, nothings in reality—is not just an occupational hazard of linguists and philosophers.[57] A hole can be the answer to brainteasers like "What's the only thing you can put into a bucket that will make it lighter?" and "The more you take from me the bigger I grow; what am I?" They are the source of trick questions, like "How much dirt is in a hole six feet wide, eight feet deep, and five feet long?" (Answer: None.) Holes have also been used to create visual illusions, such as in the face-vase on page 43, the art of Escher and of Magritte, and the "Sea of Holes" sequence in *Yellow Submarine* in which the animated Ringo slips a black oval into his pocket and retrieves it later to deflate the bubble imprisoning Sgt. Pepper's Lonely Hearts Club Band ("I've got a hole in me pocket!" he reminds himself). And then there are the little balls of fried dough sold by Dunkin' Donuts, whimsically called "donut holes."

The idea that shapes can be cognitively melted down into schematic blobs skewered on axes originally came from a theory of shape recognition by the computational neuroscientist David Marr. Marr noted how easily people recognize stick figures and animals made from pipe cleaners or twisted balloons, despite their dissimilarity from real objects in their arrangement of pixels. He proposed that we actually *represent* shapes in the mind in blob-and-axis models rather than in raw images, because such a model is stable as the object moves relative to the viewer, while the pixels in the image are all over the place.[58] Schematic models are not the only way we recognize objects—we can, for example, recognize a shirt in a hamper just by its color and texture—but they do seem to populate the interface where vision meets language and reasoning.[59] Not only do nouns for shapes (like *ribbon, layer, crust, hunk,* and *groove*) get their definitions from this world of pipe cleaners, cutouts, and balloons, but we seem to *conceive* of the objects around us in these terms. Few people think of a wire as a very, very skinny cylinder and of a CD as a very short one, though technically that's what they are. We conceive of them as having only one or two primary dimensions, respectively. Nor do we ordinarily imagine a lake as a translucent chunk with a flat top, sharp edges, and numerous bulges molded to the shape of the lake bottom. We think of it as a 2-D surface.[60]

The schematic dimensionality of objects affects not just how we recognize and visualize them but sometimes how we reason about them. We tend to think of a box, for instance, as a 3-D hollow container. In a classic experiment on problem-solving taught to every psychology undergraduate, people are given a matchbook, a box of thumbtacks, and a candle, and are asked to figure out how they can attach the candle to the wall.[61] Most people are stumped. Thinking of the box as a container, it never occurs to them to empty it of thumbtacks and tack it to the wall, providing a 2-D shelf for the candle. And when it comes to our bodies, the conception works the other way around: we think of them as solids, not containers, which gives rise to some odd intuitions. Researchers in artificial intelligence who puzzle over the nature of common sense have noticed that if there's a bag in your car, and a gallon of milk in the bag, people agree that there's a gallon of milk in the car. But if there's a person in a car, and a gallon of blood in a person, people don't think there's a gallon of *blood* in the car. Emotion researchers have noted that most people are revolted at the idea of eating from a bowl of soup that they've spat into, though no one is disgusted by the thought that his or her mouth was full of saliva in the first place.[62] There is a joke about a little girl who is filling in a hole in her garden when a neighbor looks over the fence. He politely asks, "Hi! What are you up to?" "My goldfish died," replies the girl tearfully, "and I've just buried him." The neighbor asks, "Isn't that an awfully big hole for a goldfish?" The little girl tamps down the soil and replies, "That's because he's inside your stupid cat."

Let's get back to language itself. The schematic modeling of shapes is the kind of geometry that defines most spatial terms in English and other languages.[63] A preposition, for example, locates a figure relative to a reference object, and in doing so has to specify something about the shape of the figure and something about the shape of the reference object. The most common kind of preposition, like *in, on, near,* and *at,* says nothing at all about the figure being located, treating it as a 0-D point or lump.[64] (This is the core of the holism effect we saw in chapter 2, in which a motion or change is thought to affect an entity in its entirety.) Recall that anything at all can be *in* or *on* something, whether it is a pebble, a pencil, or a pad, and it doesn't matter which way it's pointing. The reference object, in contrast, has to have a certain geometry for a preposition to apply. *In*, for example, requires a 2-D or 3-D cavity. *Along* needs a 1-D primary axis: a bug can walk *along* a pencil, but not a CD, though it can walk along the 1-D *edge* of the CD. *Through* demands a 2-D opening or an aggregate, as when a fish swims through water or a bear runs through the woods. *Inside* demands an enclosure, usually 3-D.

Actual descriptions of things in space depend on the compatibility between the pipe-cleaner geometry of the objects and the dimensional demands of the spatial terms. Because a lake is conceived as having only two primary dimensions, you can't swim *inside the lake*, though that would seem to make geometric sense. Lederer asks why we say that something can be *underwater* or *underground* even though it's surrounded by, not beneath, the water or the ground. It's because *water* and *ground* are conceived as 2-D surfaces, not 3-D volumes, geologically improbable though that is. The dimensionality of an object is also the aspect of its geometry that modifiers "see" when they combine with it in a phrase. A *big CD*, for example, has to have an above-average diameter, not an above-standard thickness (that could only be a *thick CD*), and a *big lake* has to be one with an unusually large area, regardless of its depth; it can't be a few yards wide and a mile deep.[65]

We saw in chapter 3 that once a reference object is reduced to a few sticks, sheets, or blobs, it needs to be impaled with axes that identify its directions and allow the figure to be located relative to it. The most common way for a language to assign places and directions is to superimpose a human body on the reference object and co-opt an appropriate body-part term.[66] We see this in English in generic spatial terms like *back, face,* and *head*, and even more vividly in words that carry over more of the body's geometry, such as *eye* (of a needle or a storm), *nose* (of a plane), *foot* (of a mountain or table), *mouth* (of a river), *neck, elbow, finger, groin, flank, butt,* and one that seems to baffle the alien Mr. Pi in the *Monty* comic strip on the following page.

The body metaphor, needless to say, is not the only source of words for spatial relations. In chapter 3 we saw how different languages, and different terms within a single language, choose from a menu of reference frames.

Monty © United Feature Syndicate, Inc.

These include a gravity-based frame, used for *above;* a geocentric frame, used for *north;* an object-centered frame, used in *the car's right side;* and an egocentric frame, used in *behind the pole.* The menu is a product of our visual system's dexterity in spearing objects with different coordinate systems, as we saw in the shape that flips from a square to a diamond on page 145.

One of the reasons our descriptions of space can be maddeningly ambiguous is that our spatial terms don't nail down every degree of freedom in the way a reference object can be spindled with axes. If a sunbather is lying down with her knees supporting a book, and a fly lands on her thigh, we can say that the fly is *below her knee,* if we use gravity as the reference frame, or we can say that it's *above her knee,* if we use her body as the reference frame. Actually, the possibilities for confusion are even worse than that. Talmy asks us to imagine a speaker and a hearer in the back of a church.[67] A queue of people runs from left to right, facing the right wall. John stands in the middle of the line, though he has done an about-face and is facing left. Flanking him are two people, both facing the front of the church, one standing a bit closer than John to the altar, the other a bit closer to the entrance. The following page shows a bird's-eye view of the scene.

Now, who is *in front of* John? The speaker could answer that it is Person #1, if he aligns the reference axis with John's body. He could answer that it is Person #2, if he aligns it with the queue. He could answer Person #3, if he aligns it with the church and its inherent front and back. Or he could say it is Person #4, if he aligns it with the line of sight joining himself to John. Not all spatial terms in the world's languages are as ambiguous as the English *in front of,* but all of them are ambiguous in some ways.

Why is everyday spatial language so bad? Why do we describe space—the ubiquitous medium in which all experience unfolds—with terms that

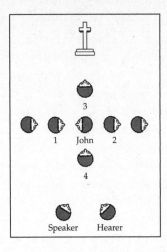

are so often ambiguous, digital, topological, schematic, and relative? As with all questions about the design of language (or anything else), the answer lies in the tradeoffs.

Recall my hypothetical ideal spatial vocabulary, whose prepositions were assembled out of six syllables, one for each of the degrees of freedom of an object's disposition relative to a reference frame. The obvious problem with this scheme is that we tend not to *like* six-syllable words, especially ones we have to use all the time, as we do spatial prepositions. (*In* and *to*, for example, are in the top-ten list of the most frequent words in English, and *on, at, by,* and *from* are close behind.)[68] Of course, each of these sesquipedalian words could be replaced by a unique short one, but that would multiply the number of distinct words children would have to memorize. Even if we made only seven gradations of length and angle, we would have to memorize more than a hundred thousand words just to master the language's spatial vocabulary, before even getting to *carburetor, doorknob,* and *trombone.* Tradeoffs between transparency, precision, word length, and vocabulary size were what doomed the various Enlightenment-era designs for a "perfect language," and in *Words and Rules* I argued that many features of real languages emerge as a compromise among these demands.[69]

Since words and syllables aren't free, languages economize when they can. One reason for the ambiguity of spatial words (and for ambiguity in general) is that many potentially ambiguous terms are clear in face-to-face conversation, where the speaker and the hearer share their surroundings and are aware at any moment of what the other one knows. But in newspapers and other texts, words are wrenched out of this intimate dialogue and have to be consumed by strangers at a distance.

Spatial terms economize in a second way. Of all the ways in which one object can hover around another, not all are equally worth distinguishing. Imagine you are in a rainstorm, ten feet from an overhanging ledge. Move one foot toward it; you still get wet. Move over another foot; you still get wet. Keep moving, and at some point you no longer get wet. Continue to move another foot in the same direction; you don't get any drier. So nature has set up a discontinuity between the segment of the path where gradual changes of position leave you equally wet and the segment where gradual changes leave you equally dry. And it is exactly at that discontinuity that one would begin to describe your position using *under* rather than *near*.[70] The digitized effects of analogue distance are the rationale behind the old saw that baseball is a game of inches, and similar sayings like "A miss is as good as a mile" and "Close only counts in horseshoes and hand grenades."

Spatial terms quantize space at the cusps where causal events play out differently on each side. As your palm gradually cups around a marble, the curvature at which you stop saying the marble is *on* the hand and start saying it's *in* it is more or less the shape that would prevent it from rolling off when you jiggle it. Likewise, a rope that is *around* a pole can do things that a rope which is merely *by* the pole cannot. A failure to attend to the qualitative semantics of a preposition can have tragic consequences:

NEWTON WOMAN RESCUED FROM FROZEN
POND IN LINCOLN DIES

A woman who fell through thin ice Sunday and was under water for at least 90 minutes died yesterday. Meanwhile, the Lincoln Fire Department said a miscommunication between the caller who reported the accident and the dispatcher significantly delayed her rescue. . . . [The] Lincoln fire department chief said the delay was due to a mix-up that led rescue workers to believe a woman had fallen on the ice, not through it, and that left the rescuers combing the woods to find the scene of the accident.[71]

A preposition, then, tends to cover a range of configurations that are similar in their manipulability, visibility, stability, and resting state. This digitization scheme is more economical than a full array of coordinates; by encoding a causal discontinuity in the world with the binary presence or absence of a symbol, you get more bang for the bit. But its bigger advantage is in making the causal powers of a configuration *explicit*. A spatial symbol, once coded in memory, lends itself to reasoning algorithms in a straightforward way. To determine whether something gets wet, you don't have do geometric calculations on a database of objects' positions; you can simply check whether the symbol "under" is present or absent.

Of course, this doesn't imply that there is a single, optimal way to carve up spatial relationships. Certainly not all languages carve them up in the same way.[72] Presumably this is because nature endows space with many causal cusps to choose among, and also because each language trades off expressiveness, precision, word length, and vocabulary size in a different way. But the quantization of spatial relations is universal, and causally important relations like contact, attachment, alignment, verticality, and proximity make their appearance in all the spatial vocabularies of the world.

When Paul Bloom and I first suggested that spatial terms align with interesting causal discontinuities, we had to go on our own intuitions.[73] Recently the idea has been corroborated in experiments by Kenny Coventry, Simon Garrod, and others. They presented people with photographs of strange arrangements of objects and asked them to rate the aptness of various prepositions in describing them.[74] The experimenters found that people's intuitions were sensitive not just to the pure geometry of the arrangements but to what the objects could *do*. A light bulb is considered to be *in* a socket when its base has been inserted, since that allows it to be illuminated, but a person is not *in* a car if only his arm extends in through a window, since that doesn't allow the car to move him or even shelter him. The position at which an umbrella is judged to be *over* a man depends on where it can best protect him against a driving rain, and a tube of toothpaste is judged to be *above* a toothbrush not when it is directly over its center of mass but when it is closer to the bristles, a compromise between geometry and functionality.

We see, then, that the conception of space expressed in language is quite different from the ubiquitous, continuous, empty Euclidean medium envisioned by Kant as the matrix of experience. It is composed of digital symbols, which idealize an object into rods, sheets, and blobs that are pierced by axes and assembled into an articulated structure. And those symbols correspond not just to expanses of matter and space but to the forces and powers that govern our use of containers, fasteners, and the business ends of tools. It's not that Kant was wrong when he implied that the mind has a pure conception of space, abstracted from its real-world contents. It's just that this conception is found at the two ends of the ladder of human cognition, skipping the level at which we habitually talk and think. At the bottom end, a spatial medium organizes our hardwired faculty of vision and imagery, as we saw at the beginning of the chapter in considering the role of space in the mind's eye. And at the top end, it can be acquired in school as part of the explicit system of knowledge about space that we call geometry.

THE DIGITAL CLOCK:
THOUGHTS ABOUT TIME

"Do not squander time," said Benjamin Franklin, "for that is the stuff life is made of." Our consciousness, even more than it is posted in space, unrolls in time. I can imagine abolishing space from my awareness—if, say, I were floating in a sensory deprivation tank or became blind and paralyzed—while still continuing to think as usual. But it's almost impossible to imagine abolishing *time* from one's awareness, leaving the last thought immobilized like a stuck car horn, while continuing to have a mind at all. For Descartes the distinction between the physical and the mental depended on this difference. Matter is extended in space, but consciousness exists in time as surely as it proceeds from "I think" to "I am."

As with every other aspect of human nature, it's been claimed that there are cultures out there that have no conception of time. The linguist Bernard Comrie examined the claims and has noted that they are not credible.[75] A person belonging to a culture with no conception of time could not generalize that people invariably are born, grow up, age, and then die, and thus would be unsurprised to meet someone who started out as a corpse, came to life as a senior citizen, grew younger and younger, and eventually disappeared into his mother's womb. Needless to say, there is no society populated by such madmen. And people in societies all over the world order the events in their autobiographies, genealogies, and histories, and their myths about such things as the creation of the world or the arrival of their ancestors.[76]

People also keep track of time in the words and constructions of their language. In many languages the ordering of events is expressed in adverbials like *yesterday* or *a long time ago*. And in about half the world's languages it is embedded in the grammar in the form of tense.[77] The semantics of time suggests that even the claim that many peoples conceive of time as cyclical should not be taken too literally. Though people are aware of the recurrence of days, years, and phases of the moon, it does not overwrite an awareness of the linear sequence of events that make up the flow of life. No language has a tense, for example, that means "at the present moment or at an equivalent point in a different cycle."[78]

But our intuitive conception of time differs from the ceaseless cosmic stream envisioned by Newton and Kant. To begin with, our experience of the present is not an infinitesimal instant. Instead it embraces some minimum duration, a moving window on life in which we apprehend not just the instantaneous "now" but a bit of the recent past and a bit of the impending future. William James called it "the specious present":

> The practically cognized present is no knife-edge, but a saddleback, with a certain breadth of its own on which we sit perched, and from which we look in two directions into time. The unit of composition of our perception of time is a duration, with a bow and a stern, as it were—a rearward- and a forward-looking end. . . . We do not first feel one end and then feel the other after it, and from the perception of the succession infer an interval of time between, but we seem to feel the interval of time as a whole, with its two ends embedded in it.[79]

How long is the specious present? The neuroscientist Ernst Pöppel has proposed an answer in a law: "We take life three seconds at a time."[80] That interval, more or less, is the duration of an intentional movement like a handshake; of the immediate planning of a precise movement, like hitting a golf ball; of the flips and flops of an ambiguous figure like those on pages 43 and 145; of the span within which we can accurately reproduce an interval; of the decay of unrehearsed short-term memory; of the time to make a quick decision, such as when we're channel-surfing; and of the duration of an utterance, a line of poetry, or a musical motif, like the opening of Beethoven's Fifth Symphony.

Time, at least as it is expressed in the grammatical machinery of language, also differs from Newtonian time in not being measurable in units. A language's tenses chop the ribbon of time into a few segments, such as the specious present, the future unto eternity, and the history of the universe prior to the moment of speaking. Sometimes the past and future are subdivided into recent and remote intervals, similar to the dichotomy between *here* and *there* or *near* and *far*. But no grammatical system reckons time from some fixed beginning point (as we do in our technical vocabulary with the traditional birth of Jesus) or uses constant numerical units like seconds or minutes.[81] This makes the location of events in time highly vague, as when Groucho told a hostess, "I've had a perfectly wonderful evening. But this wasn't it."

There is a close parallel in the degrees of precision that are available to languages in the way they express number, space, and time.[82] Using phrases composed of words, we can express quantities from the infinitesimally small to the infinitely large with any degree of precision, thanks to number phrases (*three hundred and sixty-two*), directions (*the third house on the right off Exit 23*), and dates and times (*seven forty-two P.M., May seventeenth, nineteen seventy-seven*). But if we restrict ourselves to simple words and compounds, the distinctions plummet into the dozens—with number, a few words like *one, two, twelve,* and *twenty* (or, in many languages, only "one," "two," and "many"); with space, prepositions like *across* and *along;* with time, temporal adverbs like *now, yesterday,* and *long ago*. And when we rely on the distinctions coded in grammar, the distinctions become still more schematic. In English, we distinguish only two numbers (singular and plural), and perhaps five tenses (depending on how you count); this is similar to the way that many languages dichotomize location into "here" and "there."

The imprecision in the way languages express time is related to the imprecision in the way we experience and remember it. Though no one experiences time as coarsely as the handful of distinctions in a tense system would suggest, we don't live by a mental stopwatch either.[83] There is a joke about a father who asks his son, a physicist, to explain Einstein's theory of relativity. The son says, "You see, Dad, it's like this. When you're in a dentist's chair, a minute seems like an hour. But when you have a pretty girl on your lap, an hour seems like a minute." The father ponders the explanation for a moment and says, "So tell me. For saying things like this, Mr. Einstein makes a living?"

In fairness to Mr. Einstein, his theory says that time is relative to the inertial frame in which it is measured, not that it is subjective. The human experience of time *is*, of course, subjective, and it speeds up or slows down depending on how demanding, varied, and pleasant an interval is. But one aspect of Einstein's theory does have a counterpart to the psychology of time, at least as it is expressed in language: the deep equivalence of time with space.

The similarity between space and time is limpid enough that we routinely use space to represent time in calendars, hourglasses, and other timekeeping devices. And the cognitive similarity also shows up in everyday metaphors where spatial terms are borrowed to refer to time. George Lakoff and Mark Johnson have explored a number of these "conceptual" metaphors, so called because they consist not of a single trope but of a family that share an underlying conception.[84] In the TIME ORIENTATION metaphor, an observer is located at the present, with the past behind him and the future in front, as in *That's all behind us, We're looking ahead,* and *She has a great future in front of her.* Then a metaphorical motion can be added to the scene in one of two ways. In the MOVING TIME metaphor, time is a parade that sweeps past a stationary observer: *The time will come when typewriters are obsolete; The time for action has arrived; The deadline is approaching; The summer is flying by.* But we also find a MOVING OBSERVER metaphor, in which the landscape of time is stationary and the observer proceeds through it: *There's trouble down the road; We're coming up on Christmas; She left at nine o'clock; We passed the deadline; We're halfway through the semester.* Lakoff and Johnson note that the two metaphors are incompatible, even though both use space for time. As a result, expressions like *Let's move the meeting ahead a week* are ambiguous. They can mean "make it earlier," if *ahead* is defined by the parade of time past the observer, or "make it later," if *ahead* is defined by the path of the observer through the landscape. (Note the parallel with the fly on the sunbather's thigh, which is both *above* and *below* her kneecap.)

Although the use of space to represent time appears to be universal, the way that time is aligned with a dimension of space can vary.[85] In English alone, the moving-time and moving-observer metaphors coexist with time as a pursuer, in *Old age overtook him,* and with time rotated to the vertical, in *Traditions were handed down to them from their ancestors.* Vertical metaphors for time are even more common in Chinese, with earlier events being "up" and later events being "down," presumably a legacy of their writing system.[86] And in Aymara, a language spoken in the Andes, the time-orientation metaphor is turned around 180 degrees so that the future is said to be behind one and the past in front.[87] The metaphor is unusual, but when we examine the concept of the future we will see that it is not as bizarre as it may seem.

Metaphor is not the only way in which language relates time to space. Time can be related to space and to substance in an even deeper way: in the semantics of tense and verbs. The equivalence is deeper than metaphor because it is not a mere sharing of words. It consists of a congruence in the *construal* of time, space, and substance, with no tangible linguistic thread connecting them.

Time is encoded in grammar in two ways. The familiar one is tense, which can be thought of as the "location" of an event or state in time, as in the difference between *She loves you, She loved you,* and *She will love you.* The other timekeeper is called aspect (we encountered it briefly in chapter 2); it can be thought of as the *shape* of an event in time. Aspect pertains to the difference between *swat a fly,* which is conceptualized as instantaneous (that is, within the specious present); *run around,* which is open-ended; and *draw a circle,* which culminates in an event that marks the act's completion. Aspect can also express a third kind of information related to time: the *viewpoint* on an event. An event can be described as if it is being seen from the inside (in the thick of the event as it unfolds), as in *She was climbing the tree,* or as if it is being seen from the outside (taken in as a whole), as in *She climbed the tree.*[88] (The word *aspect* is from the Latin for "to look at," and is related to *perspective, spectator,* and *spectacles.*)

Though most people have heard of tense, few have heard of aspect, because the two are often confused in language lessons and traditional grammars. Tense and aspect both have something to do with time, and both are expressed in the same vicinity, namely, on the verb or auxiliary. And as we shall see, some inflections blend a bit of tense with a bit of aspect, making it hard to keep them straight. But conceptually, they are completely different. Indeed, in theory they are independent—an event that unfolds in time in a particular way (aspect) can do so whether it takes place yesterday, today, or tomorrow (tense). And remarkably, tense and aspect each has a distinct counterpart in the realm of space and substance. We will see that the meaning of tense (location in time) is like the meaning of spatial terms, and that the meaning of aspect (shape in time and viewpoint in time) is like the meaning of words for things and stuff—complete with plurals, boundaries, and a count-mass distinction.

Tense has the reputation of being the most tortuous part of grammar. In his column-within-a-column called "Ask Mister Language Person," Dave Barry answers the following request:

> Q: Please repeat the statement that Sonda Ward of Nashville, Tenn., swears she heard made by a man expressing concern to a woman who had been unable to get a ride to a church function.
> A: He said: "Estelle, if I'd a knowed you'd a want to went, I'd a seed you'd a got to get to go."
> Q: What tense is that, grammatically?
> A: That is your pluperfect consumptive.

The horrors of tense arise from the convoluted ways that tenses can combine with verbs, aspects, adverbs, and each other (as in *Brian said that if Barbara walked home, he would walk home too*). Nonetheless, the basic meaning of tense is perfectly straightforward.

The best way to understand the language of time is to depict it, naturally enough, in space. Consider a line that runs from the past through the present moment to the future. Situations (that is, events or states) can be represented as segments along the line:

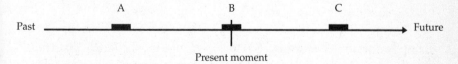

In English, the three basic tenses are child's play: the past tense is used for situation A in the diagram (the situation precedes the moment of speaking), the present tense for B (the situation overlaps the moment of speaking), and the future tense for C (the situation follows the moment of speaking). But for other English tenses, and for many tenses in other languages, we need to introduce a third moment in time: not just the event you are talking about, and the moment at which you are speaking (that is, the present moment), but also a *reference time:* an event that has been identified in the conversation and that is serving as the "now" for the actors in the narrative. (Often the "now" for the actors will be the same as the "now" for the speaker, but sometimes the two are different. For example, if it's Friday, and I'm telling a story about what Sally did on Monday, then Monday is the "now" for Sally—the reference time—even though it's no longer "now" for me.) Then we can define tense with two questions:

- Does the event occur before, after, or simultaneously with the reference time?
- Does the reference time occur before, after, or simultaneously with the moment of speaking?

With two extra wrinkles—some languages permit two or more reference times, and some languages distinguish "before" and "way before," "after" and "way after"—the answers to the two questions can, according to Comrie, capture the meaning of every tense in every language (presumably even the pluperfect consumptive).[89]

In English, the reference time plays no role in the past, present, or future tenses, but it is needed to define the other two major tenses. The pluperfect—*She had written the letter*—is shown here as pertaining to situation E:

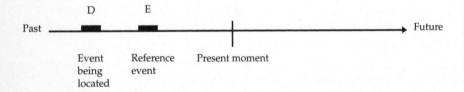

It implies that the letter-writing (D) took place prior to the "now" (E) in the story being narrated, which is prior to the moment of uttering the sentence. This is clearer when we identify the players explicitly: *Francesca had already written the fateful letter* [event being located] *when the count knocked on the door* [reference event, in the past]. The future perfect—*Francesca will have written the letter*—is similar, except that the reference event is located *later* than the present moment:

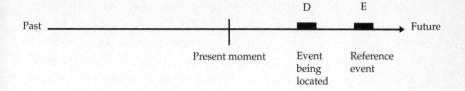

I mentioned that tenses (locations in time) work like prepositions and other spatial terms (locations in space). A tense locates a situation only relative to a reference point (the moment of speaking or a reference event), rather than in fixed coordinates such as the clock and the calendar. It cares about direction (before or after), but ignores absolute distance (days, hours, seconds). And it generally ignores the composition of the thing being located, treating it as a point or blob without visible internal parts.

But time is not identical to space, neither in reality nor in the mind, and that leads to some differences between tenses and spatial terms.[90] Most obviously, time is one-dimensional, so there are fewer tenses than there are spatial terms. And because of this one-dimensionality, the present moment ("now") intrudes between the past and the future with no detour around it, ineluctably dividing time into two noncontiguous regions. So unlike space, where we have terms like *there, far,* and *away from,* which refer to the entirety of space other than "here," no language has a tense that refers to the entirety of time other than "now," embracing the past and the future with a single marker. (There is a counterexample, but it is a word rather than a tense: *then* can refer to the past or the future, as in *She saw him then* and *She will see him then.*)

Another essential difference between time and space is that the two directions of time are very different.[91] The past is frozen and cannot be changed (except in science fiction like *Back to the Future*), whereas the future is a mere potentiality and can be altered by our choices in the present. This intuitive metaphysics is reflected in the way that many languages make only a two-way distinction between past and nonpast, the latter embracing the present *and* the future. Many languages don't express the future in the tense system at all but in a distinction between events that have actually taken place or are now taking place (*realis*) and events that are hypothetical, generic, or in the future (*irrealis*). The metaphysical and epistemological difference between past and future also underlies the Aymara metaphor in which the past is ahead and the future behind. The past has taken place and is knowable, as if it can be seen before your eyes, whereas the future is up for grabs and is inscrutable, as if it were out of view.

Even in English the future tense has a different status from the other tenses. Rather than being a form of the verb, it is expressed by the modal auxiliary *will*. It's no accident that the future shares its syntax with words for necessity (*must*), possibility (*can, may, might*), and moral obligation (*should, ought to*), because what will happen is conceptually related to what must happen, what can happen, what should happen, and what we intend to happen. The word *will* itself is ambiguous between future tense and an expression of determination (as in *Sharks or no sharks, I will swim to Alcatraz*), and its homonyms show up in *free will, strong-willed,* and *to will something to happen.* The same ambiguity between the future and the intended can be found in another marker for the future tense, *going to* or *gonna*. It's as if the language is affirming the ethos that people have the power to make their own futures. You might be wondering whether this is a product of some go-getter attitude, can-do spirit, or Protestant work ethic imbued in Anglo culture. Not so: in languages from disparate cultures all over the world, future tense markers evolve out of verbs for volition or verbs for motion, just as they did in English.[92]

The muddling of volition and futurity also plays out in the ways that the future tense is used differently for one's own actions and someone else's. Other than totalitarian despots, a person can determine his own immediate future more reliably than someone else's, so the mixture of willfulness and prediction packed into a future auxiliary can vary from the first person to the second and third. According to many language mavens, in proper English the future auxiliary is *shall* for the first person but *will* for the second and third; if you switch them around, you get a declaration of intent rather than a genuine future tense. Thus *I will drown, no one shall save me* is the defiant vow of a suicide; *I shall drown, no one will save me* is the pathetic prediction of a doomed wretch. I am skeptical that any Englishman has made this distinction in the past century; Winston Churchill seemed determined enough when he said, "We shall fight on the beaches, we shall fight on the landing grounds, we shall fight in the fields and in the streets, we shall fight in the hills; we shall never surrender." But it is true that many languages blur the future tense with notions of possibility and determination. This also explains a puzzle about tense noted by Zonker Harris in *Doonesbury:*

Doonesbury © 2002 G. B. Trudeau. Reprinted with permission of Universal Press Syndicate. All rights reserved.

The future tense is often used by flight attendants and waitstaff at fancy restaurants as a display of politeness. It pretends not to foreclose any possibilities, as if the listener's approval will be solicited at every stage, before anything is set in stone. As we shall see in chapter 8, it is an example of a common tactic of politeness in the world's languages: Pretend to give the listener options.[93]

Though native speakers of English use its tense system effortlessly, it often bewilders people who learn it as adults. While doing the research for this chapter, I came across this sentence in a paper by an Italian linguist: "It may be useful to step back and get a more general picture of what goes on." No native English speaker could have written that sentence; we would say *what is going on*. But why? The answer is that English has *two* present tenses—the simple present (*it goes*) and the present progressive (*it is going*), and they are not interchangeable. The difference hinges on the second way in which language encodes time: aspect.

Aspect, recall, is about the *shape* of an event, and one's *viewpoint* on it. By "shape" I mean how an action unfolds in time. Linguists sort verbs into classes, each called an *Aktionsart,* German for "action type," based on their temporal contour.[94] The deepest divide is between "states," in which nothing changes, like *knowing the answer* or *being in Michigan,* and "events," in which something happens. Events in turn divide into those that can go on indefinitely, like *running around* or *brushing your hair,* and those that culminate in an endpoint, like *winning a race* or *drawing a circle.* The ones with an endpoint are called "telic," a word related to *teleology,* from the Greek *telos,* "end." The endpoint is usually a change of state in the direct object that was caused by the agent. The act of *drawing a circle,* for instance, is over when the circle is complete.[95] Allegedly, Lizzie Borden took an ax and gave her mother forty whacks. If so, she *killed* her (a telic event) at the moment of whichever whack it was that caused her mother to become not alive. (Borden, as it happens, was acquitted.)

Verbs are also divided by whether they describe an event that is spread out in time, like *running* or *drawing a circle* (they are called "durative"), or instantaneous, like *winning a race* or *swatting a fly*. Of course only Superman can execute an action in no time at all; the rest of us have to raise the fly swatter, bring it down, hit the fly, and so on. But the event can be thought of as instantaneous if it lies within the specious present. Linguists sometimes call these events "momentaneous," a lovely word that was last in vogue in the seventeenth century.

To get a feel for all this, it helps, once again, to visualize time as a line.[96] Let's depict an event that lacks a fixed boundary (like *running around*) with a fuzzy border:

Past ————————————————————————————————→ Future

This is called an "activity," an event that is durative (it lasts in time) and atelic (it lacks an inherent endpoint). We can now use a pip for a momentaneous event, like *swatting a fly:*

Past ————————————————————■———————————————→ Future

A telic event has no fixed beginning, but by definition it has a terminating moment, when the agent has brought about the intended change:

Past ————————————————————■———————————————→ Future

Telic events can be described in two ways: with a durative verb, which embraces both the buildup and the climax, like *drawing a circle,* or with a momentaneous verb, which zeroes in on the climax, like *winning the race, reaching the top,* or *arriving.* (Confusingly, linguists call these *accomplishments* and *achievements.* I can never remember which is which, so I will call the latter *culminations.*) We also have iterative verbs, like *pound a nail:*

Past ————————————— |||||||||||||||| —————————————→ Future

and verbs for the inception of states, like *sit down* (as opposed to *sit*, which is an activity):

Past ————————————————————— ▮░░░ - —————————→ Future

The difference between inceptive verbs and momentaneous verbs can be illustrated by our friend Mr. Pi, the space alien whose overly literal grasp of English has already illuminated a number of subtle semantic distinctions:

Monty © United Feature Syndicate, Inc.

I mentioned that a remarkable feature of the verb action classes is that they are shaped in the same way as physical objects and substances, as if events were extruded out of some kind of time-stuff.[97] Just as in the realm of matter we beheld bounded objects (*cup*) and unbounded substances (*plastic*), in the realm of time we see bounded accomplishments (*draw a circle*) and unbounded activities (*jog*). Just as we met substance words that name homogeneous aggregates (*mud*) and plurals that name aggregates made of individuals (*pebbles*), we now meet durative verbs that name a homogeneous action (like *slide*) and iterative verbs that name a series of actions (like *pound, beat,* and *rock*). And in the same way that a huge inventory of shape nouns (*pediment, cornice, frieze,* and so on) was reduced to a skeleton of lines, sheets, and blobs, a huge repertoire of action verbs (*drumming, piping, leaping,* and so on) is reduced to a skeleton of instants and durations. The difference is that time is one-dimensional, so there are fewer skeletal "shapes" for events to assume, and thus we are left with fewer action classes than shape classes. Still, even a one-dimensional shape can be given a zero-dimensional endpoint. Lederer wonders, "Why is it called 'after dark' when it is really 'after light'?" The answer is that *dark* can refer to the instant that an interlude of darkness begins. It is a perfect mirror of the answer to his question of why we say that something is *underwater* or *underground,* where a word for a three-dimensional solid may also be used for its two-dimensional boundary.

Why look at action classes so closely? It's because they play many roles in language and reasoning.[98] The action classes determine the logical conclusions one can draw from a sentence, because the truth of a proposition depends on the stretch of time it refers to. If Ivan is *running* (atelic), we can conclude that Ivan ran, but if Ivan is *drawing a circle* (telic), we cannot conclude that Ivan drew a circle—he may have been interrupted. Note again the similarity to substances and objects—half a portion of applesauce is still applesauce, but half a horse is not a horse.

Action classes also affect the way that verbs mate with explicit expressions for time. You can say *He jogged for an hour,* but not *He swatted a fly for an hour,* because the phrase *for an hour* imposes an endpoint on an event. That works with an activity, like running, which is spread out in time and can be lopped off by a boundary, but not with a momentaneous event like swatting a fly. It's even a bit odd to say *He crossed the street for a minute* or *She wrote a paper for an hour,* since those telic accomplishments are already bounded by their culminating events and don't accept a second endpoint. But a phrase like *in an hour* works the other way around: it imposes a *beginning* boundary on an event by measuring backward from its endpoint. You can *cross the street in a minute* or *write a paper in an hour,* but you can't *jog in an hour* (other than with the meaning "an hour from now"), because it has no endpoint. Nor can you *swat a fly in an hour,* because it lacks a durative activity that can be measured out and bounded. The Doors' song "Love Me Two Times" sounds strange at first, because the temporal phrase *x times* applies only to events, not to states, and *loving someone* is a state. Of course we are meant to interpret the verb as a euphemism for having sex, which is an activity and an accomplishment (sometimes in more ways than one).

Phrases like *in an hour* and *for an hour* are part of a mental system in which stretches of time are dynamically spun out, measured, and sliced off, like the Fates in Greek mythology determining the lives of mortals. They are temporal versions of the mental packager in the noun system which can convert substances into objects, as when you order *a beer* or take out *three coffees.*[99] Another way to package events is to reach for the toolbelt of English particles like *out, up,* and *off,* which provide a culmination point to an endless activity, as in the difference between merely *shaking* something and *shaking it up.* To shake something up means to shake it until it has changed its state, sometimes metaphorically, as when Elvis Presley confessed to being "All Shook Up." What Mr. Pi shows us with his literal-mindedness Mr. Lederer shows us with his wit, and here he calls our attention to the way that many particles with spatial senses like *up, down, up,* and *out* are also used in an aspectual sense, to cap off an activity:

> Why do "slow down" and "slow up" mean the same thing? . . . You have to marvel at the unique lunacy of a language where a house can burn up as it burns down and in which you fill in a form by filling it out. English was invented by people, not computers . . . . That is why when the stars are out they are visible but when the lights are out they are invisible and why it is that when I wind up my watch it starts, but when I wind up this poem it ends.[100]

Languages have an even more powerful device for packaging durative activities or grinding telic ones: the second aspect of aspect, viewpoint. Actually a better analogy than grinding and packaging is *zooming in* to scrutinize the internal stuff of an event, with its boundaries outside the field of vision, or *stepping back,* allowing the entire event, including any fuzzy boundaries, to shrink to a smudge.[101] The first is called the *imperfective,* and can be visualized like this:

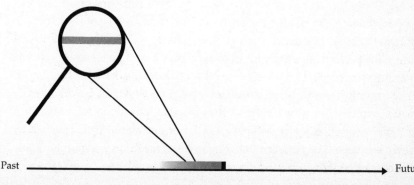

Past   Future

And the second is called the *perfective*, and can be visualized like this:

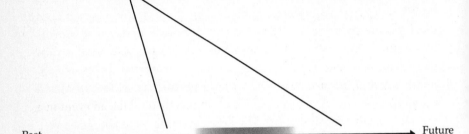

Past ⟶ Future

Why "perfective"? Because *perfect* can mean "complete," not just "flawless," as in *perfectly useless, a perfect nuisance,* and technical terms like a *perfect fifth* in music and a *perfect square* in mathematics. "Perfective" is thus a good term for a point of view that allows us to take in the whole event.

English has an imperfective aspect in the present progressive *Lisa is running* (as opposed to *Lisa runs,* the simple present). The progressive zooms in on a portion of the action making up a bounded event, turning it into a boundariless activity, just as one can mentally zoom in on the plastic composing a cup and think of it as a substance without literally having to grind the cup into bits. So while it's odd to say *Lisa drove home, but she never got there,* you can certainly say *Lisa was driving home, but she never got there*—the *-ing* in *driving* zeroes in on a portion of her driving home and excludes the endpoint from one's field of view. The imperfective is commonly used in a narrative to set the stage for an event (describing the scenery, as it were), while the past and present are used to advance the story line (as in *Lisa was driving home when suddenly a spaceship landed on the roof of her car*). Unlike many other languages, such as Russian, English doesn't have a way to mark the perfective aspect with its own suffix. But we can interpret verbs as perfective in context, as when we say *After Sarah jogged, she took a shower.* The activity of jogging, which ordinarily has no boundaries, is now taken in as a completed event, as if from a distant vantage point.

We have visited every tense in English but one, the so-called perfect, as in *I have eaten.* The perfect, confusingly, is not the same as a perfective; indeed is not really a tense at all, but a combination of a tense and an aspect.[102]

It indicates that something is currently in a state or condition that resulted from an action in the past:

Present moment

For example, *I have eaten* (perfect) suggests that one is now sated and doesn't have to eat again, whereas *I ate* can merely describe an event in a narrative at any time in the past. Unlike the state stipulated by a telic verb like *melt the butter,* the state implied by the perfect has to be interpreted in context—it is any feature of the aftermath of an action that is now deemed significant. That's why it takes some degree of chutzpah to say *I have spoken* or *I have arrived,* rather than the humbler *I spoke* or *I arrived.* ("Do not arouse the wrath of the Great and Powerful Oz! I said come back tomorrow! Oh! The Great Oz has spoken! Oh! Pay no attention to that man behind the curtain! The Great and Powerful Oz has spoken!")

In theory, tense and aspect should be completely independent. That's because the temporal contour of an event, and one's vantage point on it, should be independent of its location in time, just as the shape of an object, and whether you have zoomed in on it, are independent of its location in space. In practice it doesn't always work that way. That is because life as it unfolds is never perfectly synchronized with one's speechifying, so the relation between the events taking place in the world and the precise moment you're wagging your jaw is anything but straightforward. As a result, the interpretation of the present tense is not the same for all verbs but depends on the action class. In describing a current state, for example, you have to use the simple present—*He knows the answer; He wants a drink,* not *He is knowing the answer; He is wanting a drink.* But in describing a current activity or accomplishment, you have to use the progressive—*He is jogging; He is crossing the street,* not *He jogs; He crosses the street* (or the Italian linguist's *get a general picture of what goes on*). Presumably this is because the progressive, which turns an action into a state, is redundant with verbs like *know* and *want* that already *are* states. But it is a prerequisite for activities and accomplishments, which are perfective by default and need to be cracked open for the present moment to have a stretch of the activity to latch onto. Momentaneous events can't easily be described in the present at all—*He swats a fly* and *He is swatting a fly* both sound odd—because it's unlikely that a punctate event will just happen to take place at the exact instant one is describing it. The progressive turns a momentaneous event into an iterative one—*The light is flashing* means it's doing so repeatedly (compare *The light flashed,* which can mean only once). This is the kind of complexity that makes foreign speakers want to tear their hairs out.

As for the simple present tense, it is available for speakers to use in two different ways. One is in an ongoing narration. This is the tense of play-by-play sportscasting, as in *Lafleur skates down the ice. . . . He shoots. . . . He scores!* When the reference point of the narration is not the present moment but some point in the past, we have the "historical present," in which a writer tries to parachute the reader into the midst of an unfolding story (*Genevieve lies awake in bed. A floorboard creaks . . .*). The historical present is also often used in the setup of a joke, as in *A guy walks into a bar with a duck on his head . . .*[103] Though the you-are-there illusion forced by the historical present can be an effective narrative device, it can also feel manipulative. Recently a Canadian columnist complained about a CBC Radio news program that seemed to him to overuse the present tense, as in "UN forces open fire on protesters." The director explained to him that the show is supposed to sound "less analytic, less reflective" and "more dynamic, more hot" than the flagship nightly news show.[104]

The other use for the simple present is for actions that are habitual (*Sarah jogs every day*) or generic (*Beavers build dams*), where the verb describes a *propensity* of the subject to do something. The propensity extends over time, and hence it can be said to be in effect at the present moment, even if Sarah is at work or all the world's beavers are asleep at the precise instant you utter the sentence.

And now, dear reader, you are equipped to understand the most consequential debate about tense and aspect in human history:

QUESTION: Mr. President, I want to go into a new subject area. . . .
[Your] counsel is fully aware that Ms. Lewinsky . . . has an affidavit, which they were in possession of, saying that there was absolutely no sex of any kind in any manner, shape or form with President Clinton. That statement was made by your attorney in front of Judge Susan Webber Wright.

CLINTON: That's correct.

QUESTION: That statement is a completely false statement. Whether or not [your attorney] knew of your relationship with Ms. Lewinsky, the statement that there was no sex of any kind in any manner, shape or form with President Clinton was an utterly false statement. Is that correct?

CLINTON: It depends upon what the meaning of the word "is" is. If "is" means is and never has been, that's one thing. If it means there is none, that was a completely true statement.[105]

In August 1998 President Clinton gave this infamous testimony (since immortalized in *Bartlett's Familiar Quotations*) to a grand jury impaneled by special prosecutor Kenneth Starr. Starr was investigating perjury and obstruction of justice in a deposition Clinton had given during a sexual harassment lawsuit by Paula Jones earlier that year, in response to accusations that Clinton had had an affair with Monica Lewinsky. Clinton's lawyer had said in the deposition that "there is absolutely no sex of any kind" between Clinton and Lewinsky. In this testimony, Clinton was affirming that the statement contained the verb *is,* the present tense, and that their affair was in fact over at the moment the statement was made, so the statement was true. Note how he correctly distinguished the present-tense *is* from the perfect *has been,* which would have implied the existence of some ongoing state that held at the moment the statement was made. The prosecutor, incredulous, continued:

QUESTION: I just want to make sure I understand you correctly. Do you mean today that because you were not engaging in sexual activity with Ms. Lewinsky during the deposition that the statement Mr. Bennett made might be literally true?

CLINTON: No, sir. I mean that at the time of the deposition . . . that was well beyond any point of improper contact between me and Ms. Lewinsky. So that anyone generally speaking in the present tense saying that was not an improper relationship would be telling the truth if that person said there was not, in the present tense—the present tense encompassing many months. That's what I meant by that. . . . I wasn't trying to give you a cute answer to that.

Clinton gets full marks in this test of the semantics of tense. As we have seen, other than in an ongoing narration like a play-by-play in sports, the English present tense is used to refer to a state defined by a propensity or habit, not to a specific event. And at the time at which the verb *is* was uttered, Clinton and Lewinsky had broken up and were unlikely to have sex again, so the propensity was no longer in force. Admittedly, the termination of a durative, atelic state defined by a propensity to act is inherently fuzzy (like the boundary of an aggregate like *gravel* or *pebbles*). How much time must elapse since the last cigarette before a would-be former smoker can say, "I don't smoke"?

As to whether Clinton gave a "cute answer," this is the point at which semantics leaves off and pragmatics begins. As we shall see in chapter 8, listeners assume that speakers are conveying information relevant to what they want to know, allowing them to guess the meanings of vague expressions. This works fine when the interlocutors are cooperative and the listener's guess is the same as the speaker's intent, but not when they are adversaries, as in a legal investigation. As Clinton noted, "My goal in this deposition was to be truthful, but not particularly helpful." Given that the lawyers in the Jones trial presumably wanted to know whether Clinton had *ever* had an affair with Lewinsky, the issue then becomes whether he was legally justified in answering the question narrowly (according to its semantics) or whether a sworn commitment to "the whole truth" required answering the question as it was intended (according to its pragmatics). The Starr report reached the latter conclusion, and cited Clinton's testimony about the meaning of *is* as one of five instances in which he tried to obstruct justice and deceive the American people. The U.S. House of Representatives agreed, and in December 1998 voted to impeach him. The U.S. Senate disagreed, and in February 1999 voted to acquit him. In any case, Clinton started a trend of presidents getting into trouble because of the fine points of conceptual semantics, as we saw with regard to George W. Bush and the verb *to learn*.

The semantics of time has one last parallel with the semantics of space, and this one speaks to the Kantian project of identifying the abstract frameworks that organize our knowledge. Just as spatial language turns out to be defined not only by the geometry of objects but by how people use them, temporal language is defined not only by the way that events erupt and unreel according to a clock but by the goals and powers of the actor. The action classes were originally sketched out by Aristotle, and fit with his theory that every event has a form, a substance, an agent that brought it about, and a goal that it serves. He would not be surprised to learn that each of the four major action classes (state, activity, culmination, accomplishment) smuggles in a concept of human will in addition to its concept of temporal shape.[106]

A state is defined not just by an absence of change but by being outside the sphere of voluntary control. Generally you can't *persuade* or *force someone to know the answer,* or talk about him *deliberately* or *carefully knowing the answer;* nor can you issue the imperative *Know the answer.* The coupling of statehood and involuntariness in our language reflects a deeper coupling of the concepts in our ascription of moral responsibility. Because we construe states as involuntary, we tend not to hold people criminally responsible for them, at least not upon careful examination. Thus in 1962 the U.S. Supreme Court ruled that while a legislature can outlaw the *use* or *sale* of narcotics, it cannot outlaw *being addicted* to them. Another court decision deemed it unfair to convict someone for *being* drunk in public (the case involved a man who got drunk in his home and was dragged into a public street by the police), although someone could still be prosecuted for *getting* drunk in public or for *going out in public* drunk.[107] One exception to this generalization is the crime of *possessing* narcotics, which is indeed a state. Perhaps not coincidentally, many people feel that such laws are unfair.

Also involuntary is the momentaneous culmination that consummates an accomplishment, like *winning a race, finding a diamond, reaching Boston,* or *noticing a painting.* These verbs don't harmonize with adverbs of effort (*He deliberately won the race*), with verbs of initiating an action (*I persuaded him to notice the painting*), or with the imperative mood (*Find a diamond!*). Once one of these pursuits has been undertaken, it's the world that determines the moment of culmination, not one's intention.

Activities and accomplishments, in contrast, are generally thought of as voluntary. For that reason, accomplishment verbs, such as those in *baking a cake* and *hiding a key,* can be commanded by imperatives, and can be accompanied by volitional adverbs like *deliberately* and *carefully.* Indeed, with an accomplishment it's the actor's goal that determines the exact event that consummates it, such as causing a picture to come into existence in the case of *drawing a picture,* or being at the other side of the street in the case of *crossing the street.* Once again, this is not just a fine point of grammar but a keystone of our moral sense. Since a crime requires both a bad act and a guilty mind, criminal acts are identified by activity or accomplishment verbs: *to kill, to steal, to rape, to bribe,* and so on. If an accomplishment has not been consummated (as in the case of a would-be strangler who is interrupted by the police), we can charge the person only with a criminal attempt. And because a culmination is construed as involuntary (it is determined by the world rather than by one's intent), people are often foggy about which crime has been committed when there is a disconnect between the intended change that defines an accomplishment verb and the actual change that took place. Many hours of law-school argumentation have been spent on what to do with a man who stabs a corpse thinking it is his sleeping enemy, or whether it makes sense to charge a shooter with attempted murder if the nearest hospital is five minutes away and his victim survives, but to charge him with murder if the nearest hospital is fifteen minutes away and the victim succumbs.

So just as spatial language does not invoke an empty coordinate system, temporal language does not invoke a free-running clock. Space is reckoned with reference to objects as they are conceived by humans, including the uses to which they are put, and time is reckoned with respect to actions as they are conceived by humans, including their abilities and intentions. As central as space and time are to our language and thought, a conscious appreciation of them as universal media into which our experiences are fitted is a refined accomplishment of the science and mathematics of the early modern period.

OOMPH: THOUGHTS ABOUT CAUSALITY

Our sense of causality, Hume noted, is "the cement of the universe."[108] As we make it through the day we constantly tap our causal intuitions to understand what is going on in the world and how we should deal with it (the windows are wet, so it must have rained; if I wear a raincoat, my clothes will stay dry). When these intuitions fail, we know we are dreaming, or have projected ourselves into Wonderland or some other product of the imagination. We look to science as a purer and tougher version of our search for causes—as the best way to identify what caused an earthquake, or the arrangement of the solar system, or the appearance of the human species itself.

It's disconcerting, then, to learn that on close inspection this cement is as shoddy as the stuff used in Boston tunnels. The more you scrutinize causality, the less sense it makes, and some philosophers have suggested that science should just kiss it goodbye. At the same time, causality is deeply entrenched in our language and thought, including our moral sense, and no account of the human predicament can avoid pondering how our causal intuitions are related to the causal texture of the universe. It's no accident that the starting point for our modern understanding of causation was a book by Hume called *Treatise of Human Nature.*

Hume (and later Kant, when awakened from his dogmatic slumber) worried about how we could *justify* our inferences about unobserved events—whether we could ever elevate a deduction like "If you drop something, it will fall; I dropped a glass, therefore the glass will fall" to the level of certainty we are accustomed to in logical and mathematical deductions like "If a triangle has two equal sides, then it has two equal angles; this triangle has two equal sides, therefore it has two equal angles." He concluded that we can't, though of course we aren't being unreasonable when we expect the glass to fall. Our causal intuitions are a handy part of our psychology, even if they fall short of granting us certitude. The dubiety springs from the sad fact that our causal intuitions, deep down, are no more than expectations stamped in by experience, and these expectations are satisfied only if the universe is lawful, a brute assumption we can never prove. Here is Hume explaining why we think that one billiard ball causes a second one to move:

> Were a man, such as Adam, created in the full vigour of understanding, without experience, he would never be able to infer motion in the second ball from the motion and impulse of the first. . . . It would have been necessary, therefore, for Adam (if he was not inspired) to have had *experience* of the effect which followed upon the impulse of these two balls. He must have seen, in several instances, that when the one ball struck upon the other, the second always acquired motion. If he had seen a sufficient number of instances of this kind, whenever he saw the one ball moving towards the other, he would always conclude without hesitation that the second would acquire motion. His understanding would anticipate his sight and form a conclusion suitable to his past experience.

It follows, then, that all reasonings concerning cause and effect are founded on experience, and that all reasonings from experience are founded on the supposition that the course of nature will continue uniformly the same.[109]

Tucked into this analysis of whether we can justify our causal attributions is an offhand theory of the psychology of causality called constant conjunction: that our intuitions of cause and effect are nothing but an expectation that if one thing followed another many times in the past, it will continue to do so in the future. It's not terribly different from what happens when a dog is conditioned to anticipate food when a bell is rung, or a pigeon learns to peck a key in the expectation of food. The story that began the chapter, about the two alarms that go off in succession, raises an obvious problem for the theory. People understand (even if they don't always apply) the principle that correlation does not imply causation. The rooster's cock-a-doodle-doo does not cause the sun to rise, thunder doesn't cause forest fires, and the flashing lights on the top of a printer don't cause it to spit out a document. These are perceived to be *epiphenomena:* byproducts of the real causes.

I called Hume's theory "offhand" because he didn't consistently embrace it himself. The very example of "causation" he adduced in his summary—"when we think of the son, we are apt to carry our attention to the father"—could not be a more ruinous counterexample. We don't, of course, think that sons cause fathers, but something like the other way around, whereas in the constant-conjunction theory, the cause carries our attention to the effect. Worse, we don't need to experience a sighting of a father followed by a sighting of his son on innumerable occasions to understand the connection between them, because even the most doting father doesn't shadow his boy around the clock. People can infer an association between father and son not just by seeing one after the other but from gossip, genealogies, an upper lip that suspiciously resembles that of the mailman, or, nowadays, a DNA test. Even in the most decorous of circumstances, a full nine months must intervene between the paternal event we understand to be the cause and the filial event we understand to be the effect. And during that time, the father might desert the family or die, yet he will still be the father for all that.

Hume undoubtedly sensed this problem, because—equally offhandedly—he expanded his idea in the following passage: "We may define a cause to be an object followed by another, and where all the objects, similar to the first, are followed by objects similar to the second. Or, in other words, where, if the first object had not been, the second never had existed." But this last sentence, far from expressing the idea of constant conjunction "in other words," expresses a completely *different* idea. Indeed, in many ways it's a better idea, because it successfully rules out the embarrassment of epiphenomena. If the first alarm hadn't gone off, the second still would have; therefore the first alarm did not cause the second. Likewise, if the lights on the printer had burned out, the page would still have printed, so the lights didn't cause the printing. And it successfully rules *in* cases of causation that are separated in time or learned about circuitously. If the father had never lived, the son would not have, either, so the father is, in some sense, a cause of the son.

This *counterfactual* theory of causation—"A caused B" means "B would not have happened if not for A"—is an improvement over the constant-conjunction theory.[110] But the more you think about it, the stranger it becomes. What exactly do the "would" and the "if not for" refer to? How do we determine what is true or false in a make-believe world? You only live once, and the world unfolds whichever way it does, not some other way. It doesn't come with a PLAY OVER button that allows you to restart the game, make a different move, and see what happens. We recognize this futility in sayings like "If wishes were horses, beggars would ride" and "If my grandmother had wheels, she'd be a streetcar" (a polite version of the Yiddish saying "If my grandmother had balls, she'd be my grandfather"). Not to mention Woody Allen's remark that "my one regret in life is that I am not someone else."

Many philosophers have tried to make sense of counterfactual statements by invoking "possible worlds."[111] These are not undiscovered planets with little green men but logically consistent states of affairs: different ways that the universe could have unfolded without violating the laws of logic. To say that "A causes B" means that if A hadn't happened, B wouldn't have either, which in turn means that there are possible worlds in which A does not happen, and in every one of them B does not happen either.

Unfortunately, this still isn't enough to ground causality in counterfactual thinking. If we can be footloose and fancy free in imagining possible worlds, *any* effect can happen, even without its supposed cause: all you need to do is dream up some other circumstance that led to it. Did striking the match cause it to burn? Well, in this world, and in many possible worlds, if you don't strike it, the match doesn't burn. But what about a world in which the room suddenly heats up to 451 degrees Fahrenheit, igniting the match without anyone's having struck it? Does the existence of this possible world force us to conclude that striking matches doesn't cause them to burn in our world?

To preserve the good idea that causation depends on counterfactuals, which in turn may be defined by possible worlds, philosophers suggest that possible worlds can be lined up according to their similarity or closeness to the real world. The possible world in which I don my blue socks rather than my black socks this morning is closer to the actual world than the possible world in which I was born a woman, or the one in which World War III breaks out, or the one with an atmosphere consisting of methane and ammonia rather than nitrogen and oxygen. Getting back to causation, we can say that striking the match caused it to burn because the match does not burn in the closest possible worlds to ours in which it was not struck. And surely those worlds are ones in which the temperature of the room is room temperature rather than 451 degrees.

Why, you might ask, do philosophers bother with close and distant possible worlds rather than just saying "all things being equal" or "holding everything else constant"? It's because all else is never equal: you can't do just one thing. Consider the possible world in which New York City is in Colorado. Is New York City west of the Mississippi? Or is Colorado on the Atlantic coast?[112] Our description of a possible world has left this crucial fact unspecified. Words are cheap, so *any* description of a possible world will be mute about crucial facts that are entwined with the one you have changed with your verbal magic wand. According to one story, Nikita Khrushchev was asked how the world would have been different if he, rather than John F. Kennedy, had been assassinated in 1963. He said, "For one thing, Aristotle Onassis would probably not have married my widow." The joke is possible because nothing in "all things being equal" or "holding everything else constant" rules out one of those things being "Aristotle Onassis married the widow of the slain leader of a superpower." And according to a joke that was told many times in 1993, Bill and Hillary Clinton were being driven through her hometown when Hillary spotted an old boyfriend pumping gas. "If you hadn't married me," said Bill, "you'd be the wife of a gas station attendant." "If I hadn't married you," replied Hillary, "*he* would be president." The concept of the "closest possible world" is meant to make sense of these counterfactuals by helping us single out the coherent state of affairs that requires the fewest additional changes to the actual world in order to accommodate the single changed premise. (I doubt that this really solves the problem, but it does make us feel better.)

The definition of "A causes B" as "B does not occur in the possible worlds closest to ours in which A does not occur" is a big improvement over the idea that it means "when A occurs, B does too." For example, it fits the scientific practice by which we distinguish causation from correlation by experimental manipulation. If coffee drinkers are found to have more heart attacks, does that mean that drinking coffee causes heart disease? No, because correlation does not prove causation. Perhaps coffee drinkers tend to get less exercise, or are more likely to smoke, or eat fattier foods, and one or more of these is the real cause, while coffee drinking is an epiphenomenon. For coffee to cause heart disease, it would have to be true that in the nearest possible world in which people don't drink coffee, they get fewer heart attacks. How do you establish that? Simple: *make* such a world, by dividing a sample of people into two groups at random and having one group abstain from coffee drinking and the other enjoy their lattes. If in the first possible-made-actual world, the people have healthier hearts, we are entitled to say that coffee is the cause.[113]

Though the counterfactual theory (associated with the philosopher David Lewis) is considered among the most sophisticated analyses of causation in contemporary philosophy and jurisprudence, it is rife with problems.[114] One of them pops up whenever a *collection* of circumstances is necessary for an effect to occur. Striking the match is necessary for it to burn, but so is the dryness of the match, the presence of oxygen, and shelter from the wind. In all the possible worlds similar to ours in which the match is wet, the room is filled with carbon dioxide, or the striker is outdoors, the match does not burn. Nonetheless, if someone asks us to identify the cause of the match's burning, we single out the act of striking it, not the presence of oxygen, the dryness of the match, or the presence of four walls and a roof. For the same reason we don't consider getting married to be a cause of widowhood or stealing jewels to be a cause of the police discovering them, though in each case the later event would not have happened without the earlier one.

People somehow distinguish just one of the necessary conditions for an event as its *cause* and the others as mere *enablers* or helpers, even when all are equally necessary. The difference does not lie in the chain of physical events or in the laws they follow, but in an implicit comparison with certain other states of affairs (similar possible worlds, if you will) that we keep in the back of our minds as reasonable alternatives to the status quo.[115] Since oxygen is pretty much always around, we don't think of its presence as a cause of a match igniting. But since we spend more time not striking matches than striking them, and feel that it is up to us at any moment whether we strike or not, we do credit the striking as a cause. Change the comparison set and you change the cause. For example, if a particular kind of welding were ordinarily done in an oxygen-free chamber but one day oxygen got in and a fire ensued, we *would* identify the presence of oxygen as the cause of the fire. (The fire that killed three Apollo astronauts in 1967 is commonly blamed on the pure oxygen that filled the capsule, causing a minor spark to grow into an inferno.) Likewise, we might allow that getting married is a cause of widowhood if the woman chose to marry a man on his deathbed (an attribution that was made about Anna Nicole Smith when she married the eighty-nine-year-old oil baron J. Howard Marshall in 1994, a year before his death).[116] To label a condition as a "cause" means to identify a factor that we feel could easily have been different, or that someone could have controlled, or that someone might control in the future.

A related problem with the counterfactual theory is that causation is *transitive:* if A causes B, and B causes C, then A causes C. If smoking causes cancer, and cancer causes death, then smoking causes death. But necessary conditions (the ones behind counterfactual inferences) are not transitive. It seems reasonable to say that if Kennedy had not been president, he would not have been killed. It also seems reasonable to say that if Kennedy had not been killed, he would have been reelected. But it is quite *un*reasonable to say that if Kennedy had not been president, he would have been reelected![117]

Yet another problem with the theory is called preemption. Two snipers conspire to assassinate a dictator at a public rally. They agree that the first one to get a clear shot will fire, whereupon the other will melt into the crowd. Assassin 1 takes the dictator out with his first shot, and clearly his deed is the cause of the dictator's death. Yet it *isn't* true that if Assassin 1 had not fired, the dictator would not have died, because in that case Assassin 2 would have done the deed. (Sure enough, when people were asked about this scenario in experiments by the psychologist Barbara Spellman, they didn't exonerate either assassin.)[118] Or consider an example from the legal scholar Leo Katz: "Henri plans a trek through the desert. Alphonse, intending to kill Henri, puts poison into his canteen. Gaston also intends to kill Henri but has no idea what Alphonse has been up to. He punctures Henri's canteen, and Henri dies of thirst. Who has caused Henri's death? Was it Alphonse? Gaston? Both? Or neither?" Clearly the death was caused by someone, and most people finger Gaston, or sometimes both.[119] But the counterfactual theory predicts that they should say "neither."

The final problem is called overdetermination (or, sometimes, multiple sufficient causes). Consider a firing squad that dispatches the condemned man with perfectly synchronized shots. If the first shooter had not fired, the prisoner would still be dead, so under the counterfactual theory his shot didn't cause the death. But the same is true of the second shooter, the third, and so on, with the result that *none* of them can be said to have caused the prisoner's death. But that is just crazy.

The common denominator in all these problems is that the world is not a line of dominoes in which each event causes exactly one event and is caused by exactly one event. The world is a tissue of causes and effects that criss and cross in tangled patterns. The embarrassments for Hume's two theories of causation (conjunction and counterfactuals) can be diagrammed as a family of networks in which the lines fan in or out or loop around, as in the diagram on the following page.

One solution to the webbiness of causation is a technique in artificial intelligence called Causal Bayes Networks.[120] (They are named for Thomas Bayes, whose eponymous theorem shows how to calculate the probability of some condition from its prior plausibility and the likelihood that it led to some observed symptoms.) A modeler chooses a set of variables (amount of coffee drunk, amount of exercise, presence of heart disease, and so on), draws arrows between causes and their effects, and labels each arrow with a number representing the strength of the causal influence (the increase or decrease in the likelihood of the effect, given the presence of the cause). The arrows can be in whichever patterns of fanning or looping is necessary; there

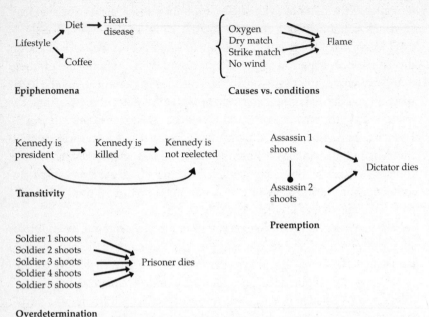

is no need to identify "the" cause of a given effect. With this diagram in hand, and measurements of the variables (such as how many cups of coffee a person drinks), a computer can crank through some arithmetic and predict the effects of a given cause (such as the increased risk of heart disease) or, in the other direction, the probability that a cause was present, given its typical effects. For example, a Causal Bayes Net would allow you to guess from a burglar alarm going off in a neighbor's house that there was probably a break-in, but if you also noticed a cat jumping around inside, you would put down the phone without calling 911. Before you use a Causal Bayes Net, you have to set it up, and that can be done from some initial assumptions about the variables and their relationships, a set of experimental interventions (such as depriving people of coffee and seeing what happens to their health), or a set of measurements of how the factors correlate in a large set of data.

Causal Bayes Nets are an optimal way to reason about causes and effects from information about their intercorrelations, and in some ways people have been shown to conform to its ideals, such as in the scenario with the cat and the burglar alarm. But it is basically Hume with a computer. It depicts causal reasoning as cranking through a huge set of correlations without caring what those variables stand for, or which mechanisms in the world allow the real-life referents of those variables to impinge on one another. It would apply perfectly well to an observer seated in front of a huge display of colored lights who kept track of whether a red light in the top right corner tended to come on a few minutes before a green light in the middle row unless a yellow square at the bottom left edge flashed twice in between. It therefore leaves out a key component of our causal thinking: the intuition that the world is made of mechanisms and forces with causal powers—some kind of push or energy or oomph that is transmitted from the cause to the effect—and that the correlations we observe are the products of these powers in action.[121]

Even a glance at human behavior suggests that people often think of causation in terms of hidden powers rather than just correlations. Many psychology experiments have shown that when people have a pet theory of how things work (such as that damp weather causes arthritis pain), they will swear that they can see those correlations in the world, even when the numbers show that the correlations don't exist and never did.[122] The habit of hallucinating causal powers and forcing experience to fit them has shaped human cultures from time immemorial, producing our species' vast compendium of voodoo, astrology, magic, prayer, idolatry, New Age nostrums, and other flimflam. Even respectable scientists don't stop at recording correlations but try to pry open nature's black boxes and identify the hidden powers at work. Sometimes the candidates don't pan out, like phlogiston or the luminiferous ether, but often they do, as with genes, atoms, and tectonic plates.

Another limitation of probabilistic theories of causation is that they apply to averages over the long run (smoking causes cancer) and have nothing to say about the causes of specific events (smoking killed Granny). But people have sharp intuitions about specific events.[123] Imagine that Uncle Irv, a two-pack-a-day smoker, is alive and well at ninety-seven. Everyone would agree that smoking did not cause him to die. Yet if "smoking causes cancer" were nothing but a statement of the odds, one couldn't say anything about its applicability to him one way or the other. Granted, people may not be *rational* in insisting that single events have identifiable causes. If many people die of heart attacks, and a drug slightly increases the risk of a heart attack, and John, who took the drug, dies of a heart attack, was the drug the cause? One could argue that the question has no answer. Yet people act as if it does have an answer. In 2005 the widow of a man who had taken the drug Vioxx was awarded $253 million by a jury in a lawsuit against the drug's manufacturer, and there are six thousand similar lawsuits pending.

People not only *apply* a causal relation to a single event but can *infer* a causal relation from a single event, without requiring the event to be repeated many times. The passengers on the *Titanic* surely believed that an iceberg caused the ship to sink, even if they had no prior experience of icebergs followed by sinking ships.[124] The simplest demonstration of the difference between recording a correlation over the long run and sensing causal powers in a single event comes from a classic experiment by the psychologist Albert Michotte.[125] Michotte showed people animations in which one dot moved along the screen until it contacted a second dot, whereupon the first stopped suddenly and the second started to move in the same direction and at the same speed. On their first viewing people had an unmistakable impression that the first one *caused* the second to move, like one billiard ball launching another. The same seems to be true for infants as young as six months and for at least one species of monkey.[126] Other displays of moving dots convey vivid impressions of related kinds of causation, such as helping, hindering, allowing, and preventing.[127]

A recent experiment by the psychologists Marc Hauser and Bailey Spaulding has shown that reasoning about causal powers without needing to see a long sequence of events beforehand is part of our primate birthright.[128] Testing rhesus monkeys who had had little to no experience with knives or paint, they found that the monkeys showed a keen appreciation of their causal powers. The monkeys were unsurprised to see a sequence in which an apple disappeared behind a screen, a hand with a knife followed it, and two apple halves emerged. They were similarly unsurprised to see a white towel and a glass of blue paint disappear behind the screen and a blue towel emerge. But they stared, as if in disbelief, when shown causally impossible events, such as a glass of water disappearing with the apple and then two apple halves emerging, or a blue knife disappearing with the white towel and then a blue towel emerging.

So causality can't be reduced to constant conjunction or to possible worlds, and our causal sensibilities aren't quite captured by statistical networks either. How can we make sense of the intuition of oomph that drives our causal instincts? The answer may be found in a look at how causation is expressed in words.

■ ■ ■

Len Talmy, the linguist who elucidated the conception of space found in language, has also elucidated the conception of causal force found in language.[129] As we saw in chapter 2, many verbs convey a notion of causality. Some verbs convey pure causation, like *begin, bring about, cause, force, get, make, produce, set,* and *start.* Others add in the nature of the effect, as in *melt, move, paint,* or *roll.* And still others convey flavors of causation that mean a lot to people but get shorter shrift in the analyses from philosophy. There are verbs of preventing, such as *avoid, block, check, hinder, hold, impede, keep, prevent, save, stop,* and *thwart.* There are verbs of enabling, such as *aid, allow, assist, enable, help, leave, let, permit,* and *support.* And there is a variety of causation expressed in connectives such as *although, but, despite, even, in spite of,* and *regardless.*

Talmy shows that all these concepts tap into a mental model of "force dynamics"—a notion of intrinsic tendencies and countervailing powers which is reminiscent of the billiard-ball animations that convey such a vivid impression of causation to our eyes. The basic player in a causal scenario is called an *agonist:* an entity conceived as having an intrinsic tendency toward motion (left) or rest (right):

The plot is joined by the appearance of an *antagonist:* an entity that exerts a force on the agonist, generally in opposition to its intrinsic tendency. If the antagonist's force is greater than the agonist's tendency (below left), the agonist will change from motion to rest or vice versa. If it is lesser (below right), the agonist will continue with whatever it ordinarily does:

If we imagine dropping in on a situation that is already in progress, we get four possibilities (a little arrow inside an agonist indicates that it is moving):

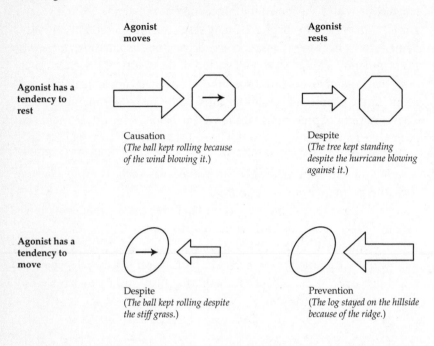

Agonist
moves

Agonist
rests

Agonist has a
tendency to
rest

Causation
(*The ball kept rolling because
of the wind blowing it.*)

Despite
(*The tree kept standing
despite the hurricane blowing
against it.*)

Agonist has a
tendency to
move

Despite
(*The ball kept rolling despite
the stiff grass.*)

Prevention
(*The log stayed on the hillside
because of the ridge.*)

And these four give us the basic meanings of causing, preventing, and two versions of keeping on: movement despite a hindrance (lower left) and stability despite a push (upper right).

To round out the types of causes that are expressed in language, we have to move from atelic activities to telic accomplishments. Imagine, now, the antagonist making an entrance or an exit, rather than being there the whole time, as in the diagram on the following page.

This gives us the dynamic version of causation, together with blocking and two kinds of allowing: letting something do its thing (bottom left) and letting it be (top right). A few other scenarios (such as the antagonist and agonist tending in the same direction, or an antagonist staying out of an agonist's way) give us the other causal relationships, such as helping, hindering, enabling, staying, keeping, and leaving alone. One final distinction among causal verbs is in whether the verb announces the effect, leaving the speaker to mention the cause as an afterthought (*The window broke because*

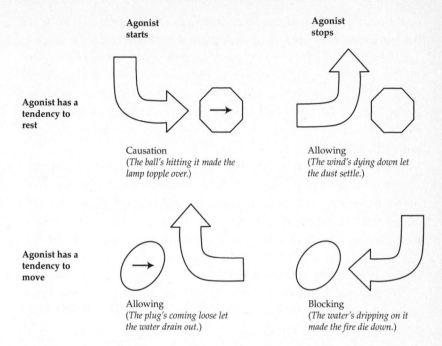

Agonist starts   Agonist stops

Agonist has a tendency to rest

Causation
(*The ball's hitting it made the lamp topple over.*)

Allowing
(*The wind's dying down let the dust settle.*)

Agonist has a tendency to move

Allowing
(*The plug's coming loose let the water drain out.*)

Blocking
(*The water's dripping on it made the fire die down.*)

a ball hit it), or announces the cause, leaving the speaker to mention the effect as an afterthought (*The ball hit the window, causing it to break*).

No doubt you've noticed that these sentences are pretty stilted. Ordinarily we say that the wind blew the ball or that the ball hit the lamp, not that the ball kept rolling because of the wind blowing on it, or that the ball's hitting it made the lamp topple. The awkward sentences are meant to lay bare the fact that only an *event* can cause another event, not an object by itself. A ball, just by existing, can't do a thing; it's only when it's thrown that things start to happen. But everyday language glosses over this logical nicety. An autonomous force, like the wind, waves, or fire, or a person exerting free will, appears as the subject of a predicate, and the predicate expresses the final event in the causal chain, with the intervening links left unspoken. Hence we typically say *Cal made the lamp break* (presumably by knocking it over) and *The wind made the tree topple* (presumably by blowing on it). Causal language can become even more compact because of a process we met in chapters 2 and 3: when an antagonist acts directly on the agonist, the act and the effect get packed into a single verb, and we say *Cal broke the lamp* or *The wind toppled the tree*. For causation to be expressed this concisely, the causal sequence, recall, has to be construed as direct, without intervening links of the same grain size: if Sybil opens the window, and the wind blows the lamp off the table, people don't ordinarily say that *Sybil broke the lamp.*[130] And with many causal verbs the antagonist must *intend* the effect. A girl who stands up and causes her balloon to touch a hot ceiling light has not *popped the balloon,* people in experiments say; nor does a man *wave a flag* when he raises it on a windy day or *dim the lights* when he pushes the switch down in his toaster.[131]

The basic script of an agonist tending, an antagonist acting, and the agonist reacting, played out in different combinations and outcomes, underlies the meaning of the causal constructions in most, perhaps all, of the world's languages. And in language after language, the prototypical force-dynamic scenario—an antagonist directly and intentionally causing a passive agonist to change from its intrinsic state—gets pride of place in the language's most concise causative construction.[132]

To confirm that force dynamics governs the way people use causal language, even when they see a scenario for the first time, Phillip Wolff brought Talmy's diagrams to life on a computer screen using a physics simulator and asked people to describe what they saw.[133] A little motorboat (an agonist) was shown proceeding across a tank of water and suddenly being buffeted by a bank of fans (the antagonist). In describing what the fans did, people used the verb *cause* when the fans deflected the boat's trajectory and pushed it toward a buoy. They used the verb *help* (a verb of enabling) when the boat was already heading toward the buoy and the fans blew it there more quickly. And they used *prevent* when the boat started out toward the buoy and the fans blew it off course. In a nifty extension, Wolff showed that the same dynamics apply to the metaphorical version of force that underlies our concept of personal influence (as when we talk about *social forces* and *peer pressure*). In these animations, a woman (the agonist) standing on a street corner indicates to a traffic cop (the antagonist) that she does or doesn't want to cross the street, and he waves his arm or holds out his palm, whereupon she crosses or doesn't cross. People who saw the various scenes described them with *cause, enable, prevent,* and *despite* according to the same calculus that applied to little boats and fans.

If people naturally conceive of causation in force-dynamic terms, one can understand why the concept should be so intimately connected to counterfactual thinking. The intrinsic tendency of the agonist is, by definition, what it would do *if* the antagonist were not acting on it (in other words, what it does in the possible worlds in which the antagonist is absent). This may be the foundation, buried deep in our cognitive makeup, on which the more rarified definition of causation in terms of counterfactuals has been erected by modern logicians. And it may be other features of that cognitive foundation that make the counterfactual edifice so rickety. The distinction between causes and conditions (striking the match versus keeping the wind out) makes little sense within the counterfactual theory, yet it has a straightforward implementation in the force-dynamic mindset: it's the difference between prototypical causing (in which an antagonist overpowers the intrinsic tendency of the agonist) and various forms of helping, enabling, and allowing (in which an antagonist joins forces with or stays out of the way of the agonist). And since the force-dynamic mindset equates our concept of causation with a metaphor based on intuitive physics, rather than a formula in formal logic, it needn't respect logical necessities such as transitivity. If Antagonist A launches Agonist B, which is then stopped by Antagonist C, there is no reason to conceive of A as impinging on C at all.

Several experiments have shown that people distinguish causal chains that exemplify different force-dynamic interactions even when they are logically equivalent.[134] In a dull but informative experiment, the psychologists Clare Walsh and Steven Sloman told their subjects about a coin that was perched on its edge and about to fall over. In one scenario, the coin is on the verge of falling forward onto heads, and Bill rolls a marble toward it and knocks it over onto tails. In another, the coin is on the verge of landing on tails, an unnamed person rolls the marble toward it, threatening to flip it to heads, but Frank catches the marble before it reaches the coin, allowing it to land on tails. Logically speaking, both guys did something that was necessary for the coin to fall on tails—if they hadn't done what they did, the coin would have landed on heads. Yet in the first scenario, people say that Bill caused the coin to land on heads, but in the second scenario, they say that Frank didn't. The difference is that Bill (who rolled a marble) was seen as an antagonist to the coin's intrinsic tendency to fall, but Frank (who caught a marble) was seen as an antagonist to the tendency of the marble to move, not as an antagonist to the tendency of the coin to fall.

Talmy points out how the mindset behind force dynamics is very different from our best understanding of force and momentum from Newtonian physics. The force-dynamic model in language singles out one entity and conceives of another as impinging on it, whereas in physics neither object in an interaction is privileged. Language conceives of the agonist as having an inner impulse toward motion or rest, whereas physics treats an object as simply continuing at its current velocity. Language distinguishes motion and rest as qualitatively distinct tendencies, whereas physics treats rest as a velocity that happens to be zero. Language treats the antagonist as exerting a force that is stronger than the intrinsic tendency of the agonist. In Newtonian physics, an action and its reaction are opposite and equal, so a pair of touching objects that are at rest, or are moving at a constant velocity, must exert equal forces on each other (if one force were stronger, the two would accelerate in that direction). In language, things can just happen, without stated causes—*The book toppled off the shelf; The sidewalk cracked*—whereas in physics every event has a lawful antecedent. And in physics, the distinction between causing, blocking, permitting, and helping plays no obvious role.

The intuitive physics embedded in language also contaminates people's physical reasoning.[135] When asked to diagram the forces that impinge on a ball when it is tossed straight up in the air, most college students say that when the ball is on its way up, the force pushing it upward is stronger than the one pulling it down; at the zenith, the two forces are equal; and on the way down, the downward force is stronger. The correct answer is that a single force, gravity, impinges on the object the whole time.

When relativity and quantum physics were accepted in the twentieth century, many physicists commented on the violence they do to common sense. Richard Feynman, for example, said, "I think I can safely say that no one understands quantum mechanics. . . . Do not keep asking yourself, if you can possibly avoid it, 'But how can it be like that?' . . . Nobody knows how it can be like that."[136] What is less appreciated is that classical Newtonian physics is *also* deeply counterintuitive. The theory in the history of physics that is closest to intuitive force dynamics is the medieval notion of impetus, in which a moving object has been imbued with some kind of vim or zest that pushes it along for a while and gradually dissipates.

So there is a big discrepancy between intuitive physics, with its discrete episodes of causing and helping and overcoming a tendency toward rest, and real physics, which is just a bunch of differential equations specifying how objects change their velocity and direction over time. Laplace's Demon, the hypothetical imp that knows the instantaneous positions and velocities of every particle in the universe, was said to be able to calculate the entire future or past by plugging these values into the equations that express the laws of mechanics and electromagnetism. The concept of a "cause," or even a discrete "event," plays no role. The discrepancy between intuitive physics and classical physics has led some philosophers to suggest that the very concept of causation is scientifically obsolete, a holdover from an evolutionary past in which we dragged branches along the ground and threw rocks at giraffes. As Bertrand Russell wrote, "The law of causation . . . is a relic of a bygone age, surviving, like the monarchy, only because it is erroneously supposed to do no harm."[137]

But this can't be right. When experts try to determine what caused the *Challenger* space shuttle to fall out of the sky or who caused John F. Kennedy's death, they are not being scientifically illiterate. Nor would they be satisfied with the advice that they look up measurements of all the atoms in the space shuttle before it disintegrated, or in Dealey Plaza on that November afternoon, and plug them into a very, very, very big set of equations. At the scales people find interesting, which are cluttered with friction and chemistry and the trillions of microscopic interactions inside other people's brains, matter in motion obeys principles of its own, and Newton's laws become hopelessly inadequate.

There's a pattern here. In summing up the language of matter, space, and time, I concluded that they are measured by human goals, not just by a scale, a clock, and a tape measure. Now we see that the fourth major category in conceptual semantics, causality, also cares about our intentions and interests. Perhaps we have been looking in the wrong place for benchmarks of the concepts expressed in language. Rather than following Kant into the halls of the physics and math departments, perhaps we should go across campus, to the schools of engineering and law.

PURE AND APPLIED

Evolutionary psychologists believe that aside from language itself, the two things that make humans stand out from other animals are a talent for tools—manipulating the physical world to our advantage—and a talent for cooperation—manipulating the social world to our advantage.[138] Engineering and law are the institutionalized versions of those talents.

When concepts of substance, space, time, and causation are applied to objects that serve human purposes, we are in the realm of engineering. And explanations in engineering freely use vernacular language, no matter how flaky the explanation may seem to the logician or theoretical physicist. When the words are chosen carefully, they can insightfully convey what is going on in a human contrivance without equations or computer simulations, at least at the executive-summary level. Let's look at part of an explanation of how a toilet flushes, from the Web site *How Stuff Works:*

> Take a bucket of water and pour it into the bowl. You will find that pouring in this amount of water causes the bowl to flush. That is, almost all of the water is sucked out of the bowl, and the bowl makes the recognizable "flush" sound and all of the water goes down the pipe. What's happened is this: You've poured enough water into the bowl fast enough to fill the siphon tube. And once the tube was filled, the rest was automatic. The siphon sucked the water out of the bowl and down the sewer pipe. As soon as the bowl emptied, air entered the siphon tube, producing that distinctive gurgling sound and stopping the siphoning process.[139]

First, let's look at how substance enters the explanation. Count nouns are used for the matter that has stable shapes and boundaries and that we keep in view during the explanation (*the bowl, the siphon tube, the sewer pipe*). Mass nouns are used for the matter that gets its shape from whatever currently contains it, and that comes and goes without our really caring about the particular portions (*water, air*). Note, too, how other count nouns are called in when the need arises to measure out portions of the mass—*a bucket of water, this amount of water*—supplemented by quantifiers, as in *all of the water* and *enough water*. Finally, note how other count nouns have reified ethereal events as if they were objects—*the flush sound, the gurgling sound, the siphoning process.*

Now, space. We have objects that may be conceptualized as 2-D boundaries of 3-D volumes—*bucket* and *bowl*—and as the boundaries of volumes with one primary and two secondary dimensions—*pipe* and *tube*. No other aspect of their shape is mentioned (the bucket or the toilet can be round, oval, or rectangular), nor are these details necessary to the explanation. The boundaries are defined not just by their geometry but by their force-dynamic powers, such as containing or directing their contents. The prepositions give us a veritable diagram of the trajectories of motion: water goes *into the bowl, out of the bowl,* and *down the pipe.* For good measure we have the verb *enter,* which also incorporates a spatial meaning ("go into").

Now, time. "What's happened is this," says the explanation. The expression *what happened* is the linguist's litmus test for a stretch of time being construed as an event rather than a state (the most basic distinction of aspect). The flushing is presented as a durative, telic event: the process unfolds over time and terminates when a new state comes into being (an empty bowl). Tense, too, is skillfully used to engage the reader's understanding. The first few sentences, which ask the reader to participate in an ongoing experiment, are in the present tense: *causes, is sucked, makes, goes.* The next two clauses use the perfect—*What [has] happened, You [have] poured*—to identify recent events that resulted in a state of current interest. Then it switches to the past tense to force the reader to review in sequence what he has wrought: *the tube was filled, sucked the water, the bowl emptied, air entered.* And the adverbial phrases—*Once the bowl filled, As soon as the bowl emptied*—give their verbs perfective aspect (the events are seen from the outside, as having been completed), setting the stage for the events that ensue.

And finally, causality. We find verbs that express causal concepts nakedly—*cause the bowl to flush, make the sound, produce the sound, stop the process*—and verbs whose causative parts are hitched to specific effects—*pour* (let flow), *suck* (cause to move by suction), *fill* (cause to be full). The three main types of causation—causing, allowing, and blocking—are all there, as is a fourth type, enabling, tucked into the time adverbials *Once* and *As soon as.* Then there is an event that is perceived as having no cause, namely, air entering the siphon. And of course we have an agonist, the water that wants to stay in the bowl, and an antagonist, the poured water that forces it out (together with smaller agonists and antagonists that appear when we crank up the microscope and look at the causal sequence at a finer grain size).

So the human renditions of Kantian concepts that seem so baffling to the physicist, geometer, and logician working in the abstract turn out to be extremely handy for the engineer working at the scales relevant to human interests and purposes. Admittedly, English is not guaranteed to provide us with clear explanations—it's easy to botch them (as we saw in the baffling descriptions I clipped out of the newspaper), and they come out somewhat differently in different languages (as we saw with the reference frames and polysemous spatial terms). But English and other languages tap an inventory of concepts that are about the right size and kind to capture our basic understanding of how things work. And while we may not spend much time talking about toilets and other products of professional engineering, we do spend a lot of time talking about the products of amateur engineering in our cooking recipes, first-aid instructions, housekeeping hints, sewing patterns, home-repair manuals, and sports tips.

The concept of causation, so essential to our ability to manipulate our physical surroundings, is equally indispensable to our ability to manipulate our social surroundings. Indeed, the concept of causation and the concept of human action are coupled. Though occasionally the first link in an interesting causal chain is a natural event like the weather or a rockslide, more often it is a human being, exercising what we think of as free will. The prototypical subject of a causative verb is a person, and as we have seen, its prototypical object is an entity that the person has directly and intentionally affected in the final link of the causal chain.

Though we conceive of voluntary actions as uncaused, that doesn't stand in the way of our trying to influence them. We influence people by holding them *responsible* for the effects they cause. When we see an event we do or don't like and attribute its cause to the intended action of a person, we shower that person with praise or opprobrium, hoping that this will lead him (and other people who hear about the policy) to act that way more or less often in the future. And the concept of causation we apply when choosing our verbs is also the concept we apply when we hold people responsible. We single out the acts that a person intentionally and directly and foreseeably caused, rather than those that just happened in his presence or that he brought about accidentally or unknowingly. Presumably it is because those are the kinds of acts that our praise and blame can affect in the future (we blame an assistant for not backing up a file, because that might make him more careful in the future, but don't blame him for a hard-disk crash, because there's nothing he can do about it). And when we mete out tangible punishment rather than verbal condemnation, and codify the policy in writing, we call it law.

"The law is a profession of words," it is said. But human actions don't come labeled with words; the movie of life has no voice-over or subtitles. To apply the words of a law to a particular happening, as lawyers must do, they must fish for examples of the *concepts* that the words represent. When our intuitive conception of causation plugs into a situation cleanly, in a way that all observers can agree upon, the case is open and shut. But when the concept must be jammed into a scenario that violates our stereotype of direct causation—something that happens more often with the behavior of people than the behavior of toilets—interested parties will argue about the best fit. Every part of the concept of causation that figures in the language of causation has served as a flashpoint in legal disputation.

Take the most basic distinction of all—between events that merely follow each other, à la Hume, and events we sense to be linked by a causal force. Norman Finkelstein, a critic of Israel, calls attention to a 1995 event in which a Palestinian man in Israeli custody was shaken by an officer and subsequently died. The event was investigated by forensic pathologists and the supreme court of Israel, all of whom agreed, according to Finkelstein, that "Harizad died from the shaking." Alan Dershowitz, a defender of Israel, pointed out that the actual quotation was that "the subject expired *after* being shaken."[140] Dershowitz noted that "the difference between 'died *from* the shaking' and 'expired *after* the shaking' is considerable." Indeed it is: it is the difference between mere succession and actual causation (conveyed in this case by the preposition *from,* which taps a force-dynamic metaphor of energy flowing from cause to effect). The semantic distinction between *after* and *from* points to a causal distinction between succession and impingement, which in turn animates a moral distinction between tragedy and evil.

Another force-dynamic distinction, the one between causing and letting, deeply penetrates our moral reasoning. The difference is exposed in the trolley problem, a famous thought experiment devised by the philosopher Philippa Foot that has long been a talking point among moral philosophers.[141] A trolley is hurtling out of control and is bearing down on five railroad workers who don't see it coming. You are standing by a switch and can divert the trolley onto another track, though it will then kill a single worker who hasn't noticed the danger either. Should you save five lives at the cost of one by switching the track? Most people say yes—not just readers of philosophy journals nodding their heads, but, in a massive experiment run by Marc Hauser, almost 90 percent of the 150,000 people in more than a hundred countries who volunteered to ponder the dilemma and share their intuitions on his Web site.[142]

Now imagine that you are on a bridge overlooking the tracks and have spotted the runaway train bearing down on the five workers. Now the only way to stop it is to throw a heavy object in its path. And the only heavy object within reach is a fat man standing next to you. Should you throw the man over the bridge? Both dilemmas present you with the option of sacrificing one life to save five, and so, by one reckoning, the two dilemmas are morally equivalent. But most people the world over disagree. Though they would pull the switch in the first dilemma, they would not heave the fat man in the second. When pressed for a reason, they can't come up with anything coherent, but then neither can most moral philosophers.

Joshua Greene, who is both a philosopher and a cognitive neuroscientist, suggests that people are equipped with an evolutionarily shaped revulsion to manhandling an innocent human being, and that this overwhelms any utilitarian calculus that would tally the lives saved and the lives lost.[143] The impulse against roughing up a person would explain other examples in which people recoil from killing one to save many, such as euthanizing a hospital patient to harvest his organs and save five dying patients in need of transplants, or smothering a baby in a wartime hideaway to prevent its cries from attracting soldiers who would kill all the occupants, baby included. In support of this idea, Greene, together with the cognitive neuroscientist Jonathan Cohen, scanned people's brains as they pondered the various dilemmas.[144] They found that the dilemmas that required killing a person with one's bare hands activated certain brain areas associated with emotion, together with other brain areas that are involved in resolving a conflict.

So we see here the unmistakable stamp of a force-dynamic mindset in thinking through a profound moral dilemma. A scenario in which the actor is an antagonist and his sacrificial victim (the fat man) is an agonist—the prototypical meaning of causative verbs—evokes an emotion that overwhelms our reckoning of lives saved and lost, whereas the alternative scenario, in which the actor is a mere *enabler* of an antagonist (the train), does not.

Does this mean that our force-dynamic mindset makes us irrational in the moral arena? Does the eye-catching difference between causing and enabling contaminate our ethics and render our intuitions untrustworthy? Not necessarily. We value people not just for what they *do* but for what they *are.* And a person who is capable of heaving a struggling man over a bridge or covering the mouth of a baby until it stops breathing is probably capable of other horrific acts that lack a redeeming reduction in the body count. Even putting aside the callousness that would be necessary to carry through these acts, the kind of person who chooses his acts only by their anticipated costs and benefits (reckoned by calculations that he arrogates to himself) might skew the sums in his favor whenever the odds and payoffs are uncertain, which in real life they always are. So the majority of people who gave the "inconsistent" answers to these thought experiments may be the victims of a trap set by moral philosophers. The philosophers have jiggered a thought experiment in which a person of good character, whose behavior tends to lead to good outcomes in typical circumstances and hence deserves our approbation, would do things that lead to a greater number of deaths. The perhaps-too-fertile imagination of the philosopher, with its hazard of baffling or entrapping our causal intuitions, has been satirized in a compendium of philosophical humor: "A brain in a vat on Twin Earth is at the wheel of a runaway trolley. On one track is a worker, Jones, who is planning to murder five men, but one of those men intends to blow up a bridge that will be crossed by a bus filled with thirty orphans . . ."

In fairness to the philosophers, any fan of the television series *Law & Order* knows that the legal system really does turn up agonizing scenarios that hinge on whether an act may be conceived as causing to die, enabling to die, or allowing to die. In fact, you don't even need to turn on the television; examples fill the newspapers and history books. We already met Charles Guiteau, the man who shot James Garfield but might have escaped the noose if only someone had thought to wash his hands before sticking his finger in the president's wound or to feed him through his mouth instead of the other end of his alimentary canal. And in a real-life puzzle of indirect causation that not even a philosopher could have made up, a Long Island widow filed a sixteen-million-dollar wrongful-death suit against the Benihana restaurant chain because one of its chefs, imitating Jackie Chan in the movie *Mr. Nice Guy,* tried to fling a grilled shrimp into her husband's mouth with a spatula. The chef had first flung a shrimp at the man's brother-in-law but missed his mouth, hitting him in the forehead instead. He threw another shrimp at the man's son, hitting him in the arm. Then the chef flung a third shrimp, this one at the victim, who tried to dodge it by jerking his head back. After the dinner the man began to experience neck pain. In the ensuing months he underwent two spinal operations. He contracted an infection after the second one and died of septicemia. According to a report in the *New York Law Journal*, the family's lawyer invoked the counterfactual theory of causation: "But for the food-flinging incident . . . [the man] would still be alive." Benihana's lawyer implicitly invoked the force-dynamic alternative: "Benihana cannot be liable for [the man's] death because of a break in the chain of causation between the first or second procedures and his death five months later." The jury, presumably showing the mind's true colors, ruled in favor of Benihana.[145]

Every other ingredient that we saw go into the semantics of causation has been a bone of contention in a court of law.[146] There is the conundrum of an intermediate link consisting of the voluntary action of another human agent. A man is ordered by a ruthless IRA gunman to drive him to a location where the gunman shoots a policeman. Was the driver an accomplice to murder? What about an alleged war criminal who says he was following orders, or a kidnap victim such as Patty Hearst who said she was brainwashed?

There is the conundrum of crimes of omission (failing to prevent, rather than causing). Should we press murder charges against a woman who failed to stop her boyfriend from fatally beating her child? The passerby who fails to prevent a homeless man from freezing to death? A man who shoots an attacker in the leg in self-defense and then waits so long to call an ambulance that the attacker dies from loss of blood?

And there is the conundrum of how to identify the goal of a causal act, which often resides in the privacy of a person's mind. This is clear enough when we distinguish accidental from intended outcomes, as in the difference between a woman who loses control of her car on an icy road and kills her husband on the sidewalk and a woman who takes aim at her husband, steps on the accelerator, and mows him down. But what about cases where there is a mismatch between the private intentions of an actor and the public result he brings about? A woman thinks she is stealing an umbrella from a stand, but it turns out to belong to her. A man has consensual sex with his stepdaughter (which is legal), thinking she's his daughter (which is not). A believer in voodoo sticks pins in a likeness of his wife, hoping to kill her.

Our concept of causation is indispensable to our attribution of credit and blame in everyday life. Yet in the full drama of human experience, it will sometimes collide with circumstances that don't meet its standard checklist. Given the endless enigmas flowing from our concept of causation, with its model of directness, intention, contact, and intrinsic tendency, it's no wonder that episodes of *Law & Order* seem to fill every channel on the cable dial morning, noon, and night.

Kant was surely right that our minds "cleave the air" with concepts of substance, space, time, and causality. They are the substrate of our conscious experience. They are the semantic contents of the major elements of syntax: noun, preposition, tense, verb. They give us the vocabulary, verbal and mental, with which we reason about the physical and social world. Because they are gadgets in the brain rather than readouts of reality, they present us with paradoxes when we push them to the frontiers of science, philosophy, and law. And as we shall see in the next chapter, they are a source of the metaphors by which we comprehend many other spheres of life.

Yet when examined through the window of language, these concepts turn out to be quite unlike the infinite aquarium, perpetual clock, or PLAY OVER button that were the best guesses about the nature of space, time, and causality in Kant's day. They are digital where the world is analogue, austere and schematic where the world is rich and textured, vague even when we crave precision, and parochial to human goals and interests even when we ought to seek the view from nowhere.

It is sometimes humbling to think that the foundations of common sense are just the design specs of one of our organs. Yet our science and reason have managed to reveal many aspects of substance, space, time, and causality that do violence to common sense but which we can see are likely to be true. No small part in this transcendence is played by scrutiny of these concepts as they appear in language and thought, allowing them to be understood as part of our makeup, and discounted accordingly. It is the closest we can come to the light dove's dream of taking wing in empty space.

5

THE METAPHOR METAPHOR

When in the Course of human events it becomes necessary for one people to dissolve the political bands which have connected them with another and to assume among the powers of the earth, the separate and equal station to which the Laws of Nature and of Nature's God entitle them, a decent respect to the opinions of mankind requires that they should declare the causes which impel them to the separation.

The United States Declaration of Independence is perhaps the best-known passage of English prose expressing an abstract political idea. Its subject matter, a challenge to power, has long been a part of the human condition. But challenges to power had hitherto been contests of sheer brawn, and here the challenge was being justified from first principles worked out by Enlightenment philosophers. Indeed, what was being articulated was not only the rationale for the challenge, but the rationale for the rationale.

At the heart of this abstract argument, though, is a string of concrete metaphors. The issue at hand was the *bands* that *connected* the colonies to England, which it was necessary to *dissolve* in order to effect a *separation*. (Though today the word *dissolve* means "absorb in a liquid," it originally meant "loosen asunder.") The four metaphors really allude to a single, unstated metaphor: ALLIANCES ARE BONDS. We see the metaphor in other expressions like *bonding, attachment,* and *family ties.*

Also palpable is the metaphor in *impel*—force to move—whose literal sense is plain in the noun *impeller,* the rotating part that pushes the water or air in a pump, and in its cousin *propeller.* The implicit metaphor is that CAUSES OF BEHAVIOR ARE FORCES. It underlies the cognates *repel* and *compel,* and analogous terms like *impetus, drive, force, push,* and *pressure.* A related metaphor may be found in *powers of the earth* (which calls to mind *horsepower* and *electric power*): A SOVEREIGN STATE IS A SOURCE OF PHYSICAL FORCE.

A bit less obvious is the metaphor for human history, *course,* which refers to a path of running or flowing, as in *the course of a river, a racecourse,* and *a headlong course.* The metaphor is that A SEQUENCE OF EVENTS IS MOTION ALONG A PATHWAY, a special case of the TIME IS MOTION metaphor we met in the previous chapter.

The very name of the document echoes two older metaphors, which we can glimpse in related words. *To declare,* like *clarify,* comes from the Latin for "make clear," an instance of the UNDERSTANDING IS SEEING metaphor, as in *I see what you mean, a murky writer,* and *shedding more heat than light.* And *independence* means "not hanging from," echoed in *suspend, pendant,* and *pendulum.* It alludes to a pair of metaphors, RELIANCE IS BEING SUPPORTED (*propped up, financial support, support group*), and SUBORDINATE IS DOWN (*control over him, under his control, decline and fall*).

If we dig even deeper to the roots of words, we unearth physical metaphors for still more abstract concepts. *Event,* from Latin *evenire,* originally meant "to come out" (compare *venture*). *Necessary* comes from "unyielding" (compare *cede*). *Assume* meant "to take up." *Station* is a standing-place, an instance of the widespread metaphor that equates status with location. *Nature* comes from the Latin for "birth" or "inborn qualities," as in *prenatal, nativity,* and *innate. Law* in the sense of "moral necessity" is based on *law* in the sense of man-made regulations, from Old Norse *lag,* "something set down." The metaphor A MORAL OBLIGATION IS A RULE also underlies e*ntitle,* from the Latin word for "inscription." *Decent* originally meant "to be fitting." *Respect* meant "to look back at" (remember *aspect*), *kind* comes from the same Germanic root as *kin, require* from "to seek in return."

Even the little grammatical words have a physical provenance. Sometimes it is evident in modern English, as in the pronoun *it* (A SITUATION IS A THING) and the prepositions *in* (TIME IS SPACE), *to* (INTENTION IS MOTION TOWARD A GOAL), and *among* (AFFILIATION IS PROXIMITY). Sometimes it is evident only in the word's ancestor, such as *of*, from a Germanic word related to "off," and *for*, from the Indo-European term for "forward."

Not much is left. *Political* comes from the Greek *politēs*, meaning "citizen," from *polis*, "city," which is a metonym rather than a metaphor, but still has an association to something tangible. *The* and *that* come from an ancient Indo-European demonstrative term (also the source of *then*, *there*, *they*, and *this*), standardly used in connection with pointing. That leaves *God*, *man*, and *people*, which mean what they mean and have for a long time, and the quasi-logical terms *and*, *equal*, and *cause*.

So if language is our guide, the lofty declaration of abstract principles is really a story with a strange and clunky plot. Some people are hanging beneath some other people, connected by cords. As stuff flows by, something forces the lower people to cut the cords and stand beside the upper people, which is what the rules require. They see some onlookers, and clear away the onlookers' view of what forced them to do the cutting.

But *should* language be our guide? It seems unlikely that anyone reading the Declaration would entertain the bizarre images in the literal meanings of these words or their roots. At the same time it's jarring to discover that even the airiest of our ideas are expressed ("pressed out") in thumpingly concrete metaphors. The explorations of language and thought in the preceding chapters have turned up these metaphors under every stone: events as objects, states as locations, knowing as having, communicating as sending, helping as giving, time as space, causation as force. What should we make of the discovery that people can't put two words together without using allusions and allegories? This chapter will try to steer a path between two extreme answers.

Maybe we should make nothing of it. All words have to be coined by a wordsmith at some point in the mists of history. The wordsmith had an idea to get across and needed a sound to express it. In principle, any sound would have done—a basic principle of linguistics is that the relation of a sound to a meaning is arbitrary—so the first coiner of a term for a political affiliation, for instance, could have used *glorg* or *schmendrick* or *mcgillicuddy*. But people are poor at conjuring sounds out of the blue, and they probably wanted to ease their listeners' understanding of the coinage rather than having to define it or illustrate it with examples. So they reached for a metaphor that reminded them of the idea and that they hoped would evoke a similar idea in the minds of their listeners, such as *band* or *bond* for a political affiliation. The metaphorical hint allowed the listeners to cotton on to the meaning more quickly than if they had had to rely on context alone, giving the word an advantage in the Darwinian competition among neologisms (the topic of the next chapter). The word spread and became endemic to the community, adding to the language's stock of apparent metaphors. But then it came to be used often enough, and in enough contexts, that speakers kicked the ladder away, and today people think not a whit about the metaphorical referent. It persists as a semantic fossil, a curiosity to amuse etymologists and wordwatchers, but with no more resonance in our minds than any other string of vowels and consonants. Call this the killjoy theory. It says that most metaphors are dead metaphors, like *coming to a head,* which people would probably stop using if they knew that it alludes to the buildup of pus in a pimple.

The other extreme is that the human mind can directly think only about concrete experiences: sights and sounds, objects and forces, and the habits of behavior and emotion in the culture we grow up in. All our other ideas are metaphorical allusions to these concrete scenarios. We can't think of political affiliations, for example, without calling to mind (perhaps unconsciously) some kind of glue or cord. And when we think about time, the parts of the brain dedicated to space light up. Human intelligence, with its capacity to think an unlimited number of abstract thoughts, evolved out of primate circuitry for coping with the physical and social world, augmented by a capacity to extend these circuits to new domains by metaphorical abstraction. And since people think in metaphors, the key to understanding human thought is to deconstruct those metaphors. People disagree because they frame a problem with different metaphors, and make a mess of their lives because of pernicious implications of these frames, which they use without awareness. A linguistically informed literary criticism is the key to resolving conflict and frustration, from psychotherapy and law to philosophy and politics. Call this the messianic theory. It is based on the idea that TO THINK IS TO GRASP A META-PHOR—the metaphor metaphor.

KILLJOYS AND MESSIAHS

The killjoy theory and the messianic theory may seem like the hundred-dollar bottles that bookend the shelves of a savvy wine dealer, but they each deserve our attention. There can be no doubt that *some* metaphors are as dead as doornails—if not *dissolving bands* and *coming to a head,* then certainly the ones composed of roots from Latin and Old Norse. After all, today's Anglophones don't have a Jungian collective memory for the metaphoric imagination of long-dead speakers. And if some metaphors can persist in the language as fossils, it puts every metaphor under a cloud of suspicion. Just spotting a metaphor in the language is no guarantee that speakers see through it as opposed to using it uncomprehendingly, the way we use *man* or *dog.* As a famous connoisseur of the symbolic once noted, sometimes a cigar is just a cigar.

One sign that metaphors are often observed in the breach is the prevalence of *mixed* metaphors, in which a speaker or writer cobbles together two metaphors that are related in their underlying meanings but ludicrous in their literal content:

> I'm not going to stick to my laurels [actress Kate Winslet, at the
> 2002 Academy Awards].[1]
> Once you open a can of worms, they always come home to roost.[2]
> Those professors tilt at the windmills of a capitalist patriarchy
> from whose teat they feed.[3]
> Once again, the Achilles' heel of the Eagles' defense has reared
> its ugly head.[4]

The opacity of everyday metaphors is also apparent in inadvertently tasteless expressions (such as the radio psychotherapist who said, "For some patients, cancer can be a growth experience"), ambiguous headlines (CHEF THROWS HIS HEART INTO HELPING FEED NEEDY), Goldwynisms ("An oral agreement isn't worth the paper it's written on"), and entry into the club called AWFUL—Americans Who Figuratively Use "Literally."[5] The charter member was Rabbi Baruch Korff, a defender of Richard Nixon during his Watergate ordeal, who at one point protested, "The American press has literally emasculated President Nixon."

But at the other end of the shelf, the ubiquity of metaphor in everyday language is truly a surprising discovery, rich with implications. Even the killjoy has to admit that metaphors were alive in the minds of the original coiners and compelling to the early adopters. And the sheer number of figures of speech that fall within a single image (itself unmentioned) suggests that the tacit metaphor must have been transparent to large numbers of coiners and adopters for a very long time. Consider just a few of the expressions that fall under the umbrella ARGUMENT IS WAR, collected by the linguist George Lakoff and the philosopher Mark Johnson:[6]

> Your claims are indefensible. He attacked every weak point in my argument. His criticisms were right on target. I demolished his argument. I've never won an argument with her. You don't agree? Okay, shoot! If you use that strategy, he'll wipe you out. She shot down all of my arguments.

Or the many variations of LOVE IS A JOURNEY:

> Our relationship has hit a dead-end street. It's stalled; we can't keep going the way we've been going. Look how far we've come. It's been a long, bumpy road. We can't turn back now. We're at a crossroads. We may have to go our separate ways. The relationship isn't going anywhere. We're spinning our wheels. Our relationship is off the track. Our marriage is on the rocks. I'm thinking of bailing out.

Because these metaphors have no poetic panache, they may be distinguished from literary metaphors like *Juliet is the sun*. Sometimes they are called conceptual metaphors, because no one ever actually had to *say* "Argument is war" or "Love is a journey"; the underlying metaphor is implicit in the family of related tropes. They are also called "generative" metaphors because people easily generate new tropes that belong to a family, like *He protected his theory in a hardened bunker,* or *Marsha told John they should step on the brakes.*[7] For the expressions to proliferate so easily, speakers and hearers must be dissecting the implied metaphor to lay bare the connections between the things named by the metaphor and the abstract concepts they are really talking about. (In literary theory these are sometimes called the "vehicle" and the "tenor" of the metaphor; cognitive scientists call them the "source" and the "target.")[8] For instance, to be fluent with the various LOVE IS A JOURNEY expressions, one has to fathom the conceptual metaphor in considerable depth. Lakoff explains:

The lovers are travelers on a journey together, with their common life goals seen as destinations to be reached. The relationship is their vehicle, and it allows them to pursue those common goals together. The relationship is seen as fulfilling its purpose as long as it allows them to make progress toward their common goals. The journey isn't easy. There are impediments, and there are places (crossroads) where a decision has to be made about which direction to go in and whether to keep traveling together.[9]

Anyone who didn't tacitly grasp this scenario might use some of the expressions by rote, but would not be able to produce or understand new ones. And a community of such literal minds would never have found the metaphors compelling enough to allow them to accumulate in the language the way they have. When you combine the productivity of expressions within a generative metaphor with the sheer number of generative metaphors—Lakoff has documented hundreds, from BIG IS IMPORTANT and THE VISUAL FIELD IS A CONTAINER to MORALITY IS CLEANLINESS and THE SELF IS A GROUP OF PEOPLE—you have to acknowledge the possibility that generative metaphors are a major phenomenon in language and an important clue to our cognitive makeup.[10] Abstract ideas are connected in a systematic way to more concrete experiences.

METAPHOR MATTERS

A lot is at stake in the question of how the mind handles conceptual metaphors. For one thing, the answer may shed a good deal of light on cognitive development and education. Children may not understand political alliances or intellectual argumentation, but they surely understand rubber bands and fistfights. Conceptual metaphors point to an obvious way in which people could learn to reason about new, abstract concepts. They would notice, or have pointed out to them, a parallel between a physical realm they already understand and a conceptual realm they don't yet understand. This would explain not only how children learn difficult ideas as they grow up but how people of any age learn them in school or from expository prose.[11] Analogies such as THE ATOM IS A SOLAR SYSTEM or AN ANTIBODY IS A LOCK FOR A KEY would be more than pedagogical devices; they would be the mechanism that the mind uses to understand otherwise inaccessible concepts.

The appeal of the metaphor metaphor runs even deeper. Ever since Darwin and Wallace proposed the theory of evolution by natural selection, people have wondered how the human mind evolved the ability to reason about abstract domains such as physics, chess, or politics, which have no relevance to reproduction and survival. The puzzle led Wallace himself to part company from Darwin and ascribe the human mind to a divine plan, foreshadowing the Intelligent Design movement in the United States more than a century later.[12] But conceptual metaphor points to a way to solve the mystery.[13]

The conceptual metaphors we met in chapters 2 and 4 were rooted in substance, space, time, and causation (itself rooted in force). These concepts were certainly within the ken of our evolutionary ancestors. In the preceding chapter we saw experiments by Marc Hauser and his colleagues showing that rhesus monkeys can reason about cause and effect (for example, they know that a hand with a knife can cut an apple but that a hand with a glass of water cannot). In other experiments Hauser has shown that tamarin monkeys have a rich understanding of the spatial and mechanical relations we express with nouns, prepositions, and verbs.[14] When given an opportunity to reach for a piece of food behind a window using objects set in front of them, the monkeys go for the sturdy hooks and canes, avoiding similar ones that are cut in two or made of string or paste, and not wasting their time if an obstruction or narrow opening would get in the way. Now imagine an evolutionary step that allowed the neural programs that carry out such reasoning to cut themselves loose from actual hunks of matter and work on symbols that can stand for just about anything. The cognitive machinery that computes relations among things, places, and causes could then be co-opted for abstract ideas. The ancestry of abstract thinking would be visible in concrete metaphors, a kind of cognitive vestige.[15]

Of course, the sources of most of the metaphors in Lakoff's vast collection aren't just objects, space, time, and causation. But many of them are other plausible obsessions for a hominid ancestor, such as conflict, plants, and disease. And even the complex ones can be built out of more basic concepts. For example, the "vehicle" in the LOVE IS A JOURNEY metaphor can be conceived as a container that moves people along a path toward a goal. If all abstract thought is metaphorical, and all metaphors are assembled out of biologically basic concepts, then we would have an explanation for the evolution of human intelligence. Human intelligence would be a product of metaphor and combinatorics. Metaphor allows the mind to use a few basic ideas—substance, location, force, goal—to understand more abstract domains. Combinatorics allows a finite set of simple ideas to give rise to an infinite set of complex ones.[16]

Another fallout of the metaphor metaphor is the phenomenon of *framing*. Many disagreements in human affairs turn not on differences in data or logic but on how a problem is framed. We see this when adversaries "talk past each other" or when understanding something requires a "paradigm shift." In the first chapter I mentioned some examples, like *invading Iraq* versus *liberating Iraq, ending a pregnancy* versus *killing an unborn child,* and *redistributing wealth* versus *confiscating earnings.* Each controversy hinges on a choice between metaphors, such as the competing force-dynamic models that underlie an invasion (an antagonist entering an area by overcoming the resistance of an agonist) and a liberation (an antagonist removes another antagonist who is impeding the free motion of an agonist). One of the reasons I explained verb constructions in chapter 2 was that they show that even our most quotidian acts can be framed in different ways, such as the difference between *spraying paint on the wall* (cause the paint to go) and *spraying the wall with paint* (cause the wall to change).

Within cognitive psychology the most famous example of the effects of framing (briefly mentioned in chapter 3) comes from an experiment by Amos Tversky and Daniel Kahneman, who posed the following problem to a sample of doctors:[17] "A new strain of flu is expected to kill 600 people. Two programs to combat the disease have been proposed." Some of the doctors were then presented with the following dilemma:

> If program A is adopted, 200 people will be saved. If program B is adopted, there is a one-third probability that 600 people will be saved and a two-thirds probability that no people will be saved. Which of the two programs would you favor?

If you're like most of the doctors who were given this choice, you will pick program A, the sure option, rather than program B, the risky one. The other set of doctors was presented with a different dilemma:

> If program C is adopted, 400 people will die. If program D is adopted, there is a one-third probability that nobody will die

and a two-thirds probability that 600 people will die. Which of
the two programs would you favor?

If you're like most of the doctors who faced *this* choice, you will avoid
program C, the sure option, and gamble with program D, the risky one.

If you reread the two dilemmas carefully, however, you will notice that
the choices are identical. If 600 people would die in the absence of treatment, then saving 200 people is the same as losing 400 people, and saving
no one is the same as losing everyone. Yet the doctors flipped their prefer-
ence depending on how the same menu of options was framed. The crucial
difference in wording alluded to a difference in metaphors. The people who
would be saved after receiving the treatment were construed as a "gain"
over what would have happened if the epidemic were left untreated, whereas
the people who would die were considered a "loss" against what would have
happened if the epidemic had never arrived. Now, it has been independently shown that people hate to lose something more than they enjoy
gaining it.[18] For example, they don't mind paying for something with a
credit card even when told there is a discount for cash, but they hate paying
the same amount if they are told there is a surcharge for using credit. As a
result, people will often refuse to gamble for an expected profit (they turn
down bets such as "Heads, you win $120; tails, you pay $100"), but they will
gamble to avoid an expected loss (such as "Heads, you no longer owe $120;
tails, you now owe an additional $100"). (This kind of behavior drives
economists *crazy,* but is avidly studied by investment firms hoping to turn
it to their advantage.) The combination of people's loss aversion with the
effects of framing explains the paradoxical result: the "gain" metaphor
made the doctors risk-averse; the "loss" metaphor made them gamblers.

Tversky and Kahneman's 1981 study, though a bit complicated, is the
gold standard for demonstrating the effects of framing on behavior: identical
events, different metaphors, flipped decision—and not just any decision, but
one that would affect hundreds of lives. Since then, the idea that framing
affects thinking has been applied to many spheres of human activity. The urban planner Donald Schön has argued that an "urban blight" metaphor led
planners to treat crowded neighborhoods as if they were diseased plants,
which had to be extirpated to prevent the spread of rot. The result was the
disastrous "urban renewal" projects of the 1960s.[19] The judge Michael Boudin
has argued that judges can be illicitly affected by metaphors such as the "fruit

of the poisonous tree" (illegally obtained evidence), the "wall of separation between church and state," and "bottleneck monopolies" (companies that control a distribution channel such as a power grid or realty listing service).[20] A book on psychotherapy called *Metaphors in Mind* calls on therapists to work with their patients' metaphors, like "I have sensitive radar for insults" and "I'm trapped behind a door."[21] A book on business leadership called *The Art of Framing* examines references to businesses as journeys, games, wars, machines, organisms, and societies.[22]

A lot follows from the idea that the mind is a metaphor-monger. Let's consider the ways in which it is and the ways in which it is not.

THE MESSIAH OF METAPHOR

If an appreciation of metaphor will bring on a messianic age, the messiah himself would be George Lakoff. Lakoff was a student of Chomsky's in the 1960s and founded the breakaway movements called Generative Semantics and Cognitive Linguistics.[23] In a series of engaging books beginning with *Metaphors We Live By* (coauthored in 1980 with Mark Johnson), Lakoff has analyzed the world of conceptual metaphors with dazzling insight and in impressive depth. And he has come to some remarkable conclusions.

Lakoff is by far the strongest advocate of the metaphor metaphor. Metaphor is not an ornamental flourish of language, he says, but an essential part of thought: "Our ordinary conceptual system, in terms of which we both think and act, is fundamentally metaphorical in nature."[24] Mental life begins with a few experiences that are not metaphorical, namely, the sensations, actions, and emotions that are built into our constitution and engage the physical world. From there, conceptual metaphors are acquired by a kind of associative conditioning. We learn that CONTROL IS UP because we experience fights in which the victor ends up on top, that GOALS ARE DESTINATIONS because we walk toward something we want, and that TIME IS A MOVING OBJECT because things that approach us get closer and closer as time elapses.

But this isn't the half of it. Since we think in metaphors grounded in physical experience rather than in logical formulas with truth values, the entire tradition of Western thought since the Greeks is fundamentally misconceived.[25] Reason is not based on abstract laws, because thinking is rooted in bodily experience. And the concept of objective or absolute truth must be rejected. There are only competing metaphors, which are more or less apt for the purposes of the people who live by them.

Western philosophy, then, is not an extended debate about knowledge, ethics, and reality, but a succession of conceptual metaphors.[26] Descartes's philosophy is based on KNOWING IS SEEING, Locke's on THE MIND IS A CONTAINER, Kant's on MORALITY IS A STRICT FATHER, and so on. Nor is mathematics about a Platonic reality of eternal truths.[27] It is a creation of the human body and senses, growing out of the activities of moving along a path and of collecting, constructing, and measuring objects. Political ideologies, too, cannot be defined in terms of assumptions or values, but only as rival versions of the metaphor that SOCIETY IS A FAMILY. The political right likens society to a family commanded by a strict father, the political left to a family cared for by a nurturant parent.[28]

Day-to-day political debates are also contests between metaphors.[29] Citizens are not rational and pay no attention to facts, except as they fit into frames, and the frames are "fixed in the neural structures of [their] brains."[30] In George W. Bush's first term, for example, he promised *tax relief,* which frames taxes as an affliction, the reliever as a hero, and anyone obstructing him as a villain. The Democrats were foolish to offer their own version of "tax relief," which accepted the Republicans' framing; it was, Lakoff said, like asking people not to think of an elephant. The Democrats should instead have reframed taxes as "membership fees" that are necessary to maintain the services and infrastructure of the society to which we belong. In 2005 Lakoff was cast as the savior of the Democratic Party in the wake of its shocking defeat in the previous year's presidential election. He conferred with party leaders and strategists, addressed caucuses, and saw his book *Don't Think of an Elephant!* become a bestseller and a liberal talisman.[31]

Now, linguistics has exported a number of big ideas to the intellectual world. They include the diversification of languages as an inspiration to Darwin for the origin of species; the analysis of contrasting sounds as a paradigm for structuralism in anthropology and literary theory; the linguistic determinism hypothesis; and Chomsky's deep structure and innate universal grammar. Even by these standards, Lakoff's theory of conceptual metaphor is a lollapalooza. If he is right, conceptual metaphor can do everything from overturning twenty-five hundred years of misguided reliance on truth and objectivity in Western thought to putting a Democrat in the White House.

Though I believe that conceptual metaphor really does have profound implications for the understanding of language and thought, I think Lakoff takes the idea a wee bit too far.

Let's start at the top, with his dismissal of truth, objectivity, and disembodied reason. To be fair, Lakoff is not a postmodernist or a radical cultural relativist. He believes there is a nonmetaphorical, physical world, and he believes that a human nature, embedded in our bodies and interacting with the world, provides universal experiences that ground many metaphors in ways that are common to humanity. Yet he also believes that many of the metaphors that ground our reason are specific to a culture, and even his universalism is a kind of *species* relativism: our knowledge is nothing but a tool suited to the interests and bodies of *Homo sapiens*. As such, Lakoff's version of relativism is vulnerable to the two standard rebuttals of relativism in general.[32]

The first is that our best science and mathematics can predict how the world will behave in ways that would be a staggering coincidence if the theories did not characterize reality. As Richard Dawkins has pointed out, even the staunchest relativist will fly to the next academic conference on a jet airplane designed according to the metaphors of modern physics rather than on a magic carpet designed according to some rival metaphor (as he put it, Show me a relativist at 30,000 feet and I will show you a hypocrite).[33] It won't do to say that scientific metaphors are merely "useful," unless one stifles all curiosity about why some metaphors are useful and others are not. The obvious answer is that some metaphors can express truths about the world. So even if language and thought use metaphors, that doesn't imply that knowledge and truth are obsolete. It may imply that metaphors can objectively and truthfully capture aspects of reality. In a later section we will see how they can do this.

The other rebuttal is that by their very effort to convince others of the truth of relativism, relativists are committed to the notion of objective truth.[34] They attract supporters by persuasion—the marshaling of facts and logic—not by bribes or threats. They confront their critics using debate and reason, not by dueling with pistols or throwing chairs like the guests on a daytime talk show. And if asked whether their brand of relativism is a pack of lies, they would deny that it is, not waffle and say that the question is meaningless. One has to look no farther than the opening lines of Lakoff and Johnson's *Philosophy in the Flesh:*

The mind is inherently embodied.

Thought is mostly unconscious.

Abstract concepts are largely metaphorical.

These are three major findings of cognitive science. More than two millennia of a priori philosophical speculation about these aspects of reason are over. Because of these discoveries, philosophy can never be the same again. When taken together and considered in detail, these three findings are inconsistent with central parts of Western philosophy.

"The mind is inherently embodied"—not "We offer the metaphor that the mind is inherently embodied." "These discoveries," "these findings"—not "these useful framings." "Inconsistent with central parts of Western philosophy"—not "a different metaphor from that of Western philosophy." Lakoff and Johnson can't help themselves. In the very act of advancing their thesis, they presuppose transcendent notions of truth, objectivity, and logical necessity that they ostensibly seek to undermine. Even if we grant Lakoff the point that abstract concepts are somehow metaphorical, the crucial next step is to show how thinking metaphorically can be rational, not to abandon rationality altogether.

Moving from philosophy to psychology, we discover a big problem with the claim that most of our thinking is metaphorical: people effortlessly transcend the metaphors implicit in their language. I have already mentioned the killjoy's point that many, if not most, conceptual metaphors are opaque to current speakers. This implies that speakers have the means to entertain the underlying concepts: the abstract idea of an approach to a climax, not the concrete idea of the head of a pimple; the abstract idea of a profusion of problems, not the concrete idea of a can of worms.

Experimental psychology is a field of killjoys, and several have shown that people don't dig down to a conceptual metaphor every time they understand a conventional one. The psychologists Boaz Keysar and Samuel Glucksberg and their collaborators showed people chains of sentences built around a conceptual metaphor:[35]

> "Love is a patient," said Lisa. "I feel that this relationship is on its last legs. How can we have a strong marriage if you keep admiring other women?" "It's your jealousy," said Tom.

The experimenters reasoned that if the readers really thought about the underlying metaphor, they should be prepared for a new example, like "You're infected with this disease." They should then recognize it more quickly than if it came out of the blue, after a bland lead-in that made no mention of the metaphor:

> "Love is a challenge," said Lisa. "I feel that this relationship is in trouble. How can we have an enduring marriage if you keep admiring other women?" "It's your jealousy," said Tom.

But in fact there was no advantage: the lead-in with the conventional metaphors left readers unprepared for the novel one, suggesting that the underlying conceptual metaphor simply didn't register. In a third condition, the readers were forced to think about the conceptual metaphor because the lead-in sentences had unconventional metaphors:

> "Love is a patient," said Lisa. "I feel that this relationship is about to flatline. How can we administer the medicine if you keep admiring other women?" "It's your jealousy," said Tom.

This time readers understood the probe sentence about being infected with a disease more quickly—as quickly as when the story was about being literally infected with a real disease. The psychologists concluded that people can read through a metaphorical expression to its underlying concepts, but only when the metaphor is fresh. When the metaphor is conventional, like most of the ones that Lakoff adduces, people go directly to the abstract meaning.

People not only can ignore metaphors, but can question and discount them, and analyze which aspects are applicable and which should be ignored. Indeed, calling attention to conventional metaphors is a common genre of humor, as when Steven Wright asked, "If all the world's a stage, where is the audience sitting?" or in the African American snap, "Your mama's so dumb, she put a ruler on the side of the bed to see how long she slept." The following page shows another example in a *Dilbert* cartoon:

Drollery aside, people could not analyze their metaphors if they didn't command an underlying medium of thought that is more abstract than the metaphors themselves. Nor, for that matter, could they use a conceptual

Dilbert © United Feature Syndicate, Inc.

metaphor to *think* with. When reasoning about a relationship, it's fine to mull over the metaphorical counterpart to a common destination, the rate at which one reaches it, and the bumps along the way. But someone would be seriously deranged if he started to wonder whether he had time to pack or where the next gas station was. In preparing for a debate, I can imagine how to shoot down someone's argument and defend my own, but I had better not worry about guarding my supply lines, issuing war bonds, or dealing with peaceniks at home. Thinking cannot trade in metaphors directly. It must trade in a more basic currency that captures the abstract concepts shared by the metaphor and its topic—progress toward a shared goal in the case of journeys and relationships, conflict in the case of argument and war—while sloughing off the irrelevant bits.[36]

For this reason, it should come as no surprise to learn that even a metaphor as ubiquitous as TIME IS SPACE does not depend on the concept of time actually camping out in the neural real estate used by the concept of space. David Kemmerer has shown that some patients with brain damage can lose their ability to understand prepositions for space, as in *She's at the corner* and *She ran through the forest,* while retaining their ability to understand the same prepositions for time, as in *She arrived at 1:30* and *She worked through the evening.*[37] Other patients showed the opposite pattern. This suggests that different circuits of the brain are responsible for understanding space and for understanding time, their metaphorical overlap notwithstanding.

The need for a stratum of thought that is deeper than metaphor also confronts us when we think about how conceptual metaphors are learned. Recall that Lakoff invokes a Pavlovian theory, in which, say, we learn MORE IS UP through experiences like watching a heap of books rising higher and higher as they are piled on a table. But this becomes preposterous when we turn to more complex metaphors. No one has to be smitten with a seatmate on a cross-country bus trip to appreciate that love is a journey, or see a pair of debaters pull out pistols to sense that arguments are like war. Lakoff thus appeals to "similarity" as a second theory of how conceptual metaphors are acquired: after we learn that A GOAL IS A JOURNEY by the Pavlovian method (that is, we arrive at the playground only after trudging over to it), we extend the metaphor to a romantic relationship because the goal of a relationship is like the goal of a physical destination, such as a playground. But then it's the abstractness of an idea like "goal" that is doing all the work! The abstract ideas define the dimensions of similarity (such as the ways in which a playground is like a romantic aspiration) that allow a conceptual metaphor to be learned and used. You can't think with a metaphor alone.

BENEATH THE METAPHOR

If learning and using a metaphor requires us to manipulate ideas in a deeper stratum of thought, do we have any idea what these ideas are? One can glimpse them in a more moderate version of the theory of conceptual metaphor originally discovered by Lakoff's MIT classmate Jeffrey Gruber and developed by another alumnus, Ray Jackendoff.[38]

The key phenomenon was introduced in chapter 2: verbs like *go, be,* and *keep* are used not just for location (*The doctor kept Pedro at home*) but also for states (*The doctor kept Pedro healthy*), possessions (*Pedro kept the house*), and times (*Pedro kept the practice session at noon*). Jackendoff noted that these verbs preserve one part of their meaning across the physical and nonphysical uses, but not other parts. The preserved part is a skeleton of spatial and force-dynamic concepts like those explored in chapters 2 and 4: thing, substance, aggregate, place, path, agonist, antagonist, goal, means. The skeleton is then labeled with a symbol for a semantic field, such as location, state, possession, or time. For example, the skeleton behind *keep* in *He kept the money* might specify an antagonist opposing an agonist's tendency to go away, with a label slapped across it that says "possession." The concept behind *keep* in *He kept the book on the shelf* would have the same skeleton but would bear the label "location."

The metaphorical flavor of language comes from the fact that skeletal concepts like "go," "place," and "agonist" maintain connections to physical reasoning. They are most easily triggered by the experience of seeing things move around; they are used by children in spatial senses before they are used in abstract senses; and they might have evolved from circuitry for physical reasoning in our primate ancestors. Yet as they take part in moment-to-moment thinking, they are abstract symbols, and need not drag with them images of hunks of matter rolling around. For that reason they are not genuinely metaphorical, at least not in Lakoff's sense.

At this point one might ask: If the concepts beneath a conceptual metaphor are ultimately just abstract symbols like x, y, and z, in what sense would *any* notion of metaphor have a role in the adult mind? And why, in the course of history, development, or evolution, would the mental machinery for location be co-opted for possession, circumstance, or time, if they just get bleached of any content having to do with real space?

The answer is that there are tools of *inference* that can be carried over from the physical to the nonphysical realms, where they can do real work. For example, it's in the nature of space, time, and causation that if A moves B over to C, then B was at not at C at a previous time, though now it is. It's also true that if A had not done the moving, B would not now be at C. Crucially, the same syllogisms apply in the realm of possession. If A *gives* B to C, then B was not *owned* by C at a previous time, though now it is. And if A had not done the giving, C would not now own B. So if an organism that is competent to reason about space can pry its rules away from spatial contents per se, it would gain the ability to reason about possession automatically, making the same kinds of predictions about the future, and the same kinds of inferences about the past.

Of course, the co-opting of spatial reasoning for possession and other abstract domains can only go so far. Physical space is three-dimensional and continuous, whereas possession is one-dimensional and all-or-nothing. For this reason the GIVING IS MOVING metaphor doesn't always make sense: you can't give something *upward* or *frontward,* or give it *partway toward Bill.*[39] But if some parts of the spatial rules are grafted onto the possession realm, while other parts are pruned away, one has the basics to reason about possession.

And so it is with times, states, and causes. Since time really is a dimension like space, we can use tools of spatial cognition to grapple with it (with provisos such as that time is one-dimensional and the future differs from the past, as we saw in chapter 4). More generally, we can use the mental mechanisms for space to deal with *any* continuous variable, from health to intelligence to the gross domestic product. By similar reasoning, the cognitive machinery for force dynamics has enough in common with the logic of counterfactuals ("If not for the antagonist, the agonist would have stayed put") that, stripped of its content from folk physics, it can be used to frame abstract causal thoughts ("If not for the scandal, Melvin would still be governor"), not just bumps and lurches. That's why the language of space and force is so ubiquitous in human discourse: few things in life cannot be characterized in terms of variables and the causation of changes in them.

On this view, metaphors are useful to think with to the extent that they are *analogies*—that they support reasoning of the form "A is to B as X is to Y."[40] Though many metaphors and similes just comment on a perceptual similarity (such as *The clouds at sunset were like boiled shrimp*), the more useful ones allude to the way the source of the metaphor is assembled out of parts. The source (for example, a journey) is stripped down to some essential components (A, B, and C). The metaphor puts these components into correspondence with the components of the target (such as a romantic relationship): A to X, B to Y, C to Z. Then some concept related to A in the source, such as B, is used to pick out an analogously related concept in the target, such as Y. In a journey, one sometimes has to pass over bumpy spots to reach a destination. The way love is like a journey is that the experience of the relationship is like the experience of travel on the road, and the couple's shared goal is like the traveler's destination. Ergo, one may deduce, for a couple to achieve a shared goal, they should be prepared to endure periods of conflict.

This is why Lakoff is right to insist that conceptual metaphors are not just literary garnishes but aids to reason—they are "metaphors we live by." And metaphors can power sophisticated inferences, not just obvious ones like "If you give something away, you no longer own it" or "If you bail out at the first sign of conflict, you'll never have a satisfying relationship." Donald Schön recounts an engineering problem faced by the designers of the first paintbrushes made with synthetic bristles.[41] Compared to brushes with natural bristles, the new ones glopped paint onto a surface unevenly, and none of the improvements the engineers could think up (varying the diameter, splitting the ends like hairs) fixed the problem. Then someone said, "You know, a paintbrush is a kind of pump!" When a painter bends the brush against a surface, he forces paint through the spaces between the bristles, which act like channels or pipes. A natural brush, it turned out, forms a gradual curve when it is bent against a wall, whereas the synthetic one formed a sharp angle, blocking the channels like a crimp in a garden hose. By varying the density of the bristles along their lengths, the engineers got the synthetic brush to curve more gently and thus to deliver paint more evenly. The crucial first step was the pump metaphor, which got them to reconstrue a brush from one big paddle that wipes paint against the surface to a set of channels that pump it out their ends. (Note, by the way, how these metaphors use the schematic geometry of nouns and prepositions that we explored in the last chapter.)

The paintbrush story shows that the power of analogy doesn't come from noticing a mere similarity of parts (synthetic bristles are like natural bristles; natural bristles have split ends; so let's give the synthetic bristles split ends). It comes from noticing *relations* among parts, even if the parts themselves are very different. A paintbrush doesn't look anything like a pump, but the relations among the parts of a brush are like the relations among the parts of a pump. A space between the bristles is like the space inside a hose; bending the brush pushes paint through those channels in the same way that compressing a chamber pushes water through the hose, and so on. The psychologist Dedre Gentner and her collaborators have shown that a focus on relationships is the key to the power of analogy as a tool of reasoning.[42] They note that many scientific theories were first stated as analogies, and often are still best explained that way: gravity is like light, heat is like a fluid, evolution is like selective breeding, the atom is like a solar system, genes are like coded messages. For an analogy to be scientifically useful, though, the correspondences can't apply to a part of one thing that merely resembles a part of the other. They have to apply to the *relationships* between the parts, and even better, to the relationships between the relationships, and to the relationships between the relationships between the relationships.

Here is one of their examples. Early in the nineteenth century the French physicist Sadi Carnot laid out the principles of thermodynamics that explain the workings of heat-driven engines, where a difference in temperature can be converted to work.[43] (For example, in one kind of steam engine, heated steam in one end of a closed chamber expands and pushes a piston toward the other end. But the piston will move only if the other end is cooler, causing its contents to condense; if that end isn't cooler, the steam on that side will just push back and the piston won't move.) Carnot's explanation used a point-by-point analogy that likened the transfer of heat between two connected bodies to the descent of water down a waterfall. The difference in height between the top and bottom of the waterfall corresponds to the difference in temperature between the hot and cold bodies. The amount of water at the top corresponds to the amount of heat in the hotter object. The maximum power available from a waterfall depends jointly on the height difference and the amount of water at the top, just as the amount of power available from a heat engine depends jointly on the temperature difference and the amount of heat in the hotter body. If one were to draw box-and-arrow diagrams of the two systems indicating what depends on what and what causes what, the geometry of the two diagrams would be the same; only the labels would be different.

Carnot was careful to pair up the entities in the two domains in a consistent and one-to-one fashion (heat in the hotter body to the volume of water at the top, and so on). He focused on the relationships among the entities (the temperature difference for heat, the height difference for water) and ignored their individual properties, like the fact that water is clear and wet or that hot things get red. And he did not get sidetracked by actual fraternization between an entity in one domain and an entity in the other, such as the fact that water itself can be hot or that steam engines use water. (That's what Dilbert's boss did when he thought about eagles using software.)

Gentner and her collaborator Michael Jeziorski point out that this mental discipline is essential to the sound use of analogy in science, but it didn't come easy.[44] Most practitioners before the modern scientific era, and most purveyors of pseudoscience today, rambunctiously mix their metaphors, crisscross the connections, and get seduced by surface similarity. The alchemists, for instance, analogized the sun to gold, because both are yellow; Jupiter to tin, because Jupiter is the god of the sky and the sky was thought to be made of tin; and Saturn to lead, because it moved slowly, as if it were heavy like lead, but also because lead is dark, like night, which in turn is like death, and Saturn is farthest from the sun, the giver of life, making Saturn the lord of death. The piling on of metaphorical and metonymic allusions was thought to make the system more compelling, whereas by modern scientific standards it makes it less compelling.

Today, woolly symbolism, shallow similarities, and mercurial mappings are the hallmarks of many kinds of quackery: the "like treats like" principle in homeopathy (such as treating hay fever with a tincture that has been in contact with onions); the use of resemblance in folk medicine (such as powdered rhinoceros horn as a treatment for erectile dysfunction); the reading of significance into the numbers matching the letters in a word in cabbalism; stabbing a doll that looks like an enemy in voodoo and other forms of sympathetic magic.

Loose and overlapping analogies are also a mark of bad science writing and teaching. The immune system is like a sentinel, except when it's a lock and key; no, wait, it's a garbage collector! The best science writers, in contrast, pinpoint the meaningful matchups in an analogy and intercept the misleading ones. In *The Blind Watchmaker,* Richard Dawkins explains how sexual selection can produce flamboyant displays like the outsize tail of a widowbird. Traits in males that are attractive to females can vary wildly over the course of evolution, Dawkins notes, because there are many stable combinations of a tail length preferred by females and an actual tail length in the population (which is itself a compromise between the length preferred by previous generations of choosy females and the length that is optimal for flight). Mathematicians call this "a line of equilibria," and to establish the conditions that produce it they require abstruse equations. But Dawkins explains the idea as follows:

> Suppose that a room has both a heating device and a cooling device, each with its own thermostat. Both thermostats are set to keep the room at the same fixed temperature, 70 degrees F. If the temperature drops below 70, the heater turns itself on and the refrigerator turns itself off. If the temperature rises above 70, the refrigerator turns itself on and the heater turns itself off. <u>The analogue of the widow bird's tail length is not the temperature (which remains constant at 70°) but the total rate of consumption of electricity.</u> The point is that there are lots of different ways in which the desired temperature can be achieved. It can be achieved by both devices working very hard, the heater belting out hot air and the refrigerator working flat out to neutralize the heat. Or it can be achieved by the heater putting out a bit less heat, and the cooler working correspondingly less hard to neutralize it. Or it can be achieved by both devices working scarcely at all. <u>Obviously, the latter is the more desirable</u>

solution from the point of view of the electricity bill but, as far as the object of maintaining the fixed temperature is concerned, every one of a large series of working rates is equally satisfactory. We have a *line* of equilibrium points, rather than a single point.[45]

In the passages I have underlined, Dawkins anticipates how his readers might misconnect the entities in the world to the entities in the analogy, and redirects their gaze to the intended points of correspondence.

Legitimate scientific analogies like those of Carnot, Schön, and Dawkins raise the question of why metaphors and analogies should be so useful in the sphere of knowledge where we feel we have the surest grip on the truth. In the case of metaphors with space and force there is little surprise, because they are simply being used to talk about variables and causal change, which are a universal language of science. But when it comes to more complex metaphors, the usefulness seems almost spooky. What is it about the world, or what is it about *us,* that allows furnaces and air conditioners to shed light on widowbirds' tails? Lakoff, recall, suggests that our scientific knowledge, like all our knowledge, is limited by our metaphors, which can be more or less apt or useful, but not accurate descriptions of an objective truth. The philosopher Richard Boyd draws the exact opposite moral.[46] He writes that "the use of metaphor is one of many devices available to the scientific community to accomplish the task of *accommodation of language to the causal structure of the world.* By this I mean the task of introducing terminology, and modifying usage of existing terminology, so that linguistic categories are available which describe the causally and explanatorily significant features of the world."

Metaphor in science, Boyd suggests, is a version of the everyday process in which a metaphor is pressed into service to fill gaps in a language's vocabulary, like *rabbit ears* to refer to the antennas that used to sprout from the tops of television sets. Scientists constantly discover new entities that lack an English name, so they often tap a metaphor to supply the needed label: *selection* in evolution, *kettle pond* in geology, *linkage* in genetics, and so on. But they aren't shackled by the content of the metaphor, because the word in its new scientific sense is distinct from the word in the vernacular (a kind of polysemy). As scientists come to understand the target phenomenon in greater depth and detail, they highlight the aspects of the metaphor that ought to be taken seriously and pare away the aspects that should be ignored, just as Dawkins did in the heater-cooler analogy. The metaphor evolves into a technical term for an abstract concept that subsumes both the target phenomenon and the source phenomenon. It's an instance of something that every philosopher of science knows about scientific language and that most laypeople misunderstand: scientists don't "carefully define their terms" before beginning an investigation. Instead they use words loosely to point to a phenomenon in the world, and the meanings of the words gradually become more precise as the scientists come to understand the phenomenon more thoroughly.[47] (More on this in the next chapter.)

This still leaves open the question of why metaphors should ever *work*. Why should so many scientific analogies allow us to reason to correct conclusions, as opposed to being mere labels, like *quark* or *Big Bang,* that are memorable but uninformative? Boyd suggests that scientific metaphors are often dispensable for things that may be characterized by a single trait or essence, like water being H_2O. But they come into their own for complex systems composed of numerous parts and properties that work in concert to keep the system stable (he calls them *homeostatic property cluster kinds*). The basic idea is that there are overarching laws of complex systems that govern diverse phenomena in the natural world.[48] One set of laws explains why solar systems, atoms, planets with their moons, and balls tethered to poles fall into stable patterns of revolution. Another explains similarities in ecosystems, bodies, and economies: in all three systems, for example, energy is taken in, internal functions differentiate, and resources get recycled. A third explains the feedback loops by which animals regulate their blood glucose, thermostats regulate house temperature, and a cruise-control device regulates the speed of a car. To the extent that these laws exist, scientists can discover their properties as they study the systems that are governed by them. And they are entitled to use a metaphor both as a label for that kind of system and as a means of generalizing from a well-understood exemplar to a less-well-understood one.

In the day-to-day conduct of science, all this leaves room for debate about whether a phenomenon really *is* an example of the system named by its metaphorical label, or whether the resemblance stops at the metaphorical terminology. No one has a problem with the idea that the lens of an eye and the lens of a telescope are two instances of the general category "lens," rather than the telescope being a "metaphor" for the eye. Nor is there any-thing metaphorical going on when we refer to "the genetic code": a code by now is an information-theoretic term for a mapping scheme, and it sub-sumes cryptograms and DNA as special cases. But do cognitive psycholo-gists use the computer as a "metaphor" for the mind, or (as I believe) can it be said that the mind *literally* engages in computation, and that the human mind and commercial digital computers are two exemplars of the category "computational system"?[49]

So the ubiquity of metaphor in language does not mean that all thought is grounded in bodily experience, nor that all ideas are merely rival frames rather than verifiable propositions. Conceptual metaphors can be learned and used only if they are analyzed into more abstract elements like "cause," "goal," and "change," which make up the real currency of thought. And the methodical use of metaphor in science shows that metaphor is a way of adapting language to reality, not the other way around, and that it can cap-ture genuine laws in the world, not just project comfortable images onto it.

A realist interpretation of conceptual metaphor casts a different light on the application of metaphor and framing to politics. True to his cognitive the-ory, Lakoff states that "frames trump facts" in the minds of citizens, and that the dominant frames are imposed by those in power to serve their in-terests. It's a condescending and cynical theory of politics, implying that average people are indiscriminately gullible and that political debate can-not and should not be about the actual merits of policies and people. But Lakoff's political theory does not follow from the nature of conceptual met-aphor any more than his theory of scientific knowledge does. Metaphors and their framings are not "fixed in the neural structures of people's brains" but can be examined, doubted, and even ridiculed (remember Woody Allen snapping his chin down on some guy's fist).

One can just imagine the howls of ridicule if a politician took Lakoff's Orwellian advice and renamed "taxes" as "membership fees." (Indeed, Or-well himself singled out *revenue enhancement* as an egregious euphemism for a tax increase in his famous 1949 essay "Politics and the English Lan-guage.") No one has to hear the metaphor *tax relief* to think of taxes as an affliction; I suspect that that sentiment has been around for as long as taxes have been around. And do frames really always trump facts? Lakoff's evidence is that people don't realize that they are really better off with higher taxes, because any savings from a federal tax cut would be offset by increases in local taxes and private services. But if that is a fact, it has to be demonstrated the old-fashioned way, as an argument backed with numbers—and in the face of people's perception, not entirely deluded, that some portion of their federal taxes goes to pork-barrel projects, corporate welfare, bureaucratic waste, and so on. Many Democrats have chided Lakoff for underestimating voters by trying to repackage 1960s-style leftism rather than coming up with compelling new ideas.[50]

But isn't it undeniable that beliefs and decisions are affected by how the facts are framed? Yes, but that is not necessarily irrational. Different ways of framing a situation may be equally consistent with the facts being described in that very sentence, but they make different commitments about *other* facts which are *not* being described. As such, rival framings can be examined and evaluated, not just spread by allure or imposed by force. To take the most obvious example, *taxes* and *membership fees* are not two ways of framing the same thing: if you choose not to pay a membership fee, the organization will cease to provide you with its services, but if you choose not to pay taxes, men with guns will put you in jail. Nor are *liberation* and *invasion* factually interchangeable. One implies that the bulk of the populace resents its current overlords and welcomes the arriving army; the other implies the opposite. Debaters who frame an event in these two ways are making competing predictions about unobserved facts, just like scientists who examine the same data and advance competing theories that can be distinguished by new empirical tests.

Even in the gold standard of framing, the Tversky-Kahneman flu problem, the frames are not truly synonymous. The description "200 people will be saved" refers to those who survive because of the causal effects of the treatment. It is consistent with the possibility that *additional* people will survive for different and unforeseen reasons—perhaps the flu may be less virulent than predicted, perhaps doctors will think up alternative remedies, and so on. So it implies that *at least* 200 people will survive. The scenario "400 people will die," on the other hand, lumps together all the deaths regardless of their cause. It implies that *no more than* 200 people will survive.

In other cases, different frames pick out the same facts but are consistent with different policies. Framing abortion as "killing an unborn child" implies that legalizing abortion is logically and morally consistent with legalizing infanticide; framing nationalized health care as "restricting choice" implies that it is consistent with allowing the government to restrict other personal choices. One can engage these framings by disputing the implied generalizations: abortion can be legal while infanticide is not because early fetuses are in crucial regards like conceptuses (whose deaths we don't criminalize) whereas newborns are like other children (whose deaths we do). Nationalized health insurance can be mandatory while other private choices are not because it falls in the category of choices with public consequences, like police protection and garbage removal, which we already force people to pay for. As with all metaphors, we can evaluate the merits of rival framings by asking which aspects of similarity we ought to take seriously and which aspects we ought to ignore.

People certainly are affected by framing, as we know from centuries of commentary on the arts of rhetoric and persuasion. And metaphors, especially conceptual metaphors, are an essential tool of rhetoric, ordinary communication, and thought itself. But this doesn't mean that people are enslaved by their metaphors or that the choice of metaphor is a matter of taste or indoctrination. Metaphors are generalizations: they subsume a particular instance in some overarching category. Different metaphors can frame the same situation for the same reason that different words can describe the same object, different grammars can generate the same corpus of sentences, and different scientific theories can account for the same set of data. Like other generalizations, metaphors can be tested on their predictions and scrutinized on their merits, including their fidelity to the structure of the world.

THE GOOD, THE BAD, AND THE UGLY

Before turning to the evidence on whether metaphorical thinking really does come naturally to people, let me close the circle by examining the source of the metaphor metaphor, namely "metaphor" itself, in the familiar sense of the literary device used in poetry, literature, and fancy speech.

Obviously a literary metaphor points out a similarity between a source and the target. Juliet, like the sun, improves my mood; in life, as on a stage, people take on roles. At first it looks as if a literary metaphor might be just a condensed simile. Yet just as obviously, turning a metaphor into a simile drains it of life and verve. "The window is like the east, and Juliet is like the sun"? "The world is like a stage and all the men and women are like players"? Bo-ring. The tanginess of a literary metaphor must come from some extra ingredients that spice up a mere overlap of traits. The ingredients must also make a literary metaphor more piquant than an everyday conceptual metaphor like *She shot down my argument* or *Our relationship isn't going anywhere.*

One of the ingredients is the metaphor's syntax. By expressing the trait as a noun phrase predicate (or by simply referring to the entity with the noun phrase), a metaphor recruits the semantics of a category or kind.[51] It's the effect we saw in chapter 4 that removes the circularity from *Boys will be boys* and prompted Jonathan Miller to say he was Jewish rather than a Jew. A noun phrase, when predicated of a subject, conveys a trait that is felt to be essential to the subject's very being. The trait defines a category that pigeonholes the subject in a way that is sensed as deeper, longer-lasting, and farther-reaching than the mere ascription of a trait. *A lawyer is a shark* says much more than *A lawyer is like a shark.*

The way that a metaphor asserts an identity rather than making a comparison is one of the reasons Lakoff says we actually think in our metaphors. It's also why he proposed that literary metaphors are essentially no different from everyday metaphors. In an insightful book on poetic metaphor with the humanities scholar Mark Turner, Lakoff notes that poetic metaphors often tap into everyday conceptual metaphors but make use of elements that are ordinarily omitted, flesh out details in unusual ways, or juxtapose related metaphors side by side.[52] For instance, when Robert Frost spoke of "the road not taken," he was elaborating the conceptual metaphor LIFE IS A JOURNEY, which we also see in everyday speech, as in *I've come a long way* and *He had to get out of the fast lane.*

In a review of the book, Jackendoff (together with the scholar David Aaron) challenges this equation, noting that people distinguish poetic metaphors both from literal assertions and from everyday metaphors.[53] Literary metaphors are special because they induce a sense of incongruity—a moment of frisson in which the listener puzzles over something that seems to make no sense. (Why is he equating a planet with a stage? Why should taking a less-traveled road make "all the difference"?) A simple way to show this is to *acknowledge* the incongruity and then point out the metaphorical similarity. In the case of a poetic metaphor, the result sounds perfectly reasonable, because it lays bare the metaphor's logic:

> Of course, the world isn't really a stage, but if it were, you
> might say that infancy is the first act.
> Of course, people aren't really heavenly bodies, but if they
> were, you might say that Juliet is the sun.
> Of course, life isn't really a road, but if it were, you might say
> that it's better to choose a less-traveled route.

But this doesn't work with literal statements. Jackendoff and Aaron consider the passage from Leviticus, "For the life of a being is in the blood." For us it is a literary metaphor, an offshoot of the conceptual metaphor LIFE IS FLUID IN THE BODY; DEATH IS LOSS OF FLUID, which also gave us *She feels drained, My juices were flowing,* and *His life ebbed away.* Yet the ancient Hebrews literally ascribed mental functions to organs and substances in the body, and the passage refers to the prohibition of consuming animal blood. Now let's apply the litmus test:

> Of course, life isn't really a fluid, but if it were, you might say
> that it is in the blood.

The ancient Hebrews would not assent to the first part, but we would. This confirms that there is a difference between taking the terms of a metaphor as a literal belief and understanding the metaphor as a literary device, and that they can be distinguished by the "not-really" test.

That test also shows that everyday conceptual metaphors work differently from literary metaphors:

> Of course, times aren't really locations, but if they were, you
> might say that we're getting close to Christmas.
> Of course, purposes aren't destinations, but if they were, you might
> say that I haven't yet reached my goal of finishing this book.
> Of course, love isn't really a journey, but if it were, you might
> say that I don't like the way our relationship is going.

These sentences are non sequiturs. The incongruity noted in the first part is true enough, but it seems to have nothing to do with the second part. That's because the expression in the second part is not interpreted as a metaphor at all; it is just the standard way of putting the thought into words.

The incongruity in a fresh literary metaphor is another ingredient that gives it its pungency.[54] The listener resolves the incongruity soon enough by spotting the underlying similarity, but the initial double take and subsequent brainwork conveys something in addition. It implies that the similarity is not apparent in the humdrum course of everyday life, and that the author is presenting real news in forcing it upon the listener's attention. We visited this effect in the discussion of Radical Pragmatics in chapter 3: the tension between a literal and an intended interpretation can convey a third message, which can be put to use in dysphemism, humor, and subtexts.

A third kind of zest in a literary metaphor comes from the emotional coloring of the source and the way it bleeds into the target.[55] Lakoff analyzes Benjamin Disraeli's remark "I have climbed to the top of the greasy pole" as meaning that he has achieved his status with effort, has endured temporary setbacks, can achieve nothing greater, and will probably lose the status before long. Missing from Lakoff's exegesis is a wry subtext: that political competition can be unpleasant, sullying, and ultimately pointless. Metaphors that are even more poetic, such as Nabokov's "shadow of the waxwing slain / By the false azure of the windowpane," are thick with these emotional carryovers: the insubstantiality of a shadow, the illusory nature of a reflection and the betrayal the illusion can inflict, the reminder of death in an innocuous print on a window, the sudden pathetic end of an unsuspecting lovely creature, and many more. These entwined allusions also mark a key difference between literary metaphors and scientific analogies.[56] Multiple, partial, and emotionally charged similarities add to the richness of poetry, but they detract from understanding in science.

And if the best way to understand something is to be shown what it is not, we can get insight into literary metaphor from the genre of humor that collects examples of bad ones. For many years the New Yorker filled out the bottom of its columns with a feature called "Block That Metaphor," which reproduced strained and ludicrous allusions from small-town publications. But my favorite example is the widely circulated list of the World's Worst Analogies, commonly attributed to high school students in various parts of the English-speaking world but actually the winning entries in a Washington Post contest. Consider these three:

> John and Mary had never met. They were like two hummingbirds
> who had also never met.

Her eyes were like two brown circles with big black dots in the
   center.
The thunder was ominous-sounding, much like the sound of a
   thin sheet of metal being shaken backstage during the
   storm scene in a play.

In these metaphors, knowing about the source adds nothing to one's knowledge of the target, so they fail the test of supporting an inference. Or consider these:

Her date was pleasant enough, but she knew that if her life was
   a movie this guy would be buried in the credits as
   something like "Second Tall Man."
Even in his last years, Granddad had a mind like a steel trap,
   only one that had been left out so long, it had rusted shut.
He spoke with the wisdom that can only come from
   experience, like a guy who went blind because he looked
   at a solar eclipse without one of those boxes with a
   pinhole in it and now goes around the country speaking
   at high schools about the dangers of looking at a solar
   eclipse without one of those boxes with a pinhole in it.

Here the writer has to work so hard to explain the metaphor's source that
he has left no incongruity for the reader to resolve, and the reader fails to
enjoy a rhetorical payoff. And what about these?

The revelation that his marriage of 30 years had disintegrated
   because of his wife's infidelity came as a rude shock, like a
   surcharge at a formerly surcharge-free ATM.
The ballerina rose gracefully en pointe and extended one
   slender leg behind her, like a dog at a fire hydrant.
McBride fell 12 stories, hitting the pavement like a Hefty Bag
   filled with vegetable soup.

In these three, the analogy is clear enough, and even informative, but the
emotional coloring of the source clashes so badly with the emotional coloring of the target that it defeats the purpose of spreading one to the other.

METAPHORS AND MINDS

Now that we have seen what metaphor can and cannot do in science, literature, and reasoning, let me return to the question that kicked off the chapter. How readily does an average person coin and grasp metaphors? Every metaphor and analogy has to come from somewhere. Perhaps they are rare pearls that drop from the pens of an elite corps of bards and scribblers and then are hoarded by a grateful populace. But given their prevalence in language, it seems more likely that they are the natural products of the way everyone's mind works. If so, we should be able to catch people in the act of sensing the deep correspondences between superficially different realms that make for a useful analogy or conceptual metaphor.

It's easy to show that people sense the connection in simple metaphors based on a single dimension of space, such as HAPPY IS UP. When experimental subjects are shown words on a screen and have to evaluate whether they are positive (like *agile, gracious,* and *sincere*) or negative (like *bitter, fickle,* and *vulgar*), they are quicker when a positive word is flashed at the top of the screen or a negative word at the bottom than vice versa.[57] People are also quicker at moving their hand toward a button near their bodies to verify a sentence like *Adam conveyed the message to you* than to verify *You conveyed the message to Adam,* and vice versa when they have to move their hand to a button away from their bodies. It's as if the physical movement of the hand was programmed in the same mental space as the metaphorical movement of the message. The same thing happens with possession, as in *You sold the land to Mike* compared with *Mike sold the land to you,* and with benefaction, as in *Tiana devoted her time to you* compared with *You devoted your time to Tiana.*[58]

A clever set of studies by the psychologist Lera Boroditsky and her collaborators has shown that people also sense the spatial allusion in the metaphors TIME IS A PROCESSION and TIME IS A LANDSCAPE.[59] Remember that the two are incompatible, making some sentences about time ambiguous. *Wednesday's meeting has been moved forward two days* can mean that it has been switched to Monday, because if one invokes the time-as-procession metaphor, "forward" is aligned with the march of time toward oneself—an event that has been moved forward in the procession of days is now closer to us. Or it could mean that the meeting has been switched to Friday, because if one invokes the time-as-landscape metaphor, "forward" is aligned with one's own marching through time—we have to walk farther, past more days, to get to an event that has been moved to a forward position. People can be tilted toward one interpretation or the other if they have recently read a sentence that is compatible with only one of the metaphors, such as *We passed the deadline two days ago* (which nudges them to the "Friday" interpretation) or *The deadline passed two days ago* (which nudges them to "Monday"). The explanation is similar to the one in the experiment showing that people analyze fresh examples of conceptual metaphors like love is a patient. The only common denominator between the preamble and the test sentence is the underlying conceptual metaphor, so that metaphor must have registered in people's minds.[60]

Boroditsky took the logic a step farther by showing that an actual *experience* of motion, not just words that use a metaphor for motion, can tip people's interpretations of the ambiguous *forward* one way or another. People tilt toward the "Friday" interpretation (where *forward* coincides with their own march through the metaphorical landscape) if they are first asked to imagine pushing an office chair. But they tilt toward the "Monday" interpretation (where *forward* coincides with time's advance toward them) if they are first asked to imagine pulling the chair toward them with a string. Similar tiltings can be induced by the experience of real motion: people are more likely to say "Friday" if they have recently advanced through a lunchroom line, gotten off a plane, or reached the end of a train trip.

But people's ability to link a single dimension of space to a single dimension of experience is a rather modest example of metaphorical thinking compared with the grandiose ambitions that we have for it. Recall that metaphors are powerful to the extent that they are like analogies, which take advantage of the *relational* structure of a complex concept. Can we show that people easily see metaphorical connections involving a set of entities that interact in particular ways? And can we find them discerning *new* metaphorical connections, rather than just picking up the hackneyed ones that already permeate the language?

One showcase for the metaphorical powers of the mind is the errors made by children in their spontaneous speech. The psychologist Melissa Bowerman, studying her two daughters in their preschool years, noticed that they occasionally used words for space and motion in unconventional ways when talking about possession, states, times, and causes:[61]

> You put me just bread and butter.
> You put the pink one to me.

> I'm taking these cracks bigger [while shelling a peanut].
> I putted part of the sleeve blue so I crossed it out with red
>   [while coloring].

> Can I have any reading behind the dinner?
> Today we'll be packing because tomorrow there won't be
>   enough space to pack.
> Friday is covering Saturday and Sunday so I can't have
>   Saturday and Sunday if I don't go through Friday.

> My dolly is scrunched from someone . . . but not from me.
> They had to stop from a red light.

When I first studied verbs in children, I found these reports tremendously exciting, because they seemed to be evidence for the metaphor metaphor out of the mouths of babes and sucklings. Of course, these are just two children, the daughters of an academic, who had her ear cocked for interesting usages almost 24/7 for years. To find out how readily children make these leaps, I recruited a student, Larry Rosen, to trawl through a computer database of about fifty thousand sentences transcribed from the speech of three other children.[62] The catch was paltry:

> I goin' put de door open.
> Now I think I take the whole crayoned [coloring in a picture].
> It's gonna stay raining.
> He put his bread and butter folded over.

Rosen and I then did an experiment to see if we could elicit the errors from live children. We asked kids to describe pictures showing changes of possession and state, such as a mother giving a ball to a girl, or a boy coloring a piece of paper. We stacked the deck by actually telling the children which verb to use—for example, "Can you tell me what she's doing, using the word *put*?"—hoping to tempt them into an occasional error like *He's putting the paper blue*. Thirty children described nineteen pictures each, for a total of 570 invitations to use a spatial term metaphorically. Once again, the net came up almost empty:

> Mother takes ball away from boy and puts it to girl.
> Square go big.
> Boy puts flowers to girl.
> Square went bigger.

The other 99.3 percent of the time, if the children used the target verb at all, they used it in its standard sense, such as *He put water on him*. Score one more point for the killjoy theory. Though children are certainly capable of seeing the parallels between changing location and changing state or possession, they show off this insight only rarely. Most of the time, they use spatial words just as their parents do.

Another tantalizing glimpse of the metaphorical mind came to me during a talk on memory by the artificial intelligence researcher Roger Schank when he recounted episodes in which one event reminded him of another.[63] Schank, of course, was not the first to reflect on the psychology of reminding. We often think of it as a remembrance of things past set off by a sensory experience, as in the famous passage in Marcel Proust's *Swann's Way:*

> My mother sent for one of those squat, plump little cakes called "petites madeleines," which look as though they had been molded in the fluted valve of a scallop shell. And soon, mechanically, dispirited after a dreary day with the prospect of a depressing morrow, I raised to my lips a spoonful of the tea in which I had soaked a morsel of the cake. No sooner had the warm liquid mixed with the crumbs touched my palate than a shudder ran through me and I stopped, intent upon the extraordinary thing that was happening to me. An exquisite pleasure had invaded my senses, something isolated, detached, with no suggestion of its origin. And at once the vicissitudes of life had become indifferent to me, its disasters innocuous, its brevity illusory—

this new sensation having had on me the effect which love has of filling me with a precious essence; or rather this essence was not in me it *was* me. I had ceased now to feel mediocre, contingent, mortal. Whence could it have come to me, this all-powerful joy?[64]

Five paragraphs later, the narrator is struck by the answer, when he is reminded of Sunday mornings in his childhood when his aunt Léonie served him madeleines soaked in lime-blossom tea.

Schank's remembrance of things past is rather different:

Someone told me about an experience of waiting on a long line at the post office and noticing that the person ahead had been waiting all that time to buy one stamp. This reminded me of people who buy a dollar or two of gas in a gas station.

Well, what do you expect from a computer scientist? But what it lacks in literary élan it makes up for in scientific importance, because the episode shows something extraordinary about human memory. Memories may be linked not just by a common thread of tastes, textures, and shapes but by a shared skeleton of abstract ideas, in this case "the inefficiency of waiting a long time to obtain a small quantity." Schank presents episodes in which the reminder has even less sensory overlap with the remembered thing:

X described how his wife would never make his steak as rare as he liked it. When this was told to Y, it reminded Y of a time, 30 years earlier, when he tried to get his hair cut in a short style in England, and the barber would not cut it as short as he wanted it.

X's daughter was diving for sand dollars. X pointed out where there were a great many sand dollars, but X's daughter continued to dive where she was. X asked why. She said that the water was shallower where she was diving. This reminded X of the joke about the drunk who was searching for his keys under the lamppost because the light was better there even though he had lost the keys elsewhere.[65]

Curious about how common this kind of bell-ringing is, I recently kept a diary of my own remindings, separating the Proustian remindings-by-shared-sensation from the Schankian remindings-by-shared-conceptual-structure. Within a couple of days I had come up with a dozen examples of the latter. For example:

> While jogging, I was listening to songs on my iPod, which were being selected at random by the "shuffle" function. I kept hitting the "skip" button until I got a song with a suitable tempo. This reminded me of how a baseball pitcher on the mound signals to the catcher at the plate what he intends to pitch: the catcher offers a series of finger-codes for different pitches, and the pitcher shakes his head until he sees the one he intends to use.

> While touching up a digital photo on my computer, I tried to burn in a light patch, but that made it conspicuously darker than a neighboring patch, so I burned that patch in, which in turn required me to burn in yet another patch, and so on. This reminded me of sawing bits off the legs of a table to stop it from wobbling.

> A colleague said she was impressed by the sophistication of a speaker because she couldn't understand a part of his talk. Another colleague replied that maybe he just wasn't a very good speaker. This reminded me of the joke about a Texan who visited his cousin's ranch in Israel. "You call this a ranch?" he said. "Back in Texas, I can set out in my car in the morning and not reach the other end of my land by sunset." The Israeli replied, "Yes, I once had a car like that."

As I was thinking about this kind of reminding, I was even reminded of a reminding. In college, a friend and I once sat through a painful performance by a singer with laryngitis. When she returned from intermission and began the first song of the set, her voice was passably clear. My friend whispered, "When you put the cap back on a dried-out felt pen, it writes again, but not for long."

A Schankian reminding must be the elusive mental act that brings a new metaphor or analogy into the world. And not a bad or superficial analogy either, like those of the alchemists or Worst Analogy contestants, but a deep analogy that sees through the veneer of sensory experience to a shared conceptual skeleton of events, states, goals, causes, and extents. These remindings came unbidden to Schank and to me, and if they are an effortless maneuver in people's mental playbook, we would have an explanation of why metaphors are so abundant in language and so useful in reasoning. We might even have an explanation for why people are so smart: Schankian reminding is the evolutionary gift that allows us to co-opt old ideas (beginning with the ones we inherited from our primate ancestors) for use in new realms.

Or maybe Schank and I and our friends are just weird. As with the childhood errors, the analogical leaps in the remindings are not so easy to reproduce when people are recruited off the street and encouraged to analogize on cue. In fact, if you asked most cognitive psychologists about how people use analogies, they would be resolute killjoys, saying that people are impressed by surface similarities and don't discern the common structure—exactly the opposite of what Schankian reminding would suggest.[66]

Beginning in the 1950s, Herbert Simon and Allen Newell, two of the founders of cognitive science and artificial intelligence, programmed computers to solve problems by first distilling a problem to a set of states (such as the possible configurations of chessmen on a board) and a set of operations that transform one state into another (such as moving a piece).[67] The states and operations could then be represented as symbols and computational steps in the computer, shorn of the sensory trappings of the world itself (such as the appearance of the chessboard or chessmen). The computer solved a problem by detecting a difference between a goal state and the current state and trying out operations known to reduce the difference. And if the computer could figure out one problem, it could automatically generalize to a new problem with a similar move-by-move logical structure, as long as the second problem was represented inside the computer in a similar form—a kind of reasoning by analogy. (Simon called the different versions of the same problem "problem isomorphs.") A famous example is the Tower of Hanoi, consisting of graduated rings stacked on poles:

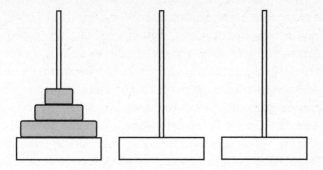

You are allowed to move one ring at a time, and you can never place a larger ring on a smaller one. The object is to get the stack of rings from the leftmost spike to the rightmost. Most people can figure it out after a while. But now consider this problem:

> Certain Himalayan inns have a tea ceremony involving a host, an elder, and a youth. The host performs three services, ranked in order of their nobility: stoking the fire, serving the tea, and reciting a poem. The ceremony must be repeated a number of times. After each performance, anyone can ask anyone else, "Honored Sir, may I perform this onerous task for you this time?" However, a person may relieve another guest only of the least noble of the tasks that the other has performed. Furthermore, if a person has just performed a task, then he may not take on another one that is nobler. Custom requires that in the last performance of the tea ceremony, all the services have been transferred to the youth. How can this be accomplished?

This is an isomorph of the Tower of Hanoi, and can be reduced to the same sequence of steps (imagine that the host is the left spike, the elder guest is the middle spike, and the youth is the right spike, and the tasks are the disks on the spikes, with "reciting a poem" on the bottom).[68] But the tea-ceremony problem seems much harder than the Tower of Hanoi. Contrary to the hope that the mind has X-ray vision for an abstract conceptual skeleton, people can find the various isomorphs of a problem radically different in difficulty, and unless they are led by the hand, they don't easily carry over their solution from one to another.

Another famous example was inspired by the Gestalt psychologist Karl Duncker (who also devised the candle-and-thumbtack problem we saw in chapter 4).[69] You are a doctor trying to destroy an inoperable stomach tumor in a patient. You can send a narrow beam of radiation at the tumor, but it destroys the tumor only at a high intensity, which would also destroy the surrounding healthy tissue. At a lower intensity it would spare the healthy tissue but also fail to kill the tumor. There is a solution, but it occurs to only one person in ten: aim several beams at the tumor from different directions, so the tumor gets a dose equal to the sum of the beams, while the surrounding tissue gets the dose of a single beam. Duncker was interested in how people get to the "Aha!" that allows them to see the solution to a problem. Decades later, the cognitive psychologists Mary Gick and Keith Holyoak wondered if one of the routes might be analogy.

In a seminal study, Gick and Holyoak gave their subjects the tumor problem, but first they gave them a kind of hint in the form of a problem isomorph together with its solution.[70] A dictator rules a small country from a fortress. Many roads radiate outward from it. The general of a liberation force knows that his army is big enough to take over the fortress. But he has just learned that the roads have been studded with mines, which allow the passage of small groups of men but would be detonated by a full army, making a full-scale invasion impossible. The general hits on a solution: he divides his army into small groups, each of which can negotiate a different road, and they can then converge at the fortress and easily overrun it. A good analogizer should spot the isomorphism and transfer the solution for the invasion problem to the tumor problem. But only about 35 percent of the participants saw the light—a threefold increase, to be sure, but still a minority. Other killjoy studies have confirmed this literal-mindedness. People rarely experience the light bulb that would tell them they can transfer the solution of one problem to an isomorph unless the problems are similar on the surface as well. For example, people do succeed at the stomach-tumor problem after reading about a doctor who treats a *brain tumor* with multiple beams of radiation.[71] Of course, when the contents of an analogy are close in their underlying structure *and* their surface qualities, it barely deserves to be called an analogy at all. It takes little analogical acumen to figure out how to order at Burger King based on your knowledge of how to order at McDonald's.

What's going on? On the one hand, analogical thinking seems to be our birthright. Metaphorical connections saturate our language, drive our science, enliven our literature, burst out (at least occasionally) in children's speech, and remind us of things past. On the other hand, when experimentalists lead the horse to water, they can't make it drink.

One factor is simply expertise. Tea ceremonies, radiation treatments, and invading armies are obscure to most students, so they don't have the needed conceptual framework at their fingertips. Subsequent studies have shown that expertise in a topic can make deep analogies come more easily. For example, when students who had taken a single physics course were shown a bunch of problems and asked which ones were similar, they lumped together the ones that had pictures of the same kinds of objects—the inclined planes in one group, the pulleys in another, and so on. But when advanced graduate students in physics did the sorting, they lumped the problems that were governed by the same *principles,* such as the conservation of energy, whether the problems involved boxes being pushed up planes or weights hanging down from springs.[72] Perhaps we are adept with conceptual metaphors in real life because we are all experts in journeys, love, arguments, and war.

But there's more to it than that. The psychologist Kevin Dunbar and his collaborators put their fingers on another way that people are at a disadvantage in the lab: the experimenters get to pick the analogy.[73] They then bury it in a story and test the subjects to see if they can find it. This treasure hunt is much harder than popping out one's *own* analogy, which is what happens in childhood errors, analogical reminding, and the coining of new metaphors in the history of a language.

Dunbar invited students to think up analogies in support of each side of a debate raging in Canada at the time over whether the federal budget should be balanced at a cost of cutting social programs.[74] Within a few minutes the average subject cooked up eleven analogies. Eighty percent of them had nothing to do with money or politics, but reached out to remote sources like farming ("If you decide it's too expensive to buy pesticide for your apples, then all your apples will die") and home life ("The deficit is like lint in your dryer. If you don't pull it out, it just accumulates and blocks, and it becomes inefficient"). He also scanned for analogies in some four hundred newspaper articles about the 1995 referendum on the political separation of Québec from Canada.[75] The search turned up more than two hundred analogies, and once again, most were based on deep rather than superficial similarities. They tapped sources as diverse as families ("It's like parents getting a divorce, and maybe the parent you don't like is getting custody"), medicine ("Separation is like a major surgery; it's important that the patient is informed and the surgeon is impartial"), and of course the national pastime ("Québec's path to sovereignty is like a hockey game. The referendum is the end of the third period").

The messiah has not come. Though metaphors are omnipresent in language, many of them are effectively dead in the minds of today's speakers, and the living ones could never be learned, understood, or used as a reasoning tool unless they were built out of more abstract concepts that capture the similarities and differences between the symbol and the symbolized. For this reason, conceptual metaphors do not render truth and objectivity obsolete, nor do they reduce philosophical, legal, and political discourse to a beauty contest between rival frames.

Still, I think that metaphor really is a key to explaining thought and language. The human mind comes equipped with an ability to penetrate the cladding of sensory appearance and discern the abstract construction underneath—not always on demand, and not infallibly, but often enough and insightfully enough to shape the human condition. Our powers of analogy allow us to apply ancient neural structures to newfound subject matter, to discover hidden laws and systems in nature, and not least, to amplify the expressive power of language itself.

Language, by its very design, would seem to be a tool with a well-defined and limited functionality. With a finite stock of arbitrary signs, and grammatical rules that arrange them in sentences, a language gives us the means to share an unlimited number of combinations of ideas about who did what to whom, and about what is where.[76] Yet by digitizing the world, language is a "lossy" medium, discarding information about the smooth multidimensional texture of experience. Language is notoriously poor, for instance, at conveying the subtlety and richness of sensations like smells and sounds. And it would seem to be just as inept at conveying other channels of sentience that are not composed out of discrete, accessible parts. Flashes of holistic insight (like those in mathematical or musical creativity), waves of consuming emotion, and moments of wistful contemplation are simply not the kinds of experience that can be captured by the beads-on-a-string we call sentences.

And yet metaphor provides us with a way to eff the ineffable. Perhaps the greatest pleasure that language affords is the act of surrendering to the metaphors of a skilled writer and thereby inhabiting the consciousness of another person. It almost lets us know what it is like to exercise mathematical genius:

> Sometimes I wander round and round in circles, going over the same ground, getting lost, sometimes for hours, or days, or even weeks. . . . But I know that if I immerse myself in it long enough, things will clarify, simplify. I can count on that. When it happens, it happens fast. Boom ba boom ba boom! One thing after the other, taking the breath away. And then, you know, I feel like I'm walking out in some remote corner of space, where no mortal's ever been, all alone with something beautiful. . . . Once, when I was in Switzerland some friends took me up in some very high cable cars, climbing up a mountain. . . . There was a restaurant on top and the view was supposed to be sublime. When we got up it was a great disappointment because the clouds were obscuring everything. But suddenly there was a rent in the clouds and there were the Jungfrau and two other peaks towering right in front of us. . . . That's what it's like.[77]

Or to compose a piece of music:

> It came as a gift. A large gray bird flew up with a loud alarm call as he approached. As it gained height and wheeled away over the valley, it gave out a piping sound on three notes, which he recognized as the inversion of a line he had already scored for a piccolo. How elegant, how simple. Turning the sequence round opened up the idea of a plain and beautiful song in common time, which he could almost hear. But not quite. An image came to him of a set of unfolding steps, sliding and descending—from the trap door of a loft, or from the door of a light plane. One note lay over and suggested the next. He heard it, he had it, and

then it was gone. There was a glow of a tantalizing afterimage
and the fading call of a sad little tune. . . . These notes were perfectly interdependent, little polished hinges swinging the melody through its perfect arc. He could almost hear it again as he
reached the top of the angled rock slab and paused to reach into
his pocket for notebook and pencil.[78]

Or to harbor an unspeakable desire:

You have to be an artist and a madman, a creature of infinite
melancholy, with a bubble of hot poison in your loins and a
super-voluptuous flame permanently aglow in your subtle spine
(oh, how you have to cringe and hide!), in order to discern at
once, by ineffable signs—the slightly feline outline of a cheek-
bone, the slenderness of a downy limb, and other indices which
despair and shame and tears of tenderness forbid me to tabu-
late—the little deadly demon among the wholesome children.[79]

Or even to reflect on the limits of language itself:

For none of us can ever give the exact measure of his needs or his
thoughts or his sorrows, and language is a cracked kettle on
which we beat out tunes for bears to dance to, while all the time
we long to move the stars to pity.[80]

6

WHAT'S IN A NAME?

All my life I have been surrounded by reminders of the commonness of my given name. Its provenance is auspicious enough: *stephanos,* the Greek word for "crown." Nonetheless, it lay in obscurity for most of the two millennia after the stoning of the first Christian martyr, who is remembered mainly in the name of the feast on which Good King Wenceslas looked out. Not until the nineteenth century do a smattering of Stephens reappear on the world stage—Austin, Douglas, Foster, Leacock, and Crane—and the first decades of twentieth century added only Benét, Spender, and Dedalus, the last of whom didn't even exist.

But between the 1930s, when it was around the seventy-fifth most popular name for an American baby boy (trailing *Clarence, Leroy,* and *Floyd*), and the 1950s, when I was born, *Stephen* (and *Steven* and *Steve*) rocketed into seventh place.[1] And it seemed to be even more popular in the demographic circles I inhabited. For as my bulky cohort made its way through the python, I repeatedly found myself surrounded by Steves. In school I was always addressed by an initial as well as a name, since every class had two or three of us, and as I furthered my education the concentration of Steveness just kept increasing. My graduate school roommate was a Steve, as was my advisor and another of his students (resulting in a three-Steve paper), and when I started my own lab, I hired two Steves in a row to run it.

But when I began to write trade science books, I became surrounded. I aspired to stand on the shoulders of Gould and Hawking, found myself debating first Gould and then Rose (the transcript of that debate is called "The

Two Steves"),[2] and at one point shared the shelves with all of them plus Budiansky and Jones. Not that it did me any harm:

Pile 'Em High cartoon by Kipper Williams, originally printed in *The Sunday Times,* reprinted with permission of the artist.

Since then we have been joined by Johnson, Landsburg, Levy, Mithen, Weinberg, Wolfram, and both authors of the bestselling *Freakonomics* (Levitt and Dubner). Steves also dominate high technology, including the CEOs of both Microsoft (Ballmer) and Apple (Jobs), both founders of Apple (Jobs and Wozniak), and the founder of AOL (Case).

The abundance of Steves in turn-of-the-century science has led to the most formidable weapon in the fight against neo-creationism today: Project Steve.[3] A brainchild of the National Center for Science Education, the initiative is a parody of the creationist tradition of publishing lists of several dozen "scientists who dissent from Darwinism." The NCSE replies: "Oh, yeah? Well, we have a list of several *hundred* scientists who *affirm* evolution—*just named Steve!*" (And Stephanie, Steffi, Stefan, and Esteban.) Part satire, part memorial to Stephen Jay Gould, the project maintains a Steve-O-Meter (now pointing past 800) and has spun off a T-shirt, a song, a mascot (Professor Steve Steve, a panda puppet), and a paper in the respected scientific journal *Annals of Improbable Research* called "The Morphology of Steve" (based on the distribution of T-shirt sizes ordered by the signatories).[4]

Hyperstevism is a phenomenon of the second half of the twentieth century. Like tulips, dot-com stocks, and other examples of the madness of crowds, the fortunes of *Steve* fell as quickly as they rose. Today it has fallen to levels not seen since the nineteen teens, and it may become as geriatric as *Elmer* or *Clem*.[5]

What makes a name rise and fall in popularity? Like most parents who give their child a fresh new name, mine had no idea they were part of a trend; they just liked the way it sounded. My grandparents tried to talk them out of it; the name, they said, made them think of a boorish laborer. I like to think it carried a hint of smoldering masculinity, as when Lauren Bacall said to Humphrey Bogart in *To Have and Have Not:*

> You know you don't have to act with me, Steve. [His name was not Steve.] You don't have to say anything and you don't have to do anything . . . not a thing. Oh, maybe just whistle. You know how to whistle, don't you, Steve? You just put your lips together and blow.

They could not have known that a generation later the name's connotations would change to the point that the slogan of the anti-gay movement in the United States would be "God made Adam and Eve, not Adam and Steve."

This chapter is about naming—naming babies, and naming things in general. Naming a baby is the only opportunity most people get to choose what something will be called. But every one of the half a million words in the *Oxford English Dictionary* had to have been thought up by a person at some point in history, accepted by a community, and perpetuated down through the eons. This is a process that enmeshes the world, the mind, and societies in surprising ways. As we shall see, one aspect of the humble act of naming has recently been found to overturn our understanding of logic, meaning, and the relation of knowledge to reality. Another aspect has been found to overturn our understanding of culture and society.

IN THE WORLD OR IN THE HEAD?

The revolution in our understanding of the logic of names began with a basic question: Where do the meanings of words live? There are two likely habitats. One is the world, where we find the things that a word refers to. The other is in the head, where we find people's understanding of how a word may be used.

For anyone interested in language as a window into the mind, the external world might seem to be an unpromising habitat. The word *cat,* for example, refers to the set of all the cats that have ever lived or will ever live.

But no mortal can be acquainted with all cats, past, present, and future. Also, many words don't have any referent in the world at all, such as *unicorn, Eliza Doolittle,* and *the Easter Bunny,* but the words are certainly meaningful to the person who knows them. Finally, people can use words with very different meanings to refer to the same thing in the world. The textbook example is *the Evening Star* and *the Morning Star,* which turned out to be two names for the planet Venus. But they certainly have different meanings to people who are innocent of astronomy and have no way of knowing that they refer to the same heavenly body. There is another well-known example of two words that refer to the same thing in the world but mean different things to a person. The words are *Jocasta* and *Mom,* and the person is Oedipus.[6] As we have seen so often, semantic niceties can make a difference.

The alternative to the idea that the meaning of a word is the set of things it refers to is that it is some kind of description, like a definition in a dictionary, or a formula in logical or conceptual symbols. Mathematics gives us an obvious model for how a finite description can stand for an infinite set of items. For example, "the set of natural numbers divisible by two without remainder" is ten words long, but it picks out the set of even numbers, which is infinite. Mathematics also shows us how two different expressions can pick out the same things: another expression for even numbers is "the set of natural numbers including zero and all the numbers that result from adding two to it any number of times." Returning to language, the meaning of *cat* might be "a small domesticated mammal that has soft fur, sharp claws, pointed ears, and, usually, a long furry tail, widely kept as a pet or to catch mice."

The two meanings of "meaning" are sometimes referred to as reference (a thing or set of things in the world) and sense (a summary formula).[7] A sense is not *necessarily* in anyone's head; it is an ideal characterization of a concept behind a word, which individuals who speak the language may know to varying degrees. But to the extent that people have *some* concept in their heads that corresponds to the sense of a word, they can be said to know what the word means. An advantage of the definition of *cat* over the set of all cats is that the definition can fit inside a person's head. Of course, somehow a word-knower must be able to identify the things that a word refers to. But at least in principle, the sense of a word can be put to use in picking out its referents in the world. In the case of *cat,* just look for small domesticated mammals with soft fur, sharp claws, pointy ears, and so on. A word's sense can thus keep a person in touch with the word's possible referents even if they are infinite in number or remote from experience.

There can't be a single answer to the question "Are meanings in the world or in the head?" because the division of labor between sense and reference is very different for different kinds of words. With a word like *this* or *that,* the sense by itself is useless in picking out the referent; it all depends on what is in the environs at the time and place that a person utters it. Logicians call words like this indexicals. The term comes from the Latin word for "forefinger," because the meaning of an indexical depends on what, in effect, you are pointing at. Linguists call them *deictic* terms (pronounced "dike-tick"), from the Greek root for the same thing, pointing. Other examples are *here, there, you, me, now,* and *then.*

At the other extreme are words that refer to whatever we say they mean when we stipulate their meanings in a system of rules. At least in theory, you don't have to go out into the world with your eyes peeled to know what a *touchdown* is, or a *member of parliament,* or a *dollar,* or an *American citizen,* or *GO* in Monopoly, because their meaning is laid down exactly by the rules and regulations of a game or system. These are sometimes called nominal kinds—kinds of things that are picked out only by how we decide to name them.

This leaves us with three not-so-clear categories: natural kinds, like *cat, water,* and *gold;* artifacts, like *pencil, oatmeal,* and *cyclotron,* and proper names, like *Aristotle, Paul McCartney,* and *Chicago.* What are the roles of the world and the mind when it comes to these kinds of entities?

Let's begin with names. At first glance, knowing a name would seem to require knowing its sense, not just its reference. It cannot be mandatory to have seen or touched the referent of a name, because we all know the meanings of names for people like Aristotle who walked the earth millennia before we were born. And we all command names that have different meanings but pick out the same referent in the world: not just *the Evening Star* and *the Morning Star,* but also *Samuel Clemens* and *Mark Twain, Clark Kent* and *Superman,* and *Puff Daddy* and *P. Diddy.* What would the meaning of a name stored in a head look like? Presumably it is like a "definite description": a characterization that picks out a single individual. The meaning of the name *George Washington,* for example, would be "the first president of the United States." Occasionally people or institutions will actually use a definite description as their sobriquet, like The Artist Formerly Known as

Prince. But we are reminded of the logical difference between a name and a definite description in Voltaire's remark that the Holy Roman Empire was neither holy nor Roman nor an empire, in Groucho's quip that Military Intelligence is a contradiction in terms, and in the bumper sticker about the American right-wing group which says THE MORAL MAJORITY IS NEITHER. According to the theory we are now entertaining, the meaning of a name would be an *abbreviation* of a definite description, though as these jokes show, not necessarily the one contained in the name itself.

Now the stage is set for an idea that, by many accounts, was one of the most surprising revelations of twentieth-century philosophy. It was independently thought up by the philosophers Saul Kripke and Hilary Putnam (and, in an earlier version, by Ruth Barcan Marcus), and it hinges on some bizarre thought experiments.[8] I will introduce it with a more plausible one from the same era, this one cooked up not by a philosopher but by a Detroit disk jockey.

Let's begin with our current assumption that a name is shorthand for a definite description. The meaning of *Paul McCartney,* for example, might be something like the definition you would find in a dictionary:

> **McCartney, Paul** (1942–), n. British musician who, as a member of the Beatles, a popular music group (1960–1971), wrote many notable songs with John Lennon, including "A Day in the Life" and "Let It Be." [Syn.: McCartney, Sir James Paul McCartney]

Now for the thought experiment—actually, a rumor, taken seriously in the fall of 1969, that McCartney was dead. According to this account, on a Wednesday morning at five o'clock in November 1966, McCartney stormed out of a recording session with the Beatles, picked up a hitchhiker named Rita, failed to notice that the lights had changed, and was killed in a horrible car accident. The Beatles were at the peak of their popularity, and the death of Paul would have meant the end of their fame and fortune. So they recruited an impostor to take his place (the winner of a Paul lookalike contest named Billy Shears) and hatched a successful plan to hide the truth from the public and the press. This explains why they stopped touring around that time (it would have been too easy to spot a fake in a live performance) and why they grew mustaches (to hide a scar on the impostor's lip). And just as you would expect in an elaborate plot to fool the world, they planted many clues to the conspiracy in their songs and album covers.

On the front cover of *Sgt. Pepper's Lonely Hearts Club Band,* the group is shown at a gravesite adorned with a floral arrangement of Paul's left-handed bass guitar. On the back cover, Paul is the only Beatle with his back to the camera. One song refers to a man who blew his mind out in a car, and another says that nothing could be done to save his life. The clues continued in other songs and albums. At the end of "Strawberry Fields Forever," John is heard singing "I buried Paul," and when "Revolution No. 9" is played backwards, you can hear him chanting "Turn me on dead man." The cover of *Abbey Road* shows a funeral procession, with John as the preacher, Ringo as the undertaker, George as the gravedigger, and the look-alike Paul barefoot (which is how they bury people in Italy). And in the opening song, John sings of the group's rebirth: "One and one and one makes three / Come together, right now, over me."

Suppose that the rumor was true. As logicians would say, there is a possible world in which it *is* true. Now think back to our definition of *Paul McCartney.* In our scenario, the man we have in mind was not, in fact, a member of the group from 1960 to 1971, did not write "A Day in the Life" or "Let It Be," and is not the man who was knighted in 1997. So if the meaning of *Paul McCartney* is the definition, then the man who was born in Liverpool in 1942 and died in London in 1966 is not Paul McCartney. And this flies in the face of people's intuitions. Most would say that the man in question was still McCartney, despite his tragic fate.

Let me push the point by expanding the thought experiment. The "Fifth Beatle," Stuart Sutcliffe, left the band in 1961 and died under mysterious circumstances in 1962. *Or so they say.* Like McCartney, Sutcliffe was handsome, played the bass guitar, and in the minds of many fans personified the group (he originated the Beatles' name, style of dress, and famous hair). Coincidence? I don't think so. Clearly there wasn't room in the band for both of them; otherwise Sutcliffe would never have left the group on the brink of their success. Perhaps McCartney engineered a compromise. It's plausible—it's conceivable—well, there's a possible world in which Sutcliffe secretly continued as a member of the group, writing the songs attributed to Paul and playing bass and singing on their albums, with Paul merely being a pretty face. When Paul died in the car crash, nothing had to be changed except the front man! Consulting our definition once again, we would have to conclude that the name *Paul McCartney* refers to Sutcliffe, the man who played with the Beatles until 1971 and composed "A Day in the Life" with Lennon. But once again, that doesn't feel right. Even if Sutcliffe turns out to meet the standard definition of *McCartney,* we feel that the name doesn't really refer to him, but rather to the man who was christened James Paul in 1942.

All this should remind you of a problem isomorph we encountered in the first chapter: the conspiracy to hide the "fact" that Shakespeare didn't write the plays attributed to him. And there, too, one has the intuition that if the rumor were true, the name *William Shakespeare* would still refer to the man who was christened as such, and not to whoever it was that wrote the plays. This is also the connection in which I discussed identity theft, where you are entitled to say that your name refers to you even when it has become associated with a description that doesn't apply to you.

Just to reassure you that these real-life rumors capture the spirit of Kripke's argument, here are a few of his own examples. Though *Aristotle* is identified as a philosopher who was a student of Plato and a teacher of Alexander the Great, we would still identify that man as Aristotle if he had decided to be a carpenter instead of a teacher, or for that matter if he had died at the age of two. Many people think that *Christopher Columbus* names the man who proved that the earth is round and that *Albert Einstein* names the man who invented the atomic bomb. These beliefs are mistaken, yet we feel that these misinformed people are still referring to the same men that we are. Similarly, most people know nothing about Cicero except that he was a Roman orator. Presumably he was not the only one, yet when people use the name *Cicero,* they intend to refer to that guy and that guy alone, not to any old Roman orator.

Kripke's conclusion was that a name is not an abbreviated description at all, but a *rigid designator*—a term that designates the same individual in every possible world. A name, in other words, refers to an individual in every conceivable circumstance in which we can sensibly talk about that individual at all, biographical facts be damned. The reference of a name is fixed when the person's parents, in effect, point to the little person whom they intend to bear the name, or at whatever later moment a name for the person sticks. It then continues to point to that person throughout his life and beyond, thanks to a chain of transmission in which a person who knows the name uses it in the presence of another person who intends to use it in the same way ("I am going to tell you about a great philosopher. His name was Aristotle . . ."). Names are, in a sense, closer to indexicals like *this* or *you* than to descriptions like "the first president of the United States" or "a small domesticated mammal that has soft fur, sharp claws, and pointed ears." When we know a name, we are implicitly pointing to someone, regardless of what we, or anyone else, know about that person.

You may find this theory easy enough to swallow in the case of proper names, which really are bestowed by a name giver at an identifiable time. But what about the other two categories we are wondering about, natural kinds and artifacts? Take natural-kind terms like *gold, atom, water,* and *whale.* Presumably each of *these* has a definition, along the lines of *gold* being the element with atomic number 79, *water* being H_2O, and *whales* being several families of the order Cetacea. And presumably it is the job of scientists to find out what these definitions are.

But as with the definitions of *McCartney* and *Shakespeare,* this seemingly innocuous idea clashes with some powerful intuitions. One of them follows from the observation that modern scientific definitions for a word may be at odds with the way that people (including the scientists themselves) used to think about its referents. Whales used to be thought of as very large fish (the animal that swallowed Jonah was called "a big fish" in the original Hebrew), but today we know they are mammals. But when scientifically illiterate people use the word *whale,* they are surely referring to the same animals that we are, not to some fish that happens to be large, like a whale shark. In the same vein, ancient alchemists and twentieth-century physicists would give very different definitions of the word *gold,* but they were surely talking about the same stuff. So were the physicists who defined an *atom* as that which could not be split and the physicists who split the atom.

When our scientific understanding of a natural kind changes, the word for the kind doesn't change its *meaning,* at least insofar as "meaning" has something to do with what the words refer to. The original referents always get grandfathered in. If that didn't happen, scientists in different eras (or scientists with different theories in the same era) could never talk about the same thing in order to hash out their differences. The meaning of a word for a natural kind, then, like the meaning of a name, is not a description or a definition, but a *pointer* to something in the world. It acquired a meaning when someone at the dawn of time dubbed a substance or object with the word (like a parent christening a child) with the intent that the word refer to that kind. People then passed on the word from generation to generation with the same intent, by saying things like "This stuff is called *gold*."

Of course, the reference of a natural-kind term can't be the exact nugget of gold or puddle of water that the first dubber pointed to, the way the reference of a name is the exact person that the parents pointed to. After all, it's not as if that nugget or puddle is preserved in a vault somewhere. The reference of a natural-kind term embraces all the things that belong to the same *kind* as the dubbed stuff—generally, those that share some hidden trait or essence with it, an essence that science can potentially discover. The dubber and subsequent users may not know what the essence is, but they sense that the kind has one, perhaps a trait that can be captured with some complicated or statistical formula. Putnam, for instance, confesses that he doesn't know the difference between an elm and a beech. Yet the words are not synonyms to him—though he can't tell the trees apart, he knows that there are experts who can, and that's good enough for him. Putnam argues that words, like goods and services, are the product of a division of labor within a society: we often have to depend on experts to distinguish the meanings, rather than doing it all ourselves.

Putnam drives home the point that natural-kind terms don't have definitions with the help of a now-famous thought experiment. Imagine a planet far away that is an exact duplicate of Earth, with people who look and think just like us, live in surroundings similar to ours, and even speak a language indistinguishable from English. The only difference is that the liquid that the natives call *water* is not H_2O but a compound with a very long and complicated chemical formula that can be abbreviated XYZ. On Twin Earth, XYZ is a colorless liquid that supports life, quenches thirst, puts out fires, falls from the sky, and fills the lakes and oceans. That means that the knowledge stored in the brains of us Earthlings and the knowledge stored in the brains of Twin Earthlings are identical—Twin Earthlings were born with the same kinds of brains as we were, and their brains were exposed to the same kinds of experiences as they were growing up. In the circumstances in which we ask for *a glass of water* (which happens to be H_2O), they also ask for *a glass of water* (which happens to be XYZ).

Now, if meaning is in the head, the meaning of *water* on Earth should be the same as the meaning of *water* on Twin Earth. Yet this doesn't sit well with most people who have thought through the scenario—they say that *water* means something different on the two planets. The difference would become blatant if a spaceship filled with our chemists ever landed on Twin Earth and analyzed the stuff coming out of the faucets—they would say, "Twin Earthlings don't drink water, they drink XYZ!" But even if this never happened, we would still feel that Earthlings and Twin Earthlings are using the words with different meanings, without, of course, realizing it. Based on the Twin Earth story and the elm-and-expert example, Putnam concludes, "Cut the pie any way you like, 'meanings' just ain't in the head!"

"All right," you might reply, "you can have *water*. But what about natural kinds that belong to more complicated conceptual systems, like kinds of animals? Surely a person couldn't be said to know the meaning of *cat* unless she knew that a cat was an animal, right?" But suppose scientists made an amazing discovery: cats are really daleks, the mutated descendants of the Kaled people of the planet Skaro, a ruthless race bent on universal conquest and domination, who travel around in mechanical casings cleverly disguised as animals.[9] Would we say that there is no such thing as a cat, since the definition of *cat* specifies a furry animal? Or would we say that, contrary to our previous beliefs, cats aren't animals? What about if we discovered small furry animals that purr and meow on some other planet—would we have been talking about *them* all along? If you answer no, yes, and no, then you agree with Putnam that natural-kind terms are, like names, rigid designators.

This leaves words for artifacts. At the very least, you might say, it's part of the meaning of *pencil* that it *is* an artifact, namely, an instrument for writing. But now suppose that scientists made an even more remarkable discovery: pencils are living organisms. When we cut them open and put them under a microscope, we see that they have nerves and blood vessels and tiny organs, and when no one is looking they spawn and have baby pencils that eventually become grown-up pencils. (Putnam notes, "It is strange, to be sure, that there is *lettering* on many of these organisms—e.g., BONDED *Grants* DELUXE made in U.S.A. No. 2—but perhaps they are intelligent organisms, and this is their form of camouflage.")[10] If you agree that these are still pencils (as opposed to concluding, "Scientists have discovered that there is no such thing as a pencil!"), then you have admitted that even the idea that a pencil is a human-made implement cannot be part of the meaning of *pencil*.

Of course, *some* part of a word meaning must be in people's heads. Not only must there be something that differentiates a person who knows a word from someone who doesn't, but as we saw earlier in the chapter, two names can point to the same thing in the world (*the Morning Star* and *the Evening Star; Jocasta* and *Mom*) but mean different things, depending on the speaker's state of knowledge. Putnam's argument, then, is not an "ain't" but an "either-or": either the meaning of a word doesn't determine its reference (the things that the word stands for), or meanings aren't in the head. Today many philosophers cut the pie a bit differently and say that there are *two* meanings of "meaning." *Narrow* meanings are in the head, in the form of definitions, conceptual structures, or stereotypes.[11] (*Water* in English and *water* in Twin English have the same narrow meaning.) *Wide* meanings, in addition, point to things in the world, and depend on many things outside a speaker's head: which people a speaker has learned the words from, which people *they* learned the words from, and, when you go far enough back, what the original coiners were pointing at when they first used the word. (*Water* in English and *water* in Twin English thus have different wide meanings.)

Why can we usually get away with ignoring the difference between narrow meanings in the head and wide meanings that bring in the world? Why do we never worry about the possibility that the ideas behind our words might mischaracterize the things we label with our words? The reason is that outside the thought experiments of philosophers and conspiracy theorists, the meanings in the head and the meanings in the world tend to pick out the same things. Our minds harmonize with the world well enough that, most of the time, what we think about corresponds to what we *think* we think about. There are, to be sure, some exceptions. There are cases of mistaken identity, such as Columbus referring to the inhabitants of Hispaniola as *Indians*. There are redrawn boundaries, such as *dolphins* being reclassified by zoologists as a kind of whale. And something must have gone terribly wrong in the chain of retelling when a rigid designator for Saint Nicholas evolved into *Santa Claus*. But most of the time we don't have to worry about these mismatches. Partly it's because there are fewer things in heaven and earth than are dreamt of in our philosophy. The real world has certain regularities, and our faculty for learning words takes them for granted. It's a pretty good bet that there is, in fact, nothing in the universe that looks like water and tastes like water but is composed of XYZ; no daleks that look like and act like cats; no organisms that look like pencils; no tragic coincidences that lead a man to unwittingly kill his father and marry his mother; and no successful Beatle cover-ups. Thanks to constraints from the way the world works, we are not easily fooled.[12]

But our reliable connection to the world requires more than the cooperation of the world itself. It also requires a tacit but deeply held conviction that words are shackled to real things, and a faith that other speakers in our community, past and present, share this conviction. This is the conviction that drives the judgments in the thought experiments that our words are locked onto certain people and things even when we learn that those people and things are very different from the way we imagined them. It is also what allows us to learn words whose referents we ourselves can't identify, trusting that there are others who can. These intuitions usually keep the chains of word-learning unbroken since the first coiners, despite long separations of time and space and considerable changes in our understanding. It is a strange and thrilling thought that every time you refer to Aristotle, you are connected through a very long chain of speakers to the man himself. And every time you use a word to refer to a thing, you fasten yourself to the end of a sinuous thread in space-time which connects you to the first people who looked on that star, or that creature, or that substance, and decided it needed a name.

The way that words can connect us to things in the world rather than to what we *think* of things in the world is not just a matter of how our intuitions get pushed around in zany thought experiments. Even putting aside practical applications like how to resolve cases of imposture and identity theft, there are major conundrums in science and in law that turn on our understanding of what in the world our words and concepts refer to.

A pivotal example in the history of biology is the meaning of terms for species. Before Darwin (and among creationists today), people used to think that every species could be defined by a set of necessary traits that characterized its essence—that there was a precise definition for *tuna, chickadee, rattlesnake,* and so on. But someone who thinks that way will have trouble wrapping his mind around the very idea of evolution, because evolution entails the appearance of intermediate forms that are literally neither fish nor fowl. According to this "essentialist" mindset, a dinosaur has a dinosaurian essence and can no more evolve into a bird than a triangle can evolve into a square. One of Darwin's conceptual breakthroughs was to treat a term for a species as a pointer to a population of organisms (a rigid designator) rather than as a type that is stipulated by a fixed set of traits. The members of that population can vary in their traits at a given time, and the distribution of the traits can shift gradually in the population's descendants. The species name, as a rigid designator, simply points to some branch of this huge genealogical tree, embracing the members that were originally dubbed with the label, their contemporaries who could breed with them, and some portion of their ancestors and descendants who are sufficiently similar to them.[13]

More recently, the nature of names has figured in a scientific controversy that is as incendiary to many laypeople as the theory of evolution. It was triggered by a discovery much like XYZ water, robot cats, and living pencils, not on some imaginary Twin Earth but in our very own solar system. It is the planet Pluto—or, I should say, the Pluto Formerly Known as a Planet. Pluto turns out to be unlike Mercury, Venus, Earth, Mars, Jupiter, Saturn, Uranus, and Neptune. Contrary to what the astronomers who discovered it first thought, Pluto is a puny iceball, smaller than our moon, poking along in a wonky orbit at the fringe of the solar system, and not terribly different from hundreds of other iceballs orbiting out there. In 2006 a panel appointed by the International Astronomical Union fiercely debated how to classify it while an anxious world held its breath. If the astronomers were to demote it from planethood, they would invalidate millions of bedroom mobiles and classroom wall charts and infuriate generations of pupils who learned mnemonics like "My very eager mother just served us nine pizzas" or "Many vile earthlings make jam sandwiches under newspaper piles." But any coherent membership rule that would have kept Pluto in the club would also have let in a motley collection of asteroids, moons, and iceballs. Nothing short of gerrymandering the definition would have allowed the word *planet* to embrace only the classic nine.

In fact, this was not a real scientific controversy but a playing out of the Kripke-Putnam argument about the logic of words. The word *planet* was, in most people's minds, a rigid designator, like a name. It referred to the collection of nine individuals that the word *planet* had come to point to in our language community—not, in this case, since a single moment of dubbing (since the word *planet* had been in use long before Pluto was discovered in 1930), but in an extended act of dubbing that had been finalized by the time we were exposed to the word. As such, people felt that the name continued to refer to the same things it always had in their collective memory, even as our knowledge about it (the sense we try to give the word) changed. The astronomers' dilemma was that they needed a technical term that captured a scientifically coherent kind (the equivalent of H_2O in chemistry or a species name in biology), and they couldn't bear to part with the vernacular word *planet*. Eventually they decided to flout the rigid-designator intuitions of the language community in favor of a scientifically defensible sense, and demoted Pluto from planetary status, to the rest of the world's chagrin.

Legal systems also have to worry about what we refer to with the meanings of our words. For a law to be fair, it has to draw clear lines that people can use to guide their behavior before the fact and that juries can use to judge it afterward. This requires that the words composing a law refer to nominal kinds, actions that are no more and no less than what the law says they are. But the concepts that enter into people's thoughts and actions are often natural kinds and artifacts. To the extent that words for these concepts are rigid designators, the law's attempt to replace them with definitions is impossible in principle. In *Bad Acts and Guilty Minds*, Leo Katz gives an example.[14] During the era of colonial rule in Africa, the British administration passed a Witchcraft Suppression Act, which included a detailed definition of witchcraft. Unfortunately, the drafters were not experts in the local customs and they botched the definition, stipulating certain rituals as witchcraft though they were in fact methods of *detecting* witches. Now what was a judge to do with a defendant who undoubtedly practiced witchcraft but didn't do what the statute *defined* as witchcraft? If the meaning of a word is its definition, the defendant should be found not guilty. But if a term like *witchcraft* is a rigid designator, it refers to acts that are of the same kind as whatever the lawmakers were referring to when they drafted the law that invokes that term, even if the description they came up with mischaracterized it. Many Americans are familiar with a parallel problem in our own legal system. For decades lawmakers and courts had been unable to enforce any consistent definition of "obscenity," leading Supreme Court Justice Potter Stewart to propose an alternative in 1964: "I know it when I see it."

Kripke draws another strange conclusion from the semantics of naming, which is captured in the second half of his title *Naming and Necessity*. At least since Kant, thinkers have distinguished two kinds of knowledge. A priori (before the fact) knowledge is the kind that can be acquired in the proverbial armchair—by divine revelation, introspection, innate ideas, or, more commonly, logical and mathematical deduction. A posteriori (after the fact) knowledge is the kind that you can acquire only by going out into the world and looking. According to a story, probably apocryphal, a seminar of medieval scholars tried to deduce from first principles how many teeth are in the mouth of a horse, and they were shocked when a young upstart suggested that they find a horse, look into its mouth, and count.[15]

"A priori knowledge" can mean a number of things to philosophers, but one of them is the set of facts that are *necessary*—facts that could not be otherwise, and therefore hold in all possible worlds.[16] A posteriori knowledge, in contrast, is about facts that are *contingent*. They depend on how the balls bounced as our world took shape, and could have come out differently if we rewound the tape and let the movie play over. After all, one couldn't very well use a set of axioms and rules to deduce something that depended on the vagaries of the way some dust swirled and settled in the primal solar system, or on which species happened to be around when the planet had the bad luck of having a comet slam into it. Conversely, if one *can* deduce something, one might think that it *has* to be that way, either because of the logical implications of the words being used in the deduction (as in *All bachelors are unmarried*), or because of the eternal and universal nature of mathematical truth which forces the deduction to come out the way it does.

Kant tried to argue for a third possibility: knowledge that is a priori, but more than just a consequence of the meanings of words—knowledge that actually characterizes the physical world as we know it. He pointed to the theorems of Euclidean geometry, which he thought characterized space, even though these theorems were deduced mathematically rather than discovered with a measuring tape and surveyor's level. Almost no one accepts this argument today, because, among other things, modern physics has shown that space is not in fact Euclidean.[17]

Kripke argued for yet another possibility, one that most philosophers had never even considered: knowledge that is a posteriori (discovered after the fact), but necessary. The discovery that the Morning Star and the Evening Star were the same thing (Venus) was a posteriori. But once it was discovered, it was a necessary truth—there is no possible world in which the Morning Star and the Evening Star refer to different things (though of course they could be *called* different things—Kripke's claim is about what *we* mean by the terms). And by the same token, if scientists are correct that water is H_2O, then water *has to be* H_2O—if something is not H_2O, then it wasn't water to begin with, according to what we mean by the term *water*. (Remember that we denied that the stuff on Twin Earth was water.) Likewise, gold *has to be* matter with atomic number 79 (if that is what its atomic number happens to be), heat *has to be* molecular motion (assuming that in fact it is molecular motion), and so on.

None of this is an attempt to do chemistry from the armchair, like the medieval scholars and the horse's teeth. The particular facts that we eventually deem to be necessary will depend on what the people in white coats have discovered. Rather, Kripke's argument is an attempt to clarify what we are logically committing ourselves to when we use proper names and names for natural kinds. We are, surprisingly, committing ourselves to a certain class of logically necessary truths (though we can't know what they are a priori). It is a major revision of our understanding of what kinds of truths there are and how we can know them—all from some intuitions about how we use names.

To be sure, when we stand this close to the concept of meaning we start to sniff an odor of paradox. What exactly am I doing when I *mean* something by a set of words—when I refer to *Aristotle,* or *Alpha Centauri,* or *water,* or *the even numbers,* or *the first baby born in 2050,* or *what the world would be like if Paul had died*? It is bracing enough to realize that just by firing some neurons or moving my lips, I can stand in a relation to a long-dead philosopher or a distant heavenly body. But at least in those cases we can glimpse a connection between the meaner and the meant in a chain of word-learning stretching back to a primeval dubber with firsthand acquaintance. The mind starts to reel, though, when we ponder what connects us to some of the other referents of our words: to water wherever it may be found in the cosmos, to an infinity of abstract entities, to a specific person who does not yet exist (but not to any of the other billions of people who do not yet exist), or to a parallel universe that has no reality but obeys certain laws. These entities spray no energy our way, and our bodies have no sense organs for them, yet somehow a diaphanous strand of semantics connects them to us. As the philosopher Colin McGinn put it, meaning seems to "enable thought to exceed the bounds of acquaintance: it can take us any distance in any direction, traveling across arbitrarily extensive portions of reality, yet keeping to fixed rails as it does so."[18] Perhaps it is not surprising that people in so many cultures think words have magical powers (as we shall see in the chapter on swearing), or that one of the gospels should begin, "In the beginning was the Word, and the Word was with God, and the Word was God." McGinn has a more prosaic explanation: the problem of meaning, like many mysteries in philosophy, may always be shrouded in enigma, because it pushes our common sense into conceptual realms that it did not evolve to think in.[19]

BLING, BLOGS, AND BLURBS: WHERE DO NEW WORDS COME FROM?

If the meaning of a name links us to an original act of dubbing, what exactly goes into this epochal event? How do people conjure up a new sound to label a concept? Which nameless concepts are deemed worthy of having a sound label them? And what sets in motion the chain of transmission that allows the word to spread through a community and get perpetuated down a line of their descendants? The rest of the chapter will take up these puzzles in turn.

The ancestry of words is distinctive among areas of human curiosity, because it is marked by (1) an astonishing amount of knowledge, and (2) an astonishing amount of codswallop. There is something about word origins that inspires people to make things up. Here are some examples from an e-mail circulating in all seriousness under the title "For Trivia Buffs: The History of Phrases":

> In Shakespeare's time, mattresses were secured on bed frames by ropes. When you pulled on the ropes the mattress tightened, making the bed firmer to sleep on. That's where the phrase, "Goodnight, sleep tight" came from.

> It was the accepted practice in Babylon 4,000 years ago that for a month after the wedding, the bride's father would supply his son-in-law with all the mead he could drink. Mead is a honey beer, and because their calendar was lunar based, this period

was called the "honey month" or what we know today as the "honeymoon."

In ancient England a person could not have sex unless you had consent of the king (unless you were in the Royal Family). When anyone wanted to have a baby, they got consent of the King & the King gave them a placard that they hung on their door while they were having sex. The placard had F.U.C.K. (Fornication Under Consent of the King) on it. Now you know where that came from.

This kind of thing inspired an etymology for the word *etymology:* "From the Latin *etus* ('eaten'), *mal* ('bad'), and *logy* ('study of'). It means 'the study of things that are hard to swallow.' "[20] Each of these tall tales could have been debunked with a glance at a dictionary, which would show that scholars have traced the real origins of just about every English word, sometimes to the original coiner, more often to a root in an ancestral or neighboring language spoken centuries or millennia ago. Generally the real etymologies are less colorful than the folk ones. *Tight* means "steadily and securely," as in *sit tight; honeymoon* alludes to a metaphorical sweetness which wanes like the moon; *fuck* is from a Scandinavian word for "beat" or "thrust." (And, by the way, *testify* does not come from a Roman practice in which men vouched for a statement by swearing on their testicles, and *shit* is not an acronym for Ship High in Transit, an advisory to keep dried manure away from the bottom of a cargo hold where it might get wet, release methane, and blow the ship to bits.) But as we shall see, sometimes they are more colorful than anything a lexical counterfeiter could dream up.

By far the most common crucible for new words is the process of assembling them out of old words or pieces of words (morphemes). Every language has a battery of combinatorial operations that crank out new words in a predictable way. In English, for instance, adding the suffix *-able* turns a verb into an adjective referring to how possible or easy it is to perform that action (*learnable, fixable, downloadable*). And gluing two nouns together creates a compound referring to a version of the second noun associated with the first (*ink cartridge, lampshade, tea strainer*). Generally we don't even notice new words like these; they are coined at the drop of a hat and understood almost effortlessly (*outdoorsiness, uncorkable, pinkness,* and so on).

If that were all there is to word coining, language would be much duller and more law-abiding than what we find. Richard Lederer, for example, would never have been able to ask whether infants commit infantry, what humanitarians eat, what goes into baby oil, or the other rhetorical questions we read on page 40. English (and other languages) get crazy because words have a habit of accumulating idiosyncrasies that can't be predicted by the logic of the rules that generated them. A *transmission* is not just an act of transmitting but a part of a car, and if a remark is *unprintable*, it's not because it will break the printer but because it is obscene. An *arrowhead* is the head of an arrow, but a *redhead* is a person with red hair, an *egghead* an intellectual, a *blackhead* a pimple, a *pothead* someone who smokes a lot of marijuana, and a *Deadhead* a fan of the Grateful Dead. One reason for this craziness is that the product of a rule can turn into just another memorized form and accumulate whatever idiosyncratic meaning might be useful to its speakers (as in *a transmission*). Another is that some rules leave open some of the semantic details of their output, which must be filled in on a word-by-word basis (as in the compounds with *head*).

Together with the more or less methodical ways of assembling new words from old ones like suffixing and compounding, speakers have a variety of more impromptu ways of retooling their words. These devices can be spotted in any list of new words, such as the "Words of the Year" announced annually by the publishers of dictionaries. The 2005 list from the *Macmillan English Dictionary*, for example, is effectively a tutorial on all the ways that people create new words:[21]

- Prefixing: *deshopping*, "to buy something with the intention of using it once and then returning it for a refund."
- Suffixing: *Whovian*, "a fan of the British science-fiction series *Doctor Who*."
- Changing the part of speech, such as turning a noun or an adjective into a verb: *supersize*, "to provide an outsize version."
- Compounding: *gripesite*, "a website devoted to making consumers aware of deficient goods and services."
- Borrowing from another language: *wiki*, "a website where users can collectively add or modify text" [from a Hawaiian word for "quick"].

- Acronyms: *ICE,* "'In Case of Emergency' contact number stored in the address book of a mobile phone."
- Truncation: *fanfic,* "new stories featuring characters and settings from a movie, book, or TV show, written by fans and not the original author."
- Portmanteau (combining the beginning of one word with the end of another): *spim* [*spam* + *I.M.*], "unwanted advertisements sent via Instant Messaging."
- Back-formation (misparsing a word and recycling one of the parts): *preheritance,* "financial support given by living parents to their children as an alternative to leaving an inheritance."
- Metaphor: *zombie,* "a personal computer infected by a virus which makes it send out spam without the user's knowledge."
- Metonym: *7/7,* "terrorist bombing" [from the attacks in the London Underground on July 7, 2005].

Of the forty words on the list, only one has a brand-new root: *dooced,* "fired because of something you posted on a weblog." In a story that outdoes the most imaginative folk etymology, the word was coined by a designer who lost her job because of a posting on her blog, www.dooce.com. She named the blog after a habitual typo she made, *dooce* for *doode.* And the intended spelling, *doode,* was her phonetic rendering of the surfer-guy pronunciation of *dude* with an exaggerated vowel.

This shaggy-dog story raises the question of how people brew up new *roots*—sounds that unlike most new coinages, aren't recycled from existing words and morphemes, but are original strings of consonants and vowels. Needless to say, most of them are not eponyms from blogs named after typos of phonetic renderings of fashionable pronunciations.

The most obvious source of a new root is onomatopoeia—a word that resembles what its referent sounds like, as in *oink, tinkle, barf, conk,* and *woofer* and *tweeter.* But onomatopoeia is pretty limited in what it can do. It applies only to noisy things, and even there the resemblance is mostly in the ear of the beholder. Onomatopoeic words are governed more by the phonological pattern of the language than the acoustic output of a noisemaker, as we see when we look at the renderings of animal sounds in other languages:

Robotman © United Feature Syndicate, Inc.

Somewhat handier than pure onomatopoeia is sound symbolism, where the pronunciation of a word merely *reminds* people of an aspect of the referent. Long words may be used for things that are big or coarse, staccato words for things that are sharp or quick, words pronounced deep in the mouth or throat for things that took place long ago or far away (compare *this* and *that, near* and *far, here* and *there*).[22] These alignments, a kind of acoustic or articulatory metaphor, are found in most of the world's languages, and experiments have shown that people are sensitive to the patterns, even in made-up words. For example, which of these is the *malooma,* and which is the *takata*?

Most people agree that the *takata* is the one on the left, because the pointy shape reminds them of the pointy sounds, and the *malooma* is the one on the right, because the blobby shape reminds them of the blobby sounds. And if I were to tell you that the Chinese words for "heavy" and "light" were pronounced qīng (with a high tone) and zhòng (with a falling tone), you would be correct if you guessed, as a majority of English speakers do, that qīng means light and zhòng means heavy rather than vice versa.[23] Sound symbolism has been "discovered" dozens of times, and every time the discoverer claims that it disproves Ferdinand de Saussure's principle that the relation of a sound and a meaning is arbitrary. It doesn't really disprove it, because you can never even come close to predicting a word's sound from its meaning or vice versa, but sound symbolism surely is part of the inspiration when a new word is coined or stretched.

Onomatopoeia and sound symbolism are the seeds of a more pervasive phenomenon in language called phonesthesia, in which families of words share a teeny snatch of sound and a teeny shred of meaning. Many words with the sound *sn-*, for example, have something to do with the nose, presumably because you can almost feel your nose wrinkle when you pronounce it. They include words for the nose itself (like *snout*), words for noselike instruments (like *snorkel* and *snoot,* a cone for directing a spotlight), words for actions and things that are associated with the nose (like *sneeze, sniff, sniffle, snivel, snore, snort, snot, snuff,* and *Snuffleupagus*), and words for looking down your nose at someone (*snarky, sneer, snicker, snide, snippy, snob, snook, snooty, snotty,* and *snub*).

Less obvious is why the *sn-* sound should be associated with quick, furtive, or acquisitive acts, as in *snack, snag, snap, snare, snatch, sneak, snip, snitch, snog,* and *snoop* (or is that a word for sticking your nose into someone's business?). Perhaps one can sniff a hint of quickness and softness in the pronunciation of *sn-*, but that seems a bit after-the-fact, and could apply to just about any English onset. It's more likely that phonesthesia grows outward from a nucleus of similar words that have coalesced for any number of reasons. Some may be products of sound symbolism. Others may be fossils of a morphological rule that was active in an earlier period of the language, or in a language from which the words were borrowed. And some might arise by sheer chance, thrown together in phonological space because the sound pattern of a language allows only so many combinations of vowels and consonants. But once these words find themselves rubbing shoulders, they can attract or spawn new members owing to the associative nature of human memory, in which like attracts like. (In *Words and Rules* I showed how this feature of memory gave rise to families of similar irregular verbs like *sing-sang, ring-rang, drink-drank,* and *wind-wound, find-found, grind-ground.*) Here are a few other clusters; you can be the judge of how symbolic their sounds are:

> *cl-* for a cohesive aggregate or a pair of surfaces in contact: *clad, clam, clamp, clan, clap, clasp, clave, cleat, cleave, cleft, clench, clinch, cling, clip, clique, cloak, clod, clog, close, clot, cloven, club, clump, cluster, clutch*

gl- for emission of light: *glare, glass, glaze, gleam, glimmer,*
    glimpse, glint, glisten, glitter, gloaming, gloss, glow

j- for sudden motion: *jab, jag, jagged, jam, jangle, jarring, jerk,*
    jibe, jig, jigger, jiggle, jimmy, jingle, jitter, jockey, jog, jostle,
    jot, jounce, judder, juggle, jumble, jump, jut

-le for aggregates of small objects, holes, or marks: *bubble, crinkle,*
    crumble, dabble, dapple, freckle, mottle, pebble, pimple, riddle,
    ripple, rubble, ruffle, spangle, speckle, sprinkle, stubble, wrinkle

Some combination of onomatopoeia, sound symbolism, and phonesthesia also gave rise to the list of words for empty speech that I provided at the end of chapter 1.

Phonesthesia is alive in the minds of children, who occasionally use it in lexical flights of fancy. The writer Lloyd Brown shared with me the following examples from his daughter Linda:

> The water was drindling down the drain.
> A mouse scuttered along the baseboard.
> I was scrumbling with the boys.
> I'm going to sloop up the gravy [with bread].
> Why is Grandma's face crimpled?
> Why does the light bulb ringle when you shake it?

And phonesthesia gives rise to a lovely puzzle for comparative linguistics: why languages seldom share a root for their word for *butterfly*.[24] In Western Europe, for example, we find *Schmetterling* in German, *vlinder* in Dutch, *somerfugl* in Danish, *papillon* in French, *mariposa* in Spanish, *farfalla* in Italian, and *borboleta* in Portuguese. The puzzle is that with just about every other kind of word, these languages share roots promiscuously. The words for *cat*, for example, are *Katze, kat, kat, chat, gato, gatto,* and *gato*. A clue may be found in the fact that while the exact word for *butterfly* in many languages is proprietary, it often has a reduplicated sound, most often *b, p, l,* or *f,* as in Hebrew *parpar,* Italian *farfalla,* and Papuan *fefe-fefe.* It's as if the words are supposed to act out the fluttering of the wings! Not all the names are phonesthetic; we also find allusions to the butterfly's properties, real or mythical. In English it's a fly with the color of butter, or that consumes butter, or whose droppings look like butter (the folk etymology that identifies *butterfly* as a spoonerism of *flutter-by* is appealing but untrue). Why the reluctance to share these metaphors and allusions? No one knows, but I am fond of a speculation by the linguist Haj Ross:

> The concept/image of butterfly is a uniquely powerful one in the group minds of the world's cultures, with its somewhat unpromising start as a caterpillar followed by its dazzling finish of visual symmetry, coupled with the motional unforgettability of the butterfly's flipzagging path through our consciousnesses. Butterflies are such perfect symbols of transformation that almost no culture is content to accept another's poetry for this mythic creature. Each language finds its own verbal beauty to celebrate the stunning salience of the butterfly's being.[25]

More recently, and far less lyrically, phonesthesia must have played a role in many of the English roots that have appeared out of the blue in recent decades, such as *bling, bonkers, bungee, crufty, dongle, dweeb, frob, glitzy, glom, gonzo, grunge, gunk, humongous, kluge, mosh, nerd, scuzzy, skank, snarf,* and *wonk.*

The loose associations of phonesthesia, applied to bigger chunks of sound, are also the source of an unignorable trend in brand names:

Zippy—Bill Griffith. King Features Syndicate.

Previously, companies named their brands after their founders (*Ford, Edison, Westinghouse*), or with a descriptor that conveyed their immensity (*General Motors, United Airlines, U.S. Steel*), or by a portmanteau that identified a new technology (*Microsoft, Instamatic, Polavision*), or with a metaphor or metonym connoting a quality they wished to ascribe (*Impala, Newport, Princess, Trailblazer, Rebel*). But today they seek to convey a je ne sais quoi using faux-Greek and Latinate neologisms built out of word fragments that are supposed to connote certain qualities without allowing people to put their finger on what they are. One can sympathize with the bemusement of Griffy, the cartoonist's alter ego. *Acura*—accurate? acute? What does that have to do with a car? *Verizon*—a veritable horizon? Does it mean that good phone service will recede into the distance forever? *Viagra*—virility? vigor? viable? Are we supposed to think it will make a man ejaculate like Niagara Falls? The most egregious example is the renaming of the Philip Morris parent company as *Altria,* presumably to switch its image from bad people who sell addictive carcinogens to a place or state marked by altruism and other lofty values.

MOTHERS WITHOUT INVENTIONS:
MYSTERIES OF THE UNNAMED AND THE UNNAMEABLE

Now that we have an inkling of where the sounds of new words come from, we arrive at the puzzle of which meanings are seen as needing a sound. Whence the urge to dub?

The most obvious reason is captured in the cliché "Necessity is the mother of invention." New words, one might guess, should materialize to fill a lexical gap: a concept that everyone wants to express, but for which *le mot juste* does not yet exist. One has only to overhear the jargon of a specialty—photography, skateboarding, hip-hop, any academic field—to appreciate that lexical suppliers will often step in to meet a demand. Even casual computer users command a lexicon that was unknown to most people a generation earlier—*modem, reboot, RAM, upload, browser,* and so on. And in an age that professes to treat women and men as equals, what would we do without *Ms.*?

But remember another cliché: If wishes were horses, beggars would ride. Many gaps in the language simply refuse to be filled. We met two of them in the first chapter: a name for the first decade of the twenty-first century, and a word for unmarried heterosexual partners. There are countless other gaps. A gender-neutral third-person pronoun to replace *he or she*—sixty proposals have been floated over the centuries (*na, shehe, thon, herm*), but none has made an inroad.[26] A term for one's adult children. A collective term for one's nieces and nephews. The parents of a child's spouse (as in the Yiddish *machetunim*). A fact that you can learn a hundred times without it sticking in memory. The lout sitting next to you on a train or in an airport lounge who screams into a cell phone the whole time. The disgusting lumps of brown snow that accumulate behind a car wheel and fall onto the garage floor. The early-morning insomnia in which your bladder is too full to allow you to fall back to sleep but you're too tired to get up to go the bathroom.[27]

English is riddled with so many gaps that an entire genre of humor has grown up to fill them. The comedian Rich Hall gave us the word *sniglet* (an example of itself) for a word that should exist but does not.[28] Here are some examples:

elbonics n. The actions of two people maneuvering for one armrest in a movie theater.
peppier n. The waiter at a fancy restaurant whose sole purpose seems to be walking around asking diners if they want ground pepper.
furbling v. Having to wander through a maze of ropes at an airport or bank even when you're the only person in line.
phonesia n. The affliction of dialing a phone number and forgetting whom you were calling just as they answer.

But sniglets were not the first sniglets; before them we had liffs. In 1983 the writer Douglas Adams (better known for *The Hitchhiker's Guide to the Galaxy*) and the television producer John Lloyd published *The Meaning of Liff*, based on the following observation: "In life (and indeed in Liff) [a town in Scotland], there are many hundreds of common experiences, feelings, situations, and even objects which we all know and recognize, but for which no words exist. On the other hand, the world is littered with thousands of spare words which spend their time doing nothing but loafing about on signposts pointing to places." Adams and Lloyd decided to label experiences that no one has a name for with names for places that no one ever needs to go to.[29] For example:

sconser n. A person who looks around while talking to you to see if there's anyone more interesting about.
lamlash n. The folder on hotel dressing-tables full of astoundingly dull information.

shoeburyness n. The vague uncomfortable feeling you get when sitting on a seat which is still warm from someone else's bottom.

hextable n. The record you find in someone else's collection which instantly tells you you could never go out with them.

The language maven Barbara Wallraff inverted the formula in *Word Fugitives,* a history of recreational word coining and a collection of her columns by that name in the *Atlantic Monthly,* in which one reader submits a lexical gap and others try to fill it:[30]

Saying something to your child and then realizing that you sound like one of your own parents: **déjà vieux, mamamorphosis, mnemomic, patterfamilias, vox pop, nagativism, parentriloquism.**

The dicey moment when you should introduce two people but can't remember one of their names: **whomnesia, persona non data, nomenclutchure, notworking, mumbleduction, introducking.**

The realization of the perfect riposte three hours after the argument: **hindser, stairwit, retrotort, afterism.**

The momentary confusion experienced by everyone in the vicinity when a cell phone rings and no one is sure if it is his/hers or not: **conphonesion, phonundrum, ringchronicity, ringxiety, fauxcellarm, pandephonium.**

The *Washington Post* Style Invitational column occasionally asks its readers to fill a lexical gap by changing a single letter of an existing word:

sarchasm n. The gulf between the author of sarcastic wit and the person who doesn't get it.

hipatitis n. Terminal coolness.

Dopeler effect n. The tendency of stupid ideas to seem smarter when they come at you rapidly.

Beelzebug n. Satan in the form of a mosquito that gets into your bedroom at three in the morning and cannot be cast out.

And then there is the frequently e-mailed list of "new words to add to your Jewish vocabulary":

yidentify v. To be able to determine the ethnic origins of celebrities even though their names might be St. John, Curtis, Davis, or Taylor.

mishpochamarks n. The assorted lipstick and make-up stains found on one's face and collar after kissing all one's aunts and cousins at a reception.

santa-shmanta n. The explanation Jewish children get for why they celebrate Hanukkah while the rest of the neighbors celebrate Christmas.

meinstein n. My son, the genius.

Despite the itch they scratch, most recreationally coined words seldom become permanent members of the language. Nor do most of the "Words of the Year," like *deshopping* or *preheritance*. I mentioned in chapter 1 that the American Dialect Society annually designates the new words that are most notable, most useful, or most likely to succeed.[31] A follow-up of the picks from the 1990s shows that the experts in the society are about as accurate as tabloid psychics. Some of the words were political barbs that died with the careers of their targets, such as *to newt* and *to gingrich*. Others depended on a topical cuteness that wore thin once the inspiring event faded from memory, like *-razzi* (an aggressive pursuer, from 1997, the year that Princess Diana died while fleeing paparazzi) and *drive-by* (cursory, from 1996, the year Bill Clinton campaigned against *drive-by deliveries* in maternity hospitals, a play on *drive-by shootings*). And some consisted of bets on the wrong name for an innovation: *notebook PC* (in conversation they're still called *laptops*), *s-mail* (overshadowed by its source, *snail mail*), *W3* (that's the World Wide Web), *information superhighway* (too Al Gore), and *Infobahn* (yuck).

The fortunes of new words are a mystery. Filling a lexical gap is no guarantee of success; nor are two other features that one might think would be selling points, brevity and transparency. *WWW* takes longer to say than what it abbreviates, the *World Wide Web,* and despite the number of times we end up saying those nine syllables, it has resisted shorter replacements like *triple-dub, wuh-wuh-wuh,* and *sextuple-u*.[32]

As for transparency, the verbs *boot up* and *reboot* stand their ground against the crystal-clear *start up* and *restart* (the terms used in the operating system menus), despite the fact that most people have no idea what *boot* alludes to. It has nothing to do with giving your computer a good swift kick, but alludes to the way computers started up in the Mesozoic era of computing, when I did my doctoral research. The mini-computers of the time were blank slates, not even smart enough to read in their operating system from a tape or disk. You had to spoon-feed programs and data into them one byte at a time by toggling a bank of switches on the front panel, one for each bit in the byte, which stood for the 1s and 0s that made up a program instruction or piece of data. Since this was too tedious even for a graduate student, you toggled in a very short program that gave the computer just enough wherewithal to read in a few bytes on a punched paper tape. Those bytes made up a slightly larger program, which told the computer how to load in the rest of the tape, which contained the computer's operating system. The little program at the beginning of the tape was called a "bootstrap loader" because its self-loading magic reminded someone of the expression "lift yourself up by your bootstraps." The entire procedure was referred to as *booting up* your computer. This might sound like Fornication Under the Consent of the King, but I was there, and can swear it's how the word caught on.

Alan Metcalfe, a past president of the American Dialect Society and the author of *Predicting New Words,* has tried to identify why some words succeed and others fail. He summarizes his guesses in the acronym FUDGE: Frequency, Unobtrusiveness, Diversity of users and situations, Generation of other forms and meanings, Endurance of the concept. It's a good start, but it raises more questions than it answers. All words originate with a single coiner, so they all begin life with a Frequency and Diversity of 1. The fact that some words increase their frequency and user diversity is the phenomenon we are trying to explain, rather than a prior cause that can explain it. The same circularity threatens Generation of new forms (polysemy), as in the spread of *blockbuster* from a large bomb to something that is commercially successful. Frequent words are more polysemous (chapter 3), so it may be that the success of a word causes it to be a better generator of new meanings, rather than that the generative power of a word causes it to become more successful. Endurance of the concept isn't a good predictor of a word's survival either. Although we have less occasion nowadays to talk about cabooses, flappers, zoot suits, and the Cold War (Metcalfe's examples), when we do have to talk about them the words are there for the using.

This leaves Unobtrusiveness, a chastening reminder to recreational word coiners and word-of-the-year anointers that most coinages that call attention to themselves end up in the dustbin of history, while the real winners slip into the language by stealth. Certainly most of the sniglets, liffs, and word fugitives are too clever for their own good, destined to amuse rather than last. The same is true of most of the nonce coinages by witty journalists that catch the eye of the assemblers of the year's-best lists, like *Brown-out,* the poor handling of an emergency reminiscent of the director of FEMA after Hurricane Katrina, and *flee-ancée,* a bride who flees her own wedding in the fashion of Jennifer Wilbanks, who was briefly famous in 2005. The humorist Gelett Burgess (1866–1951) had better luck with *blurb* and *bromide* in 1907, but not with *alibosh* (a blatant lie), *quisty* (useful but not beautiful), *cowcat* (a person who does nothing but occupy space), or *skyscrimble* (to go off-topic).

Even when a coiner pulls out all the stops to disseminate a new word to fill a lexical gap, an ungrateful populace will usually ignore him. In the year 2000, the conceptual artist Miltos Manetos noticed that the English language lacks a word for the high-tech aesthetic in product design, as in *The new iPod Nano is really X,* and a word for the genre of technologically driven artistic media such as video art, computer graphics, and digital animation, as in *Our gallery showcases new artists working in X.* Manetos suggested that a single word, used as an adjective and as a noun, could fill both gaps. In the spirit of the movement he was naming, he hired Lexicon Branding (the company that dreamed up *Pentium, Celeron, Zima, Vibrance, Optima,* and *Alero*) to generate candidates with their computer algorithm and staff of linguists. From the list they provided him, Manetos chose *neen,* which means "now" in old Greek. He rolled out the word at a packed event in a major New York art gallery, complete with journalists, critics, and a panel of commentators, including me. I predicted that the word would not catch on because it had the wrong phonesthetics, resonating too much with the *sn-* words and with childhood taunts like *nyah-nyah* and *neener-neener.* I proved to be right (just try googling *neen*), but it was an easy prediction to make, because most conspicuous coinages fail no matter what.

But it isn't even safe to predict that a coinage will fail because it is jocular or self-conscious. The language has recently welcomed *podcast,* "an audio show downloaded to a digital music player," a pun on *iPod* and *broadcast,* and *blog,* "Web log," which taps the amorphousness of *blob* and *bog* and insouciantly cuts a word against its syllabic grain in the style of 1970s campus slang like *shroom* (mushroom), *strawb* (strawberry), *burb* (suburb), and *rents* (parents). Canadians refer to their one-dollar coins as *loonies,* an irreverent allusion to the loon on its reverse, and when the two-dollar coin was introduced, it immediately became known as a *toonie.* Earlier decades gave us *Yuppie* (Young Urban Professional, a play on *hippie, Yippie,* and *preppy*), *couch potato, palimony, qwerty* (technological inertia, from the keys on the top row of a typewriter), and of course the silliest of all, *spam.* Nor is this a new phenomenon. *Soap opera* is from the 1930s, *hot dog* from the 1890s (from a campus joke about its alleged ingredient), and *gerrymander* from 1812. The verb *razz* and noun *raspberry* (the rude noise you make with your tongue stuck out) are not onomatopoeic. They are a product of Cockney rhyming slang, in which a word is replaced by a phrase that rhymes with it and the rhyming part is dropped, as in *loaf* for "head" (*head → loaf of bread → loaf*) or *apples* for "stairs" (*stairs → apples and pears → apples*). By the same logic, you can figure out the inspiration for *raspberry* when you are told that it was shortened from *raspberry tart.*

Despite the occasional success of a facetious coinage, Metcalfe is on to something when he points to unobtrusiveness as one of the usual entry requirements for admission of a new word to the language. That something, though, is not ordinariness per se, but an ability to satisfy the cognitive requirements of wordhood. Not everything that flits through our minds has the right coherence and stability to stand as the meaning of a word.[33] For a concept to be nameable, it usually has to refer to an orderly kind of thing, or to an event that plays out in a stereotyped way on every occasion. Also, other than with proper nouns, a nameable concept has to be generic rather than particular. For example, a new common noun like *latte-drinker* refers to the kind of person who typically drinks lattes (say, a young urban sophisticate), not to a person who happens at the moment to be drinking a latte. It can thus be used without contradiction in the sentence *Craig is a latte-drinker in every sense except that he doesn't actually drink lattes.* (In that regard it differs from the corresponding *phrase;* you can't say *Craig drinks lattes in every sense except that he doesn't actually drink lattes.*) Words tend to be reserved for whole entities ("rabbit," not "undetached rabbit parts"), for stable qualities ("green," not "green until 2020 and blue thereafter"), for natural kinds, for events that are terminated by a single change of state or a single goal, for artifacts with a function, and for actions with a salient cause, effect, means, or manner. And words are given to the players that have roles in the events we make assertions about, not to the assertions themselves. A sentence can be true or false, but a word cannot.

By these lights, interesting neologisms tend to fail precisely because they *are* interesting—not because their construction is clever but because the coiner is really *commenting* on something rather than *naming* something. Take the words on a typical end-of-year list, such as *egosurfing* (googling oneself), *celanthropist* (celebrity philanthropist), *infomania* (obsessively checking for e-mail and text messages), *security mom* (a voter concerned with terrorism), *ubersexual* (a heterosexual male who is masculine yet sensitive and socially aware), and *greenwashing* (a public-relations gesture intended to make a company look environmentally conscientious). These words are really news reports on social trends; you can almost feel the compiler nudging you in the ribs as if to say, "Look how our lives are being revolutionized by the Internet!"—or by changing sex roles, or technology, or terrorism, or environmental awareness.

The sniglets and their relatives, for their part, are really saying, "Don't you hate it when . . . ," "Isn't it silly that . . . ," or "Have you ever noticed how . . . ?" Indeed, by packing one of these aperçus into a single word, the coiners are commenting on their comments. They are saying, "This phenomenon is so common and recognizable that it should have a word of its own!" I think it is this exploitation of the psychology of wordhood, as much as the punning, that gives sniglets their drollness. But it also makes them bad as words. You may feel silly when you furble through a bank or have a bout of shoeburyness on a bus, but there aren't many occasions to talk about these events other than to comment on how silly you feel. And the large number of sniglets devoted to blunders and follies (such as *purpitation,* "to take something off the grocery shelf, decide you don't want it, and then put it in another section") may be comforting to fellow bumblers, but they are hardly the kinds of events demarcated by a goal that make for a good verb meaning.

But even with all this knowledge—the tools of word formation, the lexical gaps, the phonesthetic landscape, and the conceptual requirements of wordhood—we still can't predict when a new word will take root. The remaining parts of the mystery will lead us to a new way of thinking about culture and society—but first we must return to the mystery of Steve.

PROJECT STEVE REVISITED

The rise and fall of Steve and other baby names tells us that something more than a pleasing sound and a nameable concept goes into the acceptance of words. Naming a child ought to be the most straightforward instance imaginable of adding a new word to the language. The parents are free to pick the sound, and the community pretty much has to respect their choice. Yet naming a baby implicates enigmatic forces that determine the spread of a word and perhaps can shed light on why other words succeed or fail.[34]

In some ways, naming a child is different from coining other words. We never come across a human sniglet—an unfortunate child whom parents never got around to naming—nor do we get to reject the parents' choice, other than with an occasional nickname that sticks. And most of the time, parents draw on a common pool of sounds in naming their newborns, rather than minting a new one for the occasion. But because naming a child is our purest and most democratic act of dubbing, and because data on baby names are precise and abundant (the Social Security Administration maintains a database of all American forenames since the 1880s), they are a gold mine of information on how words spread.[35]

One doesn't have to be a Steve to appreciate that names can undergo boom-or-bust cycles. Given a person's first name, most observers can guess her approximate age with an accuracy well above chance.[36] An *Edna, Ethel,* or *Bertha* is a senior citizen; a *Susan, Nancy,* or *Debra* is an aging baby boomer; a *Jennifer, Amanda,* or *Heather* is a thirty-something; and an *Isabella, Madison,* or *Olivia* is a child. I wrote "guess *her* age" because girls' names turn over more quickly than boys': *Robert, David, Michael, William, John,* and *James* just won't go away. But even with males, you will do better than throwing darts if you try to guess the age of an *Ethan,* a *Clarence,* a *Jason,* or of course a *Steve.* Names don't invariably have fashion cycles—in many societies babies must be named after saints or ancestors, and many parents continue to invest their sons' names with a heavier burden of carrying on the lineage than they do their daughters'—but there is always some turnover, and the rate of churning shot up in Western countries in the twentieth century.[37]

Like my mother and father, most parents have no recollection of hearing about a baby with the name they ended up giving their own. They often say that they had a favorite relative or character in mind or that they just liked the sound. But when they get to the day-care center and call for their Tyler or Zoë, they are startled to hear three little voices respond. Their carefully considered choice, it turns out, was also the carefully considered choice of thousands of other parents. Leibniz wrote that if you see a pair of clocks that show the same time, there are three explanations. They might be synchronized by a shaft or wire connecting them. They might be adjusted continually by a skilled clockmaker who keeps them in sync. Or the two clocks may run so similarly that they just stay synchronized on their own. If parents don't coordinate their choice of names by directly copying each other, what about the other two alternatives: an external influence, and an independently developed similarity in tastes?

We can immediately set aside the kinds of influence that are most often adduced for other tastes and fashions. The sociologist Stanley Lieberson was prompted to write his book *A Matter of Taste* after he and his wife named their daughter Rebecca and were surprised to learn that so many of their contemporaries had had the same idea. He knew that he couldn't explain the trend by appealing to the usual suspects:

> There was no advertising campaign sponsored by the NRA—the National Rebecca Association—let alone an effort to demean those who preferred a competitor. The ascent of *Rebecca* and the decline of another name was not the same as the intense competition between Pepsi and Coca-Cola. Neither Wal-Mart nor Neiman Marcus were promoting the name as part of a newborn daughter's fashion ensemble. And there was no factory rebate for naming your daughter *Rebecca*.[38]

There are other possible outside influences. The most popular folk theory about trends in baby naming is that parents are influenced by heroes, leaders, actors, or the characters the actors play. Hillary Clinton used to say that her parents named her after the man who first climbed Mt. Everest. But she was born in 1947, when Edmund Hillary was an obscure New Zealand beekeeper; he would not scale the peak until 1953, when she was six. Lieberson examined the correlation between the rise and fall of baby names

and the appearance and disappearance of famous people, real and fictitious, in the public eye. In almost every case he saw an uptick in the name
before the celebrity burst on the scene. Often the famous moniker then
boosts the name to new heights—Lieberson calls it "riding the curve"—but
rarely does it light the fuse.

Marilyn, for instance, was a fairly popular name in the 1950s, and most
people would point to an obvious explanation: the fame of Marilyn Monroe.
Unfortunately for this theory, the name had begun shooting up decades
earlier and was already popular when Norma Jeane Baker adopted it as her
stage name in 1946. In fact, the popularity of *Marilyn* fell somewhat after
Monroe became famous. At this point people assume that she must have
driven it down: parents didn't want to name their girls after a sex kitten
during the suburban prudishness of the 50s or the nascent feminism of the
60s. Wrong again—the name had peaked in the 1930s and was already in
decline when Monroe burst on the scene.

In a few cases, a real or fictitious celebrity really does spawn a name.
Darren was unknown in England until Britons were introduced to a witch's
husband with that name in the 1960s sitcom *Bewitched.* And *Madison,* currently the third most popular name for a baby girl, did not even exist as a girls'
forename until a mermaid played by Daryl Hannah in the 1984 movie *Splash*
climbed out of the East River, spotted it on a street sign, and took it as her
name.[39] Also, Lieberson spotted some blips and wiggles that could be attributed to famous names. In the 1930s *Herbert* went down and *Franklin* went
up, and since then *Adolf* has vanished, all for obvious reasons. But in general
the catchiness of famous names is a cognitive illusion. People recall an instance or two in which a name is made famous by a celebrity and the same
name is given to a lot of babies, and assume that the first phenomenon caused
the second. But of course they have no way of mentally sorting the names by
year to make sure the timing is right, nor can they dredge up the counterexamples, such as the millions of babies who could have been named (but
weren't named) *Humphrey, Bing, Cary, Hedy, Greta, Elvis, Ringo,* and so on.
Nor can people easily distinguish cause and effect. Scriptwriters have to give
their characters plausible names, and aspiring actors in search of stage names
have to choose attractive ones. It stands to reason that they would be buffeted
by the same forces that affect the expectant parents of the day.

What about larger social trends, like changes in sentiments toward na-
tionalism, religion, or gender roles? Again, it's not what you think. Recent

decades have seen a resurgence of biblical names, like *Jacob, Joshua, Rachel,* and *Sarah*. Everyone's first guess is that it reflects a resurgence of religion in American life. But Lieberson shows that during this period the trends for biblical names and the trends for religious observance go in opposite directions, and also that people who give their babies biblical names are no more religious than those who don't. Feminism would seem to be a more promising spark, but even here the effects are equivocal. Some girls' names based on flowers, for instance, have declined since the 1970s (*Rose, Violet, Daisy*), but others have become more popular (*Lily, Jasmine*).[40] You can also pick whichever trend you want when you plot the popularity of girls' names that are diminutives of boys' names: some go down, like *Roberta, Paula,* and *Freda,* while others go up, like *Erica, Michaela, Brianna,* and of course *Stephanie*.[41]

The reason for the gap between conventional wisdom and the facts is that most people have the wrong theory of cultural change. They think that the changes are the predictable effects of external causes: governments, advertisers, celebrities, the economy, wars, cars, technology, and so on. They also think that cultural changes are *meaningful*—that one can give a purposeful explanation of why a society does something in the same way that one can give a purposeful explanation of why a person does something.[42]

The writer Edward Tenner has documented an example of this fallacy.[43] Until the 1960s most men wore fedoras in public; today almost none do. What happened? There is no shortage of explanations. John F. Kennedy set the trend by going hatless after his inauguration. People moved to the suburbs and started spending a lot of time in cars, so their heads didn't get as cold, and also it's awkward to get in and out of a car wearing a hat. Men grew their hair longer as a form of self-expression and didn't want to hide it or, worse, suffer from an embarrassing case of hathead. There was a greater emphasis on the natural, and hats represented the incompleteness of nature. Hats were associated with the political establishment, and the younger generation rebelled against it. The culture began to idolize youth, and hats were associated with old men.

This kind of potted sociology can be found in any newspaper article on a social trend. But all the explanations are wrong. If you measure the popularity of dress hats for men over the decades, you find that it declined steadily since the 1920s (the 1960s just saw the coup de grâce, when hat wearing fell below a critical threshold). And this decline exactly paralleled a falloff in *women's* hats and in women's *gloves* over the same period.[44] Not a single one of the pop-social-science explanations, referring as they do to hats, men, the 1950s, or the 1960s, is compatible with this chronology. There *is* something afoot during this era—a decline in formality in all spheres of life, including dress, grooming, public comportment, and terms of address (as in the use of first names rather than *Mr.* or *Mrs. So-and-So*). It's hard to imagine any external or purposeful explanation of this trend (war, politics, the economy, technology) that could push it steadily in one direction from the 1920s through the first decade of the twenty-first century and beyond. The same conclusion emerges from quantitative studies of women's fashions. Contrary to popular beliefs, skirt lengths are not correlated with the stock market, fabric shortages, high-concept advertising campaigns, or anything else, though they do exhibit slow, smooth rises and falls over the decades.[45]

Lieberson argues that we have to rethink how we explain cultural change. A "trend" is shorthand for the aggregate effects of millions of men and women making personal decisions while anticipating and reacting to the personal decisions made by others. This gives rise to an *internal* dynamic of change—hat wearing in one year affects hat wearing in the next—and to trends that have a logic of their own, rather than fitting in to some meaningful narrative about the choices of society as a whole.

Many fashions—skirt lengths, lapel widths, tail fins, facial hair, and of course baby names—undergo smooth waves of swelling and subsidence, rather than a sudden jump from one level to another or the spiky chaos of the stock market. It's tempting to invoke a physical metaphor like momentum or pendular motion, but we'd still need an explanation of why the metaphor would be apt. The economist Thorstein Veblen and the art critic Quentin Bell noted that cycles of fashion could be explained by the psychology of status.[46] The elite want to differentiate themselves from the rabble in their visible accoutrements, but then the folks in the next stratum down will copy them, and then those in the stratum below that, until the style has trickled down to the hoi polloi. When it has, the elite move to a new look, which prompts the bourgeoisie to imitate them, then the lower middle class and so on, in a never-ending, internally fueled rocking and rolling of fashion. (It calls to mind Dawkins's analogy of the heater and cooler for the evolution of another kind of status symbol, the widowbird's tail.) The status hierarchy was traditionally reckoned by wealth and class, but it can also be reckoned by other metrics of prestige—honor, hipness, sophistication—within different coteries and cliques.

Lieberson adds that the cutting edge of fashion will continue to push in the same direction for as long as it can, because any backtracking would defeat the purpose of looking different from everyone else. Hence the metaphorical momentum. But at some point it reaches a practical limit—skirts can't shrink to the size of garter belts, nor can they grow into twenty-foot trains—and the elite is forced to yank the style in the opposite direction. Hence the metaphorical pendulum. This sounds like it could explain anything and hence nothing, but Lieberson notes that whenever a fashion switches direction, the trendsetters introduce some *other* change at the same time, so the skirts or beards or fenders can't be confused with last decade's model.

And this brings us back to names. Names don't grow and shrink along a single dimension like lapels, but they do have many traits that vary: the sounds of their onsets or codas, their etymology (Hebrew, Latin, Greek, Celtic, Anglo-Saxon), their literal meanings (flowers, jewels, weapons, months), their associations with famous people.[47] The pool of names is enormous, and it can be supplemented from fiction (*Miranda* comes from *The Tempest, Wendy* from *Peter Pan*), by promoting surnames to middle names and then to forenames (like *Morgan* and *McKenzie*), by borrowing from other languages (*Siobhan, Natalia, Diego*), and by adding prefixes and suffixes like *-a, -ene, -elle,* or, among African Americans, *La-, Sha-,* and *-eesha* (*Latonya, Latoya, Lakeesha,* and so on). Different classes and ethnic groups sample from different pools, not surprisingly.

Most parents want to give their child a name that is distinctive enough that it won't be shared by all the other children they meet, in accord with Sam Goldwyn's advice to an employee, "Don't name your son William. Every Tom, Dick, and Harry is called William." On the other hand, they don't want to saddle their child with a name that is *so* distinctive that the child is marked as coming from a family of greenhorns or misfits. At one extreme we find celebrities like the actress Rachel Griffiths, who named her son Banjo, the magician Penn Jillette, who named his daughter Moxie Crime-Fighter, and the rock star Bob Geldof, who named his daughters Little Pixie, Fifi Trixibelle, and Peaches Honeyblossom. At the other we have the boxer George Foreman, who named all five of his sons George. Most parents are somewhere in between. The problem with everyone trying to be moderately distinctive is that they are in danger of being moderately distinctive in the same way. Hence we get a school full of Susans and Steves in the 1960s and of Chloës and Dylans today. But then the next generation of parents will re-act to the hyperstevism or hyperdylanism by looking for a new new thing, sending the curves careening in yet another direction. The dynamic is rem-iniscent of Yogi Berra's restaurant review: "No one goes there anymore. It's too crowded." Baby-name advisors like Pamela Satran try to advise parents on categories of names on the horizon that are neither too common nor too outré—perhaps heroes' last names, like *Monet* or *Koufax,* or color words, like *Taupe* or *Cerulean.*[48] (*Yogi,* anyone?)

A powerful force in the quest for moderately distinctive names is phon-esthesia. A popular name spreads its allure to existing names, or buds off new names, that share an onset, syllable, or coda. Earlier in the twentieth century, *Jane* begot *Janice, Janet, Jan,* and *Janelle,* each with numerous spellings. *Carol* spawned *Carolyn, Karen, Carrie, Cara,* and *Carina.* More recently, *Jennifer* gave a boost to *Jessica* and *Jenna* at one end of the word and to *Heather* and *Amber* at the other. Lieberson notes that many popular names attributed to the influence of celebrities (Janet Gaynor, Jessica Lange) are really just riding the wave of a popular *sound.* Phonesthetics can thus help explain why only some celebrity names fall on fertile ground. *Darren* shot up in England in the 60s, for example, while other television names did not (like *Ricky* and *Maxwell*), because at the time more than a third of British boys' names ended in *n.*[49]

Gender plays a role beyond the obvious fact that in almost all cultures, boys' names can be distinguished from girls'. Parents occasionally give a boys' name to a girl, perhaps because they were wishing for a boy, but more likely to endow the girl with a suggestion of independence and strength. For some reason these names seem to land on girls who are destined to be-come sexy actresses and supermodels (or perhaps the names are chosen by them), like Drew Barrymore, Blair Brown, Glenn Close, Jamie Lee Curtis, Cameron Diaz, Jerry Hall, Daryl Hannah, Mel Harris, James King, and Sean Young (all of whose forenames are rare among women of their age).[50] Feminism would be an obvious explanation were it not for the fact that the process has been going on for a hundred years or more. At the beginning of the twentieth century, *Beverly, Dana, Evelyn, Gail, Leslie, Meredith, Robin,* and *Shirley* were all primarily names for men. And during this entire time, androgynous naming has been a one-way street.[51] Once a boys' name is given to too many girls, it is ruined as a boys' name, presumably because parents are more squeamish about conferring feminine traits on their sons than masculine traits on their daughters. As Johnny Cash noted, "Life ain't easy for a boy named 'Sue.' "

Age is another impeller of trends. Many names and sounds get stooped and wrinkled along with their bearers and are pensioned off because parents don't want to think of their children as graybeards or little old ladies. This was the fate, for example, of *Ethel, Dorothy,* and *Mildred,* and of many *s*-final names like *Gladys, Florence, Lois, Doris, Frances,* and *Agnes.* But once the bearers are dead or out of sight, the name can lose its geriatric connotations and be recycled, at least if it harmonizes with the phonesthetics of the new era. *Elizabeth, Christina,* and *Joseph* enjoyed these rebirths, for example, and if you find yourself among Max, Rose, Sam, Sophie, Jake, and Sadie, you are either in an old-age home or a day-care center.[52]

What does this have to do with the acceptance of new words? As I mentioned, names differ in some ways from other kinds of words. When it comes to children, everyone gets named, the pool of names is limited, and the name usually sticks, whereas when it comes to concepts, many remain nameless, those that are named usually get new combinations of sounds, and most words-of-the-year are famous for fifteen minutes. Nonetheless, the internal dynamics that drive the cycling of fashions for names apply in part to the coining and acceptance of other kinds of words.

We have already seen the fashions in company and product names, from *Buick* to *Mustang* to *Elantra.* And teenage, campus, and hipoisie slang can be dated as accurately as names (*the cat's pajamas, hep, groovy, far out, way cool, phat, da bomb*). Another fashion cycle affects terms for superlatives. Speakers always want to impress their listeners with how great something in their experience was, so they describe it with a superlative, but that reduces the value of the superlative, so subsequent speakers reach for a new superlative by co-opting some other word for an extreme experience, and so on, in a spiral of semantic inflation. Long ago our linguistic ancestors diluted the original sense of *terrific* (causing terror), *fantastic* (worthy of fantasy), *tremendous* (causing one to tremble), *wonderful* (inspiring wonder), and *fabulous* (celebrated in fable). Speakers in recent decades have similarly taken the punch out of *awesome, excellent, outstanding,* and, in Britain, *brilliant.*

Another example. Terms for concepts in emotionally charged spheres of life such as sex, excretion, aging, and disease tend to run on what I call a euphemism treadmill. They become tainted by their connection to a fraught concept, prompting people to reach for an unspoiled term, which only gets sullied in its turn. For instance, *toilet,* originally a term for bodily care (as in *toilet kit* and *eau de toilette*), came to be applied to the device and room in which we excrete. It was replaced by *bathroom,* leading to absurdities like *The dog went to the bathroom on the rug* and *In Elbonia, people go to the bathroom on the street.* As *bathroom* became tainted (as in *bathroom humor*), it was replaced in successive waves by *lavatory, WC, gents', restroom, powder room,* and *comfort station.* A similar treadmill cycles terms for the handicapped (*lame, crippled, handicapped, disabled, challenged*), names of disfavored professions (*garbage collection, sanitation, environmental services*), and academic activities (*gym, physical education, human biodynamics*), and names for oppressed minorities (*colored, Negro, Afro-American, black, African American*).[53]

Science is not immune to these waves of fashion. Originally scientists were expected to dub their discoveries with Greek and Latin jargon like *ligand, apoptosis,* and *heteroskedasticity.* This gave way to a tolerance for English circumlocutions (*frequency-dependent selection, secondary messenger*), then to an indulgence of whimsical allusions (such as *quark, Big Bang,* and *Sonic Hedgehog,* a gene named after a video game character), and now to hip truncations like *brane theory* in physics, short for *membrane.*

The surging and cycling of fashion is not the only internal dynamic that decides the fate of words. Even if tastes remained stable, the success of a new word would depend on lexical epidemiology—the way a neologism spreads from its coiner to a new speaker, who then may infect others, and so on. Eventually the neologism either dies out or becomes endemic, depending on how many people a word-knower talks to in a day, and on how readily they notice and remember it. As with real epidemics, it's hard to predict what will happen. Depending on small differences in how catchy the word is, and on how well-connected, trusted, or charismatic its first adopters are, it may or may not reach the tipping point that would lead it to become entrenched in the community and perpetuated down the generations.[54] This is a way to make sense of the Frequency and Diversity components of the FUDGE factors that Metcalfe suggests are the secrets to a word's success.

So a look at the spread of words and names upends the conventional wisdom of where culture comes from and how it changes. In the twentieth century, a culture came to be thought of as a superorganism that pursues goals, finds meaning, responds to stimuli, and can be the victim of manipulation or the beneficiary of intervention. But the fortunes of names, a cultural practice par excellence, doesn't fit that model. Names change with the times, yet they don't reflect other social trends, they aren't driven by role models, and the only influence from Madison Avenue is the wacky chain of events that made *Madison* the third most popular name for girls. To make sense of the phenomenon we have to look at the features of human nature that enter into a naming decision—the psychology of status, of parents, and of language—together with the products of the decisions of previous namers and the epidemiology of ideas, a field that barely exists.[55]

In his 1978 book *Micromotives and Macrobehavior,* the economist Thomas Schelling called attention to many social phenomena that are unplanned and often unwanted but that emerge from people making individual choices that affect the choices of others. One example is the way that a city can become segregated, not from an apartheid policy and not because anyone wants to live exclusively with people of their own race, but because no one wants to be in too small a minority in their neighborhood. As each family moves in or out to avoid this marginality, they become part of the neighborhood of others, affecting their decisions, and so on. Eventually all-black neighborhoods and all-white neighborhoods can emerge, neither of which anyone planned or wanted. Another example is the way that a lane of traffic can get tied up when each driver slows down for a few seconds to rubberneck at an accident scene—a bargain that none would have agreed to if they could have coordinated their behavior beforehand. Schelling notes how these patterns can arise whenever individual decisions are interdependent:

> If your problem is that there is too much traffic, you are part of the problem. If you join a crowd because you like crowds, you add to the crowd. If you withdraw your child from school because of the pupils he goes to school with, you remove a pupil that *they* go to school with. If you raise your voice to make yourself heard, you add to the noise that other people are raising their voices to be heard above. When you cut your hair short you change, ever so slightly, other people's impressions of how long people are wearing their hair.[56]

In his recent bestseller *The Tipping Point,* the journalist Malcolm Gladwell applies Schelling's idea to recent social trends such as changes in rates of literacy, crime, suicide, and teenage smoking. In each case the conventional wisdom attributed the trend to external social forces such as advertising, government programs, or role models. And in each case the trend was really driven by an internal dynamic of personal choices and influences and their feedback. The naming of babies, and of things in general, is another example in which a large-scale social phenomenon—the composition of a language—emerges unpredictably out of many individual choices that impinge on one another.

A name seems like such a simple thing—a link between a sound and a meaning, shared in a community. Yet the more we concentrate on how names work, the wider the sphere of human experience that opens up before us. In previous books I have written about how names are represented in the head of a single person. I marveled at how many words a person knows, how quickly children learn them, how elegantly they are composed out of parts, and how easily the brain recognizes them. In earlier chapters of this book we saw the precision and abstractness of the conceptual structure that captures a word's sense. Now we see how names entangle us in the world *outside* our heads. A name gets its meaning from a namer who chooses a sound to point to an individual or kind, and who passes it down a chain of speakers who intend to use it in the same way. Each of these steps embroils us in an irony. The act of pointing and the intent to replicate it connect us to reality, not just to our ideas about reality, though one might have thought we have no way of telling them apart. And the choice of a sound connects us to society in a way that encapsulates the great contradiction in human social life: between the desire to fit in and the desire to be unique.

7

THE SEVEN WORDS YOU CAN'T SAY
ON TELEVISION

Freedom of speech is a foundation of democracy, because without it citizens can't share their observations on folly and injustice or collectively challenge the authority that maintains them. It's no coincidence that freedom of speech is enshrined in the first of the ten amendments to the Constitution that make up the Bill of Rights, and is given pride of place in other statements of basic freedoms such as the Universal Declaration of Human Rights and the European Convention on Human Rights.

Just as clearly, freedom of speech cannot be guaranteed in every circumstance. The U.S. Supreme Court recognizes five kinds of unprotected speech, and four of the exclusions are compatible with the rationale for enshrining free speech as a fundamental liberty. Fraud and libel are not protected, because they subvert the essence of speech that makes it worthy of protection, namely, to seek and share the truth. Also unprotected are advocacy of imminent lawless behavior and "fighting words," because they are intended to trigger behavior reflexively (as when someone shouts "Fire!" in a crowded theater) rather than to exchange ideas.

Yet the fifth category of unprotected speech—obscenity—seems to defy justification. Though some prurient words and images are protected, others cross a vague and contested boundary into the category of "obscenity," and the government is free to outlaw them. And in broadcast media, the state is granted even broader powers, and may ban sexual and scatological language that it classifies as mere "indecency." But why would a democracy sanction the use of government force to deter the uttering of words for two activities—sex and excretion—that harm no one and are inescapable parts of the human condition?

In practice as well as in theory, the prosecution of obscene speech is a puzzle. Throughout history people have been tortured and killed for criticizing their leadership, and that is the fate of freethinkers in many parts of the world today. But in liberal democracies the battle for free speech has mostly been won. Every night millions of people watch talk-show hosts freely ridiculing the intelligence and honesty of the leaders of their nation. Of course, eternal vigilance is the price of liberty, and civil libertarians are rightly concerned with potential abridgments of speech such as those in copyright law, university speech codes, and the USA Patriot Act. Yet for the past century the most famous legal battles over free speech have been joined not where history would lead us to expect them—in efforts to speak truth to power—but in the use of certain words for copulation, pudenda, orifices, and effluvia. Here are some prominent cases:

- In 1921, a magazine excerpt from James Joyce's *Ulysses* was declared obscene by an American court, and the book was banned in the United States until 1933.
- D. H. Lawrence's *Lady Chatterley's Lover,* written in 1928, was not published in the United Kingdom until 1960, whereupon Penguin Books was prosecuted (unsuccessfully) under the Obscene Publications Act of 1959.
- *Lady Chatterley* was also banned from the United States, together with Henry Miller's *Tropic of Cancer* and John Cleland's *Fanny Hill.* In a series of court decisions reflecting the changing sexual mores of the 1960s, the bans were overturned, culminating in a Supreme Court ruling in 1973.
- Between 1961 and 1964, the comedian Lenny Bruce was repeatedly arrested for obscenity and banned from performing in many cities. Bruce died in 1966 while appealing a four-month sentence imposed by a New York court, and was finally pardoned by Governor George Pataki thirty-seven years after his death.
- The Pacifica Radio Network was fined in 1973 by the Federal Communications Commission for broadcasting George Carlin's monologue "Seven Words You Can Never Say on Television." The Supreme Court upheld the action, ruling that

the FCC could prohibit "indecent" language during hours when children might stumble upon a broadcast.

- The FCC fined Howard Stern's popular radio program repeatedly, prompting Stern to leave broadcast radio in 2006 for the freedom of satellite radio. Many media experts predicted that it would be a tipping point in the popularity of that medium.

Other targets of sanctions include Kenneth Tynan, John Lennon, Bono, 2 Live Crew, Bernard Malamud, Eldridge Cleaver, Kurt Vonnegut, Eric Idle, and the producers of *Hair* and *M*A*S*H*.[1]

The persecution of swearers has a long history. The third commandment states, "Thou shalt not take the name of the Lord thy God in vain," and Leviticus 24:16 spells out the consequences: "He that blasphemeth the name of the Lord shall be put to death." To be sure, the past century has expanded the arenas in which people can swear. As early as 1934, Cole Porter could pen the lyric "Good authors, too, who once knew better words / Now only use four-letter words / Writing prose. Anything goes." Most of the celebrity swearers of the twentieth century prevailed (if only posthumously), and many recent entertainers, such as Richard Pryor, Eve Ensler, and the cast of *South Park,* have cussed with impunity. Yet it's still not the case that anything goes. In 2006 George W. Bush signed into law the Broadcast Decency Enforcement Act, which increased the fines for indecent language tenfold and threatened repeat offenders with the loss of their license.

Taboo language, then, enters into a startling array of human concerns, from capital crimes in the Bible to the future of electronic media. It stakes out the frontier of free speech in liberal democracies, not only in government control of the media but in debates over hate speech, fighting words, and sexual harassment. And of course it figures in our everyday judgments of people's character and intentions.

Whether they are referred to as swearing, cursing, cussing, profanity, obscenity, indecency, vulgarity, blasphemy, expletives, oaths, or epithets; as dirty, four-letter, or taboo words; or as bad, coarse, crude, foul, salty, earthy, raunchy, or off-color language, these expressions raise many puzzles for anyone interested in language as a window into human nature. The fear and loathing are not triggered by the concepts themselves, because the organs and activities they name have hundreds of polite synonyms. Nor are they triggered by the words' sounds, since many of them have respectable homonyms in names for animals, actions, and even people. The unprintable can become printable with a hyphen or asterisk, and the unsayable sayable with the flip of a vowel or consonant. Something about the *pairing* of certain meanings and sounds has a potent effect on people's emotions.

Shakespeare wrote, "But words are words. I never yet did hear / That the bruised heart was pierced through the ear." Yet most people don't see it that way. The FCC and network censors are not inveterate prudes; they are responding to a huge constituency of listeners who light up a station switchboard like a Christmas tree when an actor or guest lets slip an obscenity. To these guardians of decency, profanity is self-evidently corrupting, especially to the young. This argument is made in spite of the fact that everyone is familiar with the words, including most children, and that no one has ever spelled out how the mere hearing of a word could corrupt one's morals.

To the libertines, what's self-evident is that linguistic taboos are absurd. A true moralist, they say, should hold that it's violence and inequality that are "obscene," not sex and excretion. And the suppression of plain speaking about sex only leads to teen pregnancy, sexually transmitted disease, and the displacement of healthy sexual energy into destructive behavior. This air of progressiveness helped to make Bruce a martyr among artists and intellectuals: "A moral conscience second to none," wrote the critic Ralph J. Gleason; "Saint Lenny, I should call him; he died for our sins," wrote the performance artist Eric Bogosian.[2]

Yet since the 1970s, some of the progressive constituencies that most admired Bruce have imposed linguistic taboos of their own. During the O. J. Simpson trial, the prosecutor Christopher Darden referred to the *n*-word as "the dirtiest, filthiest, nastiest word in the English language, and it has no place in a courtroom." Yet it has repeatedly found its way into the courtroom, most famously in the Simpson trial to prove that a police officer was a racist, and in other trials to determine whether a person can be fired for using it, or excused for assaulting someone else who uses it.[3] And in "the new Victorianism," casual allusions to sex, even without identifiable sexism, may be treated as forms of sexual harassment, as in Clarence Thomas's remarks about porn stars and pubic hair.[4] So even people who revile the usual bluenoses can become gravely offended when they hear words on their own lists of taboos.

Another puzzle about swearing is the range of topics that are the targets of taboo.[5] The seven words you can never say on television refer to sexuality and excretion: they are names for feces, urine, intercourse, the vagina, breasts, a person who engages in fellatio, and a person who acts out an Oedipal desire. But the capital crime in the Ten Commandments comes from a different subject, theology, and the taboo words in many languages refer to perdition, deities, messiahs, and their associated relics and body parts. Another semantic field that spawns taboo words across the world's languages is death and disease, and still another is disfavored classes of people such as infidels, enemies, and subordinate ethnic groups. But what could these concepts—from mammaries to messiahs to maladies to minorities—possibly have in common?

A final puzzle about swearing is the crazy range of circumstances in which we do it. There is cathartic swearing, as when we hit our thumb with a hammer or knock over a glass of beer. There are imprecations, as when we suggest a label or offer advice to someone who has cut us off in traffic. There are vulgar terms for everyday things and activities, as when Bess Truman was asked to get the president to say *fertilizer* instead of *manure* and she replied, "You have no idea how long it took me to get him to say *manure*." There are figures of speech that put obscene words to other uses, such as the barnyard epithet for insincerity, the army acronym *snafu*, and the gynecological-flagellative term for uxorial dominance. And then there are the adjective-like expletives that salt the speech and split the words of soldiers, teenagers, Australians, and others affecting a breezy speech style.

This chapter is about the puzzle of swearing—the strange shock and appeal of words like *fuck, screw,* and *come; shit, piss,* and *fart; cunt, pussy, tits, prick, cock, dick,* and *asshole; bitch, slut,* and *whore; bastard, wanker, cocksucker,* and *motherfucker; hell, damn,* and *Jesus Christ; faggot, queer,* and *dyke;* and *spick, dago, kike, wog, mick, gook, kaffir,* and *nigger.* We will explore the biological roots of swearing, the areas of experience that spawn taboo words, and the occasions on which people put them to use. Finally I will ask why these words are felt to be not just unpleasant but taboo—why merely hearing or reading them is felt to be corrupting—before offering some reflections on what we should do about swearing.

POTTYMOUTHS

As with the rest of language, swearing can be called universal, though only with qualifications.[6] Certainly the exact words and concepts considered taboo can vary across times and places. During the history of a language, we often see clean words turning dirty and dirty words turning clean.[7] Most English speakers today would be surprised to read in a medical textbook that "in women the neck of the bladder is short, and is made fast to the cunt," yet the *Oxford English Dictionary* cites this from a fifteenth-century source. In documenting such changes the historian Geoffrey Hughes has noted, "The days when the dandelion could be called the *pissabed,* a heron could be called a *shitecrow* and the windhover could be called the *windfucker* have passed away with the exuberant phallic advertisement of the codpiece."[8] The changing fortunes of taboo words can buffet the reception of a work of literature. *Huckleberry Finn,* for example, has been the target of repeated bans in American schools because *nigger,* though never a respectful term, is far more incendiary today than it was in the time and place in which Mark Twain wrote.

Words can shed their taboos over time, too. When Eliza Doolittle chirped "Not bloody likely" at an upper-class tea in *Pygmalion,* she scandalized not only her fictional companions but the audiences who saw the play when it opened in 1914. Yet by the time it was adapted into the musical *My Fair Lady* in 1956, *bloody* had become so unexceptionable that the scriptwriters worried that the humor would be lost on the audience, and added the scene in which Eliza is taken to the Ascot races and shouts at a horse, "Move your bloomin' arse!" Many parents today are horrified when their children come home from school innocently using the verbs *suck, bite,* and *blow,* unaware of their origin as words for fellatio. But those parents probably gave just as little thought to their own use of the now-innocuous *sucker* (from *cocksucker*), *jerk* (from *jerk off*), and *scumbag* (a condom). Progressive comedians have tried to help this process along by repeating obscenities to the point of desensitization (a process that psycholinguists call semantic satiation) or by momentarily turning into linguistics professors and calling attention to the principle of the arbitrariness of the sign. Here is an excerpt from one of Lenny Bruce's best-known routines:

> Tooooooo is a preposition. To is a preposition. Commmmmme is a verb. To is a preposition. Come is a verb. To is a preposition. Come is a verb, the verb intransitive. To come. To come. . . . It's been like a big drum solo. To come to come, come too come too, to come to come uh uh uh uh uh um um um um um uh uh uh uh uh uh—TO COME! TO COME! TO COME! TO COME! Did you come?

Did you come? Good. Did you come good? Did you come good? Did you come? Good. To. Come. To. Come—Didyoucomegood? Didyoucomegooddidyoucomegood?⁹

And this is from Carlin's monologue on the "Seven Words":

> Shit, Piss, Fuck, Cunt, Cocksucker, Motherfucker, and Tits, wow. Tits doesn't even belong on the list, you know. It's such a friendly sounding word. It sounds like a nickname. "Hey, Tits, come here. Tits, meet Toots, Toots, Tits, Tits, Toots." It sounds like a snack, doesn't it? Yes, I know, it is, right. But I don't mean the sexist snack, I mean, New Nabisco Tits. The new Cheese Tits, and Corn Tits and Pizza Tits, Sesame Tits Onion Tits, Tater Tits, yeah.

Tits is now clean enough to have been left out of the Clean Airwaves Act and to be printable in "The Gray Lady," the *New York Times*. But many words stay taboo for centuries, and which words become cleaner or dirtier is as capricious as the rise and fall of *Steve*.¹⁰

Similar desensitization campaigns have been aimed at epithets for women and minorities, who often try to "reclaim" the words by using them conspicuously among themselves. Thus we have NWA (Niggaz With Attitude, a hiphop group); Queer Nation, queer studies, and *Queer Eye for the Straight Guy;* Dykes on Bikes (a cycling group for lesbians) and www.classicdykes.com; and the Phunky Bitches, a "real-time community of women (and men) who are into live music, travel, and a host of other interests." I have never heard of a temple brotherhood meeting at which the attendees greet each other with "What's happenin', kike!" but in the 1970s the novelist Kinky Friedman led a country band called the Texas Jewboys, and there is a hip magazine for young Jewish readers called *Heeb*. At the same time, these terms have not been neutralized so much as flaunted as a sign of defiance and solidarity, precisely because they *are* still offensive in the language community at large. Woe betide the outsider who misunderstands this, like the Hong Kong detective played by Jackie Chan in *Rush Hour* who innocently follows the lead of his African American partner and greets the black patrons of a Los Angeles bar with "Whassup, my nigger," thereby starting a small riot.

The punch of specific words can vary even more from one language to another.¹¹ In Québecois French, *merde* (shit) is far milder than its English equivalent, a bit closer to *crap,* and most speakers are at best dimly aware that *con* (idiot) originally meant *cunt*. But some of the worst things you can say to someone are *Tabernac!* (tabernacle), *Calisse!* (chalice), and *Sacrement!* (sacrament). In 2006 the Catholic Church tried to reclaim these words by splashing them on billboards with their original religious definitions underneath. (One columnist sighed, "Is nothing sacred?") Religious profanity is common in other Catholic regions, as it was in England before the Reformation, when sexual and scatological terms started to take over.[12]

But despite the variation across time and space, it's safe to say that most languages, probably all, have emotionally laden words that may not be used in polite conversation. Perhaps the most extreme example is Djirbal, an Aboriginal language of Australia, in which *every* word is taboo when spoken in the presence of mothers-in-law and certain cousins. Speakers have to use an entirely different vocabulary (though the same grammar) when those relatives are around. In most other languages, the taboo words are drawn from the same short list of topics from which English and French get their curses: sex, excretion, religion, death and infirmity, and disfavored groups.[13]

Claims that profanity is lacking altogether in a particular language have to be taken with a grain of salt. It's true that in many places if you ask speakers to list their profanities, they may demur. But swearing and hypocrisy go hand in hand, to the extent that some personality questionnaires include items like "I sometimes swear" as a check for lying. In *Expletive Deleted: A Good Look at Bad Language,* the linguist Ruth Wajnryb reports:

> One of my informants, an Englishman married to a Japanese woman, asked his wife the questions I was using to elicit data about Japanese. She told him she couldn't help because she didn't know any Japanese swear words. This she said, mind you, in wide-eyed innocence to a husband who was fully aware, as she was aware that he was, from firsthand experience of her skills in that department.[14]

A review in a periodical called *Maledicta: The International Journal of Verbal Aggression* contains an extensive list of Japanese sexual insults and vulgar terms, and the other cross-cultural surveys appearing in that journal also have a familiar ring to them.[15]

Taboo speech is part of a larger phenomenon known as word magic.[16] Though one of the foundations of linguistics is that the pairing between a sound and a meaning is arbitrary, most humans intuitively believe otherwise. They treat the name for an entity as part of its essence, so that the mere act of uttering a name is seen as a way to impinge on its referent. Incantations, spells, prayers, and curses are ways that people try to affect the world through words, and taboos and euphemisms are ways that people try *not* to affect it. Even hardheaded materialists find themselves knocking wood after mentioning a hoped-for event, or inserting *God forbid* after mentioning a feared one, perhaps for the same reason that Niels Bohr hung a horseshoe above his office door: "I hear that it works even if you don't believe in it."

THE BLASPHEMING BRAIN

The ubiquity and power of swearing suggest that taboo words may tap into deep and ancient parts of the emotional brain. We saw in chapter 1 that words have not just a denotation but a connotation: an emotional coloring distinct from what the word literally refers to, as in *principled* versus *stubborn* and *slender* versus *scrawny*. The difference is reminiscent of the way that taboo words and their synonyms differ, such as *shit* and *feces, cunt* and *vagina,* or *fucking* and *making love.* Long ago psycholinguistics identified the three main ways in which words' connotations vary: good versus bad, weak versus strong, and active versus passive.[17] *Hero,* for example, is good, strong, and active; *coward* is bad, weak, and passive; and *traitor* is bad, weak, and active. Taboo words cluster at the very bad and very strong edges of the space, though there are surely other dimensions to connotation as well.

Are connotations and denotations stored in different parts of the brain? It's not implausible. The mammalian brain contains, among other things, the limbic system, an ancient network that regulates motivation and emotion, and the neocortex, the crinkled surface of the brain, which ballooned in human evolution and which is the seat of perception, knowledge, reason, and planning. The two systems are interconnected and work together, but it's not far-fetched to suppose that words' denotations are concentrated in the neocortex, especially in the left hemisphere, whereas their connotations are spread across connections between the neocortex and the limbic system, especially in the right hemisphere.[18]

A likely suspect within the limbic system is the amygdala, an almond-shaped organ buried at the front of the temporal lobe of the brain (one on each side), which helps invest memories with emotion.[19] A monkey whose amygdalas have been removed can learn to recognize a new shape, like a striped triangle, but has trouble learning that the shape foreshadows an unpleasant event like an electric shock. In humans the amygdala "lights up"—it shows greater metabolic activity in brain scans—when the person sees an angry face or an unpleasant word, especially a taboo word.[20] Well before psychologists could scan the working brain, they could measure the emotional jolt from a fraught word by strapping an electrode on a person's finger and measuring the change in the skin conductance caused by the sudden wave of sweat. The skin response accompanies activity in the amygdala, and like the activity recorded from the amygdala itself, it can be triggered by taboo words.[21] The emotional flavoring of words seems to be picked up in childhood: bilingual people often feel that their second language is not as piquant as their first, and their skin reacts more to hearing taboo words and reprimands in their first language than in their second.[22]

The involuntary shudder set off by hearing or reading a taboo word comes from a basic feature of the language system: understanding the meaning of a word is automatic. It's not just that we don't have earlids to shut out unwanted sounds, but that once a word is seen or heard we are incapable of treating it as a squiggle or noise but reflexively look it up in memory and respond to its meaning, including its connotation. The classic demonstration is the Stroop effect, found in every introductory psychology textbook and the topic of more than four thousand scientific papers. People are asked to look through a list of letter strings and to say aloud the color of the ink in which each one is printed. Try it with this list, saying "black," "white," or "gray" for each item in turn from left to right:

word word word **word** **word** word

It should be pretty easy. Now this is even easier:

gray white **black** white **black** gray

But this is much, much harder:

white black gray black gray **white**

The explanation is that among literate adults, reading a word is such an overlearned skill that it has become mandatory: you can't will the process "off," even when you're trying to ignore the words so you can pay attention to the ink. That's why you're helped along when the experimenters arrange the ink into a word that also names its color, and slowed down when they arrange it into a name for a different color. A similar thing happens with *spoken* words. When people have to name color patches like this:

the task becomes much harder when a voice over headphones recites a sequence of distracting color words like "black, white, gray, white, gray, black."[23]

Now, taboo words are especially effective at snatching a reader's attention. You can feel the effect yourself in a Stroop test. Try naming the color of the ink in each of these words:

cunt shit **fuck** **tits** piss asshole

The psychologist Don MacKay has done the experiment, and found that people are indeed slowed down by an involuntary boggle as soon as the eyes alight on each word.[24] The upshot is that a speaker or writer can use a taboo word to evoke an emotional response in an audience quite against their wishes.

Some companies have exploited this effect by giving their products names that are similar enough to a taboo word to grab people's attention, such as the restaurant chain called Fuddruckers, the clothing brand called FCUK (French Connection UK), and the movie called *Meet the Fokkers*. Involuntary responses to taboo words can actually shape a language over the course of its history because of a linguistic version of Gresham's Law: bad words drive good words out of circulation. People often avoid using innocent terms that they fear might be misheard as profanity. *Coney,* an old name for "rabbit" that rhymes with *honey,* dropped out of use in the late nineteenth century, probably because it sounded too much like *cunt.*[25] The same is happening to the polite senses of words like *cock, prick, pussy, booty,* and *ass* (at least in America; in Britain the rude word is still *arse*). People named *Koch, Fuchs,* and *Lipschitz* often change their surnames, as did the family of Louisa May Alcott, formerly Alcox. In 1999, an aide to the mayor of Washington, D.C., resigned after describing his budget as *niggardly* at a staff meeting. A staffer had taken umbrage, even though *niggard* is a Middle English word meaning "miser" and has nothing to do with the epithet based on *negro,* the Spanish word for black, which came into English centuries later.[26] Unfair though that may be, both to the aide and to the word, *niggardly* is doomed. So are the original senses of *queer* and *gay.*

Swearing aloud, like hearing the swear words of others, taps the deeper and older parts of the brain. Aphasia, a loss of articulate language, is typically caused by damage to the cortex and the underlying white matter along the horizontal cleft (the Sylvian fissure) in the brain's left hemisphere.[27] For almost as long as neurologists have studied aphasia, they have noticed that patients can retain the ability to swear.[28] A case study of a British aphasic recorded him as repeatedly saying "Bloody hell," "Fuck off," "Fucking fucking hell cor blimey," and "Oh you bugger." The neurologist Norman Geschwind studied an American patient whose entire left hemisphere had been surgically removed because of brain cancer. The patient couldn't name pictures, produce or understand sentences, or repeat polysyllabic words, yet in the course of a five-minute interview he said "Goddammit" seven times, and "God!" and "Shit" once apiece.[29]

The survival of swearing in aphasia suggests that taboo epithets are stored as prefabricated formulas in the right hemisphere.[30] Such formulas lie at the opposite end of a continuum from propositional speech, in which combinations of words express combinations of ideas according to grammatical rules. It's not that the right hemisphere contains a profanity module, but that its linguistic abilities are confined to memorized formulas rather than rule-governed combinations. A word is the quintessential memorized chunk, and in many people the right hemisphere has a respectable vocabulary of words, at least in comprehension. The right hemisphere also can sometimes store idiosyncratic counterparts to rule-governed forms such as irregular verbs.[31] Often it commands longer memorized formulas as well, such as song lyrics, prayers, conversation fillers like *um, boy,* and *well yes,* and sentence starters like *I think* and *You can't.*

The right hemisphere may be implicated in swearing for another reason: it is more heavily involved in emotion, especially negative emotion.[32] Yet it may not be the cerebral cortex in the right hemisphere that initiates epithets but an evolutionarily older brain structure, the basal ganglia.[33] The basal ganglia are a set of clusters of neurons buried deep in the front half of the brain. Their circuitry receives inputs from many other parts of the brain, including the amygdala and other parts of the limbic system, and loops back to the cortex, primarily the frontal lobes. One of their functions is to package sequences of movements, or sequences of reasoning steps, into chunks that are available for further combining when we're learning a skill. Another is to inhibit the execution of the actions packaged into these chunks.[34] Components of the basal ganglia inhibit one another, so damage to different parts can have opposite effects. Degeneration of one part of the basal ganglia can cause Parkinson's disease, marked by tremors, rigidity, and difficulty initiating movement. Degeneration of another part can cause Huntington's disease, resulting in chorea or uncontrolled movements.

The basal ganglia, with their role as packagers and inhibitors of behavior, have been implicated in swearing by two trails of evidence. One is a case study of a man who suffered a stroke in the right basal ganglia, leaving him with a syndrome that is the mirror image of classic aphasia.[35] He could converse fluently in grammatical sentences, but couldn't sing familiar songs, recite well-practiced prayers and blessings, or swear— even when the beginning of a curse was given to him and he only had to complete it.

The basal ganglia have a far more famous role in swearing, thanks to a syndrome that was obscure to most people until the 1980s, when it suddenly was featured in dozens of television plots: Gilles de la Tourette Syndrome, Tourette syndrome or Tourette's for short.[36] Tourette syndrome is a poorly understood neurological condition linked to abnormalities, partly hereditary, in the basal ganglia. As any couch potato knows, its most florid symptom is a vocal tic consisting of shouted obscenities, taboo ethnic terms, and other kinds of verbal abuse.[37] This symptom is called *coprolalia* (dung speech), from a Greek root also found in *coprophilous* (living in dung), *coprophagy* (feeding on dung), and *coprolite* (fossilized dinosaur poop). In fact coprolalia occurs in only a minority of people with Tourette syndrome; the more common tics are blinks, twitches, throat-clearing sounds, and repeated words or syllables.

Coprolalia shows off the full range of taboo terms, and embraces similar meanings in different languages, suggesting that swearing really is a coherent neurobiological phenomenon. A recent literature review lists the following words from American Tourette's patients, from most to least frequent:[38]

> fuck, shit, cunt, motherfucker, prick, dick, cocksucker, nigger, cockey, bitch, pregnant-mother, bastard, tits, whore, doody, penis, queer, pussy, coitus, cock, ass, bowel movement, fangu (fuck in Italian), homosexual, screw, fag, faggot, schmuck, blow me, wop

Patients may also produce longer expressions like *Goddammit, You fucking idiot, Shit on you,* and *Fuck your fucking fucking cunt.* A list from Spanish-speaking patients includes *puta* (whore), *mierda* (shit), *cono* (cunt), *joder* (fuck), *maricon* (fag), *cojones* (balls), *hijo de puta* (son of a whore), and *hostia* (host, the wafer in a communion ceremony). A list from Japan includes *sukebe* (lecherous), *chin chin* (cock), *bakatara* (stupid), *dobusu* (ugly), *kusobaba* (shitty old woman), *chikusho* (son of a whore), and an empty space in the list discreetly identified as "female sexual parts." There has even been a report of a deaf sufferer of Tourette's who produced "fuck" and "shit" in American Sign Language.

People with Tourette's experience their outbursts not as literally involuntary but as a response to an overpowering urge, much like an irresistible itch or a mounting desire to blink or yawn. This tug-of-war between an unwanted impulse and the forces of self-control is reminiscent of one of the symptoms of obsessive-compulsive disorder (OCD) called horrific temptations—the obsessive fear that one might do something awful such as shouting "Fire!" in a crowded theater or pushing someone off a subway platform. Like Tourette's, which it often accompanies, OCD seems to involve an imbalance between the brake pedal and accelerator circuits in the basal ganglia. It suggests that one of the roles of the basal ganglia is to designate certain thoughts and desires as unthinkable—taboo—in order to keep them in check. By tagging, encapsulating, and inhibiting these thoughts, the basal ganglia solve the paradox that you have to think the unthinkable in order to know what you're not supposed to be thinking—the reason that people have trouble following the instruction "Don't think of a polar bear."[39] Ordinarily the basal ganglia can hide our bad thoughts and actions with a Don't-Go-There designation, but when they are weakened, the lock-boxes and safety catches can break down, and the thoughts we tag as unthinkable or unsayable assert themselves.

In unimpaired people, the so-called executive systems of the brain (comprising the prefrontal cortex and another part of the limbic system, the anterior cingulate cortex) can monitor behavior emanating from the rest of the brain and override it in midstream. This may be the origin of the truncated profanities that we use in polite company and which serve as the strongest epithets that pass the lips of vicars and maiden aunts when they stub their toes. Every one of the standard obscenities offers a choice of bowdlerized alternatives:[40]

For *God:* egad, gad, gadzooks, golly, good grief, goodness gracious, gosh, Great Caesar's ghost, Great Scott

For *Jesus:* gee, gee whiz, gee willikers, geez, jeepers creepers, Jiminy Cricket, Judas Priest, Jumpin' Jehoshaphat

For *Christ:* crikes, crikey, criminy, cripes, crumb

For *damn:* dang, darn, dash, dear, drat, tarnation (from *eternal damnation*)

For *goddam:* consarn, dadburn, dadgum, doggone, goldarn

For *shit:* shame, sheesh, shivers, shoot, shucks, squat, sugar

For *fuck* and *fucking:* fiddlesticks, fiddledeedee, foo, fudge, fug, fuzz; effing, flaming, flipping, freaking, frigging

For *bugger:* bother, boy, brother

For *bloody:* blanking, blasted, blazing, bleeding, bleeping, blessed, blighter, blinding, blinking, blooming, blow

In *Pygmalion,* Henry Higgins is admonished by his housekeeper not to swear in Eliza's presence:

MRS. PEARCE: . . . there is a certain word I must ask you not to use. The girl has just used it herself because the bath was

too hot. It begins with the same letter as *bath*. She knows
no better: she learnt it at her mother's knee. But she must
not hear it from your lips.

HIGGINS [*loftily*]: I cannot charge myself with having ever
uttered it, Mrs. Pearce. [*She looks at him steadfastly. He
adds, hiding an uneasy conscience with a judicial air*]
Except perhaps in a moment of extreme and justifiable
excitement.

MRS. PEARCE: Only this morning, sir, you applied it to your
boots, to the butter, and to the brown bread.

HIGGINS: Oh, that! Mere alliteration, Mrs. Pearce, natural to a
poet.

The devices that are natural to a poet are the source of most of the euphemisms for taboo words. Alliteration and assonance figure in the rerouted
profanities in the list we just saw. Rhyme gives us *ruddy* for *bloody, son of a
gun* for *son of a bitch,* and the dozens of substitutions for taboo words in
Cockney slang, like *raspberry* for *fart* (from *raspberry tart*) and *Friar* for
fuck (from *Friar Tuck*). It also led to the stereotypical French expletive *Sacre
bleu!* from *Sacre Dieu.*

Poetic devices generally repeat one of the mental structures that organize words in our minds, such as onsets, rimes, and codas.[41] Phonologists
have also identified structures that are more abstract than these. The syllables making up a word are attached to a skeleton that defines the word's
rhythmic meter and its decomposition into morphemes.[42] When parts of
a linguistic skeleton are repeated in poetry or rhetoric, we have the device
called structural parallelism (as in the Twenty-third Psalm's "He maketh
me to lie down in green pastures / He leadeth me beside the still waters").
In the realm of swearing, we see structural parallelism in the numerous
euphemisms for *bullshit* that share only its metrical and morphological
structure. Many terms for insincerity are compounds made of two stressed
words, either monosyllables or trochees, with primary stress on the first
one:

applesauce, balderdash, blatherskite, claptrap, codswallop, flap-
doodle, hogwash, horsefeathers, humbug, moonshine,
poppycock, tommyrot

Another fertile ground for terms of abuse is phonetic symbolism. Imprecations tend to use sounds that are perceived as quick and harsh.[43] They tend to be monosyllables or trochees, and contain short vowels and stop consonants, especially *k* and *g:*

> fuck, cock, prick, dick, dyke, suck, schmuck, dork, punk, spick,
>   mick, chink, kike, gook, wog, frog, fag
> pecker, honky, cracker, nigger, bugger, faggot, dago, paki

(In the 1970s a friend of mine saw a bumper sticker reading NO NUKES, then an unfamiliar term, and thought it was a racist slogan!) Hughes notes, "While it may be objected, quite validly, that most swearing makes no attempt at originality, . . . certain affinities with poetry can be observed. In both fields the language used is highly charged and very metaphorical; extreme, pointed effects are created by alliteration or by playing off different registers of the word-hoard against each other, and rhythm is very important."[44]

THE SEMANTICS OF SWEARING:
THOUGHTS ABOUT GODS, DISEASE, FILTH, AND SEX

Now that we have taken a tour of the linguistic, psychological, and neurological underpinnings of swearing, can we identify a common thread in its meaning and use? The most obvious thread is strong negative emotion. Thanks to the automatic nature of speech perception, a taboo word kidnaps our attention and forces us to consider its unpleasant connotations. That makes all of us vulnerable to a mental assault whenever we are in earshot of other speakers, as if we were strapped to a chair and could be given a punch or a shock at any time. To understand swearing, then, we have to examine what kinds of thoughts are upsetting to people, and why one person might want to inflict these thoughts on another.

The historical root of swearing in English and many other languages is, oddly enough, religion.[45] We see this in the third commandment, in the popularity of *hell, damn, God,* and *Jesus Christ,* and in many of the terms for taboo language itself: *profanity* (that which is not sacred), *blasphemy* (literally "evil speech" but in practice disrespect toward a deity), and *swearing, cursing,* and *oaths,* which were originally secured by the invocation of a deity or one of his symbols, like the tabernacle, chalice, and wafer incongruously found in Catholic maledicta.

In English-speaking countries today, religious swearing barely raises an eyebrow. Gone with the wind are the days when people could be titillated by a character in a movie saying, "Frankly, my dear, I don't give a damn." If a character today is offended by such language, it's only to depict him as an old-fashioned prude. The defanging of religious taboo words is an obvious consequence of the secularization of Western culture. As G. K. Chesterton remarked, "Blasphemy itself could not survive religion; if anyone doubts that let him try to blaspheme Odin." To understand religious vulgarity, then, we have to put ourselves in the shoes of our linguistic ancestors, to whom God and Hell were real presences.

Swearing and oaths, in the literal sense of guarantees of one's promises, take us into the Strangelovian world of paradoxical tactics, where voluntary self-handicapping can work to one's advantage.[46] Say you need to make a promise. You may want to borrow money, and so must promise to return it. You may want someone to bear or support your child and forsake all others, and so must promise to be faithful in kind. You may want to do business with someone, and so must promise to deliver goods or services in the future in exchange for something you receive today. Why should the promisee believe you, knowing that it may be to your advantage to renege? The answer is that you can submit to a contingency that would impose a penalty on you if you did renege, ideally one so certain and severe that you would always do better to keep the promise than to back out. That way your partner no longer has to take you at your word; he can rely on your self-interest.

Nowadays we secure our promises with legal contracts that make us liable if we back out. We mortgage our house, giving the bank permission to repossess it if we fail to repay the loan. We submit to marriage laws, giving our spouses the right to alimony and a division of property if we desert or mistreat them. We post a bond, which we forfeit if we fail to come through on our obligations. But before we could count on a commercial and legal apparatus to enforce our contracts, we had to do our own self-handicapping. Children still bind their oaths by saying, "I hope to die if I tell a lie." Adults used to do the same by invoking the wrath of God, as in *May God strike me dead if I'm lying* and variations like *As God is my witness, Blow me down!, Shiver me timbers!,* and *God blind me!*—the source of the British *blimey.*[47]

Such oaths, of course, would have been more credible in an era in which people thought that God listened to their entreaties and had the power to carry them out. At the same time, every time someone reneges on an oath and is not punished by the big guy upstairs, it casts doubt on his existence, his potency, or at the very least how carefully he's paying attention. The earthly representatives of God would just as soon preserve the belief that he does listen and act in matters of importance, and so are unhappy about people diluting the brand by invoking God as the muscle behind their small-time deals. Hence the proscriptions against taking the name of the Lord in vain.

Short of literally asking God to serve as one's escrow agent, one can sanctify one's promises in a more tactful way, by bringing God into the discussion obliquely. (As we shall see in the next chapter, some of the most effective threats are veiled ones.) One can link one's credibility to appurtenances of God in which he presumably takes a continuing interest, such as his name, his symbols, his writings, and his body parts. Thus we have the phenomenon of "swearing by" and "swearing on." Even today, witnesses in American court proceedings have to swear on the Bible, as if an act of perjury undetected by the legal system would be punished by an eavesdropping and easily offended God. In earlier times Englishmen swore by gruesome reminders of the crucifixion: God's blood (*'sblood*), his nails, his wounds (hence *zounds*), his hooks (*gadzooks*), and his body (*odsbodikins*).[48] They also swore by the cross, the source of children's "Cross my heart." Perhaps the most creative was Oliver Cromwell, who wrote to the Church of Scotland, "I beseech you, in the bowels of Christ, think it possible you may be mistaken."

Even if these oaths aren't seen as literally having the power to bring down divine penalties for noncompliance, they signal a distinction between everyday assurances on minor favors and solemn pledges on weightier matters. The holiness of a religious relic is a social construction that depends on its being treated with awe and reverence by everyone in a community. This requires a collective mind control in which one doesn't look at, think about, or talk about a sacred thing casually. To bring the sacred into the discussion when making a promise is to force listeners to think about something they don't casually think about and hence to indicate that one means business. By the same token, if people swear by a sacred entity too freely, its sacredness is threatened by semantic inflation, and authorities who base their power on that sacredness will take steps to prevent that from happening. Laws against "swearing" may even have popular support, since every individual wants to keep the linguistic powder dry for occasions on which *he* wants to bind an oath, and not allow others to spoil it through overuse.

Though the invocation of blood and bowels to bind an oath may seem archaic, the psychology behind it is still with us. Even a parent without an iota of superstition would not say "I swear on the life of my child" lightly. The mere *thought* of murdering one's child for ulterior gain is not just unpleasant; it should be unthinkable if one is a true parent, and every neuron of one's brain should be programmed against it. Voluntarily thinking the thought is no small matter, and it's a kind of self-threat that can enhance the credibility of a promise. The literal unthinkability of betraying an intimate or ally is the basis of the psychology of taboo in general, and this is the mindset that is tapped in swearing on something sacred, whether it be a religious trapping or a child's life.[49] And thanks to the automatic nature of speech processing, the same sacred words that consecrate promises—the "oath binding" sense of *swearing*—may be used to attract attention, to shock, or to inflict psychic pain on a listener—the "dirty word" sense of *swearing*.

Religion also figures in the other ambiguous verb for taboo language, *cursing.* As we shall see, just about any misfortune or indignity can be wished upon someone in a curse, but Christianity has furnished execrators with a particularly disagreeable thought to inflict on their targets: the possibility that they might spend eternity in Hell. Today, *Go to hell!* and *Damn you!* are among our milder epithets, but they would have packed more of a wallop in an era in which people actually feared they might be sentenced forever to searing flames, agonizing thirst, terrifying ghouls, and bloodcurdling shrieks and groans. Perhaps the closest we can come to appreciating the original impact of wishes of damnation is to imagine someone looking us in the eye and saying, "I hope you are convicted of tax fraud and sentenced to twenty years in prison. I hope your cell is hot and humid and is crawling with roaches and reeks of urine and excrement. I hope you have three vicious cellmates who beat and sodomize you every night." And so on. When we consider how brutal cursing *could* be, and how brutal it must have been when most people believed in Hell, we should be grateful that most hotheads today confine themselves to a small lexicon of hackneyed scatological and sexual imprecations that were drained of their imagery long ago.

Another semantic field that has lost its sting is disease and pestilence, as in *A plague on both your houses!* (from *Romeo and Juliet*), *A pox on you!*, and the Polish-Yiddish *Cholerya!* (cholera). In an era of sanitation and antibiotics, it's hard to appreciate the power of these allusions. It helps to visualize the "Bring out your dead!" scene in *Monty Python and the Holy Grail,* or to read in a medical textbook about the pustules, hemorrhaging, eye ulcers, diarrhea, and other grisly symptoms of these diseases. The equivalent today might be "I hope you are trapped in a fire and get third-degree burns all over your body. I hope you suffer a stroke and spend your life drooling and twisted in a wheelchair. I hope you get bone cancer and waste away in front of your loved ones." Once again, cultural critics who see swearing as a sign of the coarsening of our culture should consider how mild our curses are by the standards of history. Tellingly, there is a hint of taboo in the name of our most dreaded malady, *cancer*. It has spawned euphemisms like *the big C, malignancy, neoplasm, mitotic figure,* and one that is still seen in many obituaries, *a long illness*.

Though we no longer swear about disease, we do swear about bodily effluvia and their orifices and acts of excretion. *Shit, piss,* and *asshole* are still unspeakable on network television and unprintable in most newspapers. The *New York Times,* for example, currently identifies a bestseller by the philosopher Harry Frankfurt as "*On Bull----*." *Fart* is barely more acceptable: the *Times* will print it as part of the ageist epithet *old fart* but not as the vernacular term for flatulence. *Ass* (or *arse*), *bum, snot,* and *turd* are also on the border of respectability.

Bloody is another word that calls to mind a bodily fluid. As with many taboo terms, no one really knows where it came from, because people tend not to set down their profanities in print. That has not stood in the way of people concocting various folk etymologies, as we saw in chapter 6 with "Fornication Under Consent of the King" and "Ship High in Transit." Hughes notes, "I am sure that I am not the first logophile to have been informed (on several occasions and with complete assurance) that the origin of *bloody* lies in the religious ejaculation *By our lady!*"[50] Not bloody likely, say the historians. Nor is *God's blood* the source. *Bloody* is probably another word that became taboo because it refers to an icky bodily substance, perhaps the blood that oozes from a wound, perhaps menstrual blood. Menstruation is the target of several Judeo-Christian taboos. An Orthodox Jew, for example, may not shake hands with a woman on the off chance that she is "unclean."

Some people have been puzzled about why *cunt* should be taboo. It is not just an unprintable word for the vagina but the most offensive epithet for a woman in America and a not-too-polite term for a man in Britain and the Commonwealth. One might have thought that in the male-dominated world of swearing the vagina would be revered, not reviled. After all, it has been said that no sooner does a boy come out of it than he spends the rest of his life trying to get back in. The puzzle becomes less mysterious if one imagines the connotations in an age before tampons, toilet paper, regular bathing, and antifungal drugs.

On the whole, the acceptability of taboo words is only loosely tied to the acceptability of what they refer to, but in the case of taboo terms for effluvia the correlation is fairly good. *Shit* is less acceptable than *piss,* which in turn is less acceptable than *fart,* which is less acceptable than *snot,* which is less acceptable than *spit* (which is not taboo at all). That's the same order as the acceptability of eliminating these substances from the body in public.[51]

The linguists Keith Allan and Kate Burridge tried to expand this observation by administering a Revoltingness Questionnaire to staff and students at their Australian universities.[52] Tied for first place were feces and vomit. Menstrual blood (among men) came next, followed by urine and semen. Then, in decreasing order of revoltingness, there was a three-way tie among flatulence, pus, and nasal mucus, followed by menstrual blood (among women), belched breath, skin parings, sweat, nail parings, breath, blood from a wound, hair clippings, breast milk, and tears. The correlation with vulgarity is far from perfect: though vomit and pus are decidedly revolting, they have no taboo terms in English. Nonetheless, the vulgar words for effluvia do cluster at the top end of the scale, including the taboo terms for semen such as *cum, spunk, gizzum, jizz,* and *cream.*

Words for effluvia are taboo in many cultures, and so are the effluvia themselves. The biologists Valerie Curtis and Adam Biran summarize the results of questionnaires given in Europe, India, and Africa: "Bodily secretions are the most widely reported elicitors of the disgust emotion. Feces appear on all of the lists, while vomit, sweat, spittle, blood, pus, and sexual fluids appear frequently."[53] Effluvia have an emotional charge that makes them figure prominently in voodoo, sorcery, and other kinds of sympathetic magic.[54] People in many cultures believe that a person can be harmed by mutilating or casting spells on his feces, saliva, blood, nails, and hair, and that a person can be protected from harm if those substances are cursed, buried, drowned, or otherwise ostentatiously discarded. The potency of these substances in people's minds also leads them to be used in medicines or charms, often in homeopathic or purified doses. The emotion of disgust and the psychology of sympathetic magic are entwined. The psychologists Paul Rozin and April Fallon have shown that modern Westerners respect the laws of voodoo in their own disgust reactions, such as recoiling from an object if it merely looks like a disgusting substance or has been in contact with one in the past.[55] Word magic simply extends this chain of associations by one link, and gives the *words* for effluvia a dreadful power as well.

The dread of effluvia, of course, can also be modulated, as it must be in sex, medicine, nursing, and the care of animals and babies. As we shall see, this desensitization is sometimes helped along with the use of euphemisms that play down the repellence of the effluvia.

The big deal that people ordinarily make out of effluvia—both the words and the substances—has puzzled many observers. As the religion scholar A. K. Reinhart puts it, "Pus, vomit, urination, menstruation, sexual fluids, and so on [are] all substances and acts that, for some reason, many cultures tend to see as repellent and, despite their constant presence in human life, as abnormal."[56] Curtis and Biran identify the reason.[57] It can't be a coincidence, they note, that the most disgusting substances are also the most dangerous vectors for disease. Feces are a route of transmission for the viruses, bacteria, and protozoans that cause at least twenty intestinal diseases, as well as ascariasis, hepatitis A and E, polio, amoebiasis, hookworm, pinworm, whipworm, cholera, and tetanus. Blood, vomit, mucus, pus, and sexual fluids are also attractive to pathogens as vehicles for getting from one body into another. In modern countries, flush toilets and garbage removal quickly separate us from our effluvia, but in the rest of the world they transmit millions of cases of disease every year. Even citizens of industrial countries may be quickly threatened with cholera and typhoid in times of war or natural disasters, such as the flooding in New Orleans in the wake of Hurricane Katrina in 2005.

The strongest component of the disgust reaction is a desire not to eat or touch the offending substance.[58] But it's also disgusting to *think* about effluvia, together with the body parts and activities that excrete them, and because of the involuntariness of speech perception, it's unpleasant to hear the words for them. The effluvia that evoke the strongest disgust reaction are viscous ones, but urine is also mildly disgusting, and the word *piss* is mildly taboo. Urine is not generally infectious, but it is, of course, a waste product that carries away metabolites and toxins that the body doesn't want, and thus it should not be appealing. Vermin make up a major class of disease vectors, and are widely considered disgusting.[59] Not surprisingly, they lend their names in English to verbal imprecations such as *rat, louse, worm, cockroach, insect,* and *slug,* though the words don't rise to the level of taboo. Why some of the words for unpleasant things are taboo in a particular culture and era, while others are not, is something of a mystery. Perhaps taboo terms have to be acquired in emotion-tinged settings in childhood. Or perhaps they are self-perpetuating, and remain taboo for as long as people treat them as taboo.

The other major source of taboo words is sexuality. Since the 1960s, many progressive thinkers have found these taboos to be utterly risible. Sex is a source of mutual pleasure, they reason, and should be cleansed of stigma and shame. Prudery about sexual language could only be a superstition, an anachronism, perhaps a product of spite, as in H. L. Mencken's definition of *puritanism* as "the haunting fear that someone, somewhere may be happy." Lenny Bruce ended his "Did you come?" routine by saying, "If anyone in this room finds that verb intransitive, *to come,* obscene, vile, vulgar—if it's really a hang-up to hear it and you think I'm the rankest for saying it—*you* probably can't come."
 Bruce was also puzzled by our most common sexual imprecation:

> What's the worst thing you can say to anybody? "Fuck you, Mister." It's really weird, because if I really wanted to hurt you I should say "Unfuck you, Mister." Because "Fuck you" is really *nice!* "Hello, Ma, it's me. Yeah, I just got back. Aw, fuck you, Ma! Sure, I mean it. Is Pop there? Aw, fuck you, Pop!"[60]

Part of the puzzlement comes from the strange syntax of *Fuck you,* which, as we shall see, does not in fact mean "Have sex." But it also comes from a modern myopia (particularly in young men) for how incendiary sexuality can be in the full sweep of human experience.

Consider two consenting adults who have just had sex. Has everyone had fun? Not necessarily. One partner might see the act as the beginning of a lifelong relationship, the other as a one-night stand. One may be infecting the other with a disease. A baby may have been conceived, whose welfare was not planned for in the heat of passion. If the couple is related, the baby may inherit two copies of a deleterious recessive gene and be susceptible to a genetic defect. There may be romantic rivals in the wings who would be enraged with jealousy if they found out, or a cuckolded husband in danger of raising another man's child, or a two-timed wife in danger of losing support for her own children. Parents may have marriage plans for one of the participants, involving large sums of money or an important alliance with another clan. And on other occasions the participants may not both be adults, or may not both be consenting.

Sex has high stakes, including exploitation, disease, illegitimacy, incest, jealousy, spousal abuse, cuckoldry, desertion, feuding, child abuse, and rape. These hazards have been around for a long time and have left their mark on our customs and our emotions. Thoughts about sex are likely to be fraught, and not entertained lightly. *Words* for sex can be even touchier, because they not only evoke the charged thoughts but implicate a sharing of those thoughts between two people. The thoughts, moreover, are shared "on the record," each party knowing that the other knows that he or she has been thinking about the sex under discussion. As we shall see in the next chapter, this embroils the dialogue in an extra layer of intrigue.

Evolutionary psychology has laid out the conflicts of interest that are inherent to human sexuality, and some of these conflicts play themselves out in the linguistic arena.[61] Plain speaking about sex conveys an attitude that sex is a casual matter, like tennis or philately, and so it may seem to the partners at the time. But the long-term implications may be more keenly felt by a wider circle of interested parties. Parents and other senior kin may be concerned with the thwarting of their own plans for the family lineage, and the community may take an interest in the illegitimate children appearing in their midst, and in the posturing and competition, sometimes violent, that can accompany sexual freedom. The ideal of sex as a sacred communion between a monogamous couple may be old-fashioned and even unrealistic, but it sure is convenient for the elders of a family and a society. It's not surprising to find tensions between individuals and guardians of the community over casual talk about sex (accompanied by hypocrisy among the guardians when it comes to their own casual sex).

Even keener than the sexual conflicts between young and old and between individual and society is the conflict between men and women. We are mammals, and have inherited the asymmetry that runs through that class: in every act of reproduction, females are committed to long stretches of pregnancy and lactation, while males can get away with a few minutes of copulation. A male can have more progeny if he mates with many females, whereas a female will not have more progeny if she mates with many males—though her offspring will do better if she has chosen a mate who is willing to invest in them or can endow them with good genes. Not surprisingly, in all cultures men pursue sex more eagerly, are more willing to have casual sex, and are more likely to seduce, deceive, or coerce to get sex.[62] All things being equal, casual sex works to the advantage of men, both genetically and emotionally. We might expect casual *talk* about sex to show the same asymmetry, and so it does. Men swear more, on average, and many taboo sexual terms are felt to be especially demeaning to women—hence the old prohibition of swearing "in mixed company."[63]

A male-female difference in tolerance for sexual language may call to mind the stereotype in which a Victorian woman who heard a coarse remark would raise her wrist to her forehead and swoon onto the fainting couch. But an unanticipated consequence of the second wave of feminism in the 1970s was a revived sense of offense at swearing, the linguistic companion to the campaign against pornography. Groucho Marx might be surprised to learn that today's universities and businesses have implemented his platform for running Freedonia in *Duck Soup:* No one's allowed to smoke, or tell a dirty joke. Many published guidelines on sexual harassment include "telling sexual jokes" in their definitions, and in 1993 the veteran *Boston Globe* journalist David Nyhan was forced to apologize and to donate $1,250 to a women's organization when a female staffer overheard him in the newsroom using the word *pussy-whipped* with a male colleague who declined his invitation to play basketball after work.[64] The feminist writer Andrea Dworkin, famous for her activism against pornography and her suggestion that all intercourse is rape, explicitly connected coarse sexual language to the oppression of women:

> Fucking requires that the male act on one who has less power and this valuation is so deep, so completely implicit in the act, that the one who is fucked is stigmatized. . . . In the male system, sex is the penis, the penis is sexual power, its use in fucking is manhood.[65]

While it's tempting to ridicule the backlash against sexual swearing as a throwback to Victorian daintiness, it remains true that an atmosphere of licentiousness may be less conducive to women's interests than to men's. In the decade between the sexual revolution of the early 1960s and the feminist revolution of the early 1970s, many works of popular culture celebrated the overthrow of puritanism with sympathetic portrayals of lascivious men (examples include works by Joe Orton, Tom Lehrer, Woody Allen, and the Rolling Stones, the James Bond movies, and *Rowan and Martin's Laugh-In*). Revisiting these works can make for painful watching and listening. Their exuberant leering, thought to be sophisticated and risqué at the time, seems misogynistic today, with depictions of women as bimbos and an amused tolerance of rape, harassment, and spousal abuse. (A song from the musical *Hair* began, "Sodomy, fellatio, cunnilingus, pederasty. Father, why do these words sound so nasty?"—showing an indulgence toward pedophilia that would be unthinkable today.) The brief glorification of lechery in middle-class culture, bookended at one end by a challenge of youth to age and the individual to society, and at the other by a challenge of women to men, exposes some of the conflicts of interest that charge the language of sex.

Though people are seeing, talking about, and having sex more readily today than they did in the past, the topic is still not free of taboo. Most people still don't copulate in public, swap spouses at the end of a dinner party, have sex with their siblings and children, or openly trade favors for sex. Even after the sexual revolution, we have a long way to go before "exploring our sexuality" to the fullest, and that means that people still set up barriers in their minds to block certain trains of thought. The language of sex can tug at those barriers.

FIVE WAYS TO CUSS

Now that we have visited the content of taboo language (its semantics), we can turn to the ways in which it is used (its pragmatics). Recall that the common denominator of the content of swearing is an emotional charge that people would rather not have running through their minds at the drop of a hat—a sense of awe (for God and his trappings), fear (for Hell and disease), disgust (for bodily effluvia), hatred (for traitors, heretics, and minorities), or depravity (for sexuality). Because speech perception is automatic, uttering a taboo word can force a listener's mind to go in a direction it ordinarily prevents itself from going in. This helps us to focus the question of how profanity is used. Why do speakers try to impose their wills on their listeners' minds in this way? There is no single answer, because people swear in at least five different ways: descriptively (*Let's fuck*), idiomatically (*It's fucked up*), abusively (*Fuck you, motherfucker!*), emphatically (*This is fucking amazing*), and cathartically (*Fuck!!!*). Let's look at each in turn.

Many of the puzzles around profanity come down to a single question: what is it about a taboo word that makes it different from a genteel synonym that refers to the same thing? What are people responding to so strongly, for example, when they choose *feces* over *shit, penis* over *prick, vagina* over *cunt, have sex* over *fuck*?

The major difference is that the taboo term is dysphemistic—it calls to mind the most disagreeable aspects of the referent, rather than just pointing to it. Now, people don't like to think about feces any more than they like to see it, smell it, or touch it. Yet we are incarnate beings, for whom feces is part of life, and there are occasions on which we have no choice but to confer on what to do with it. The solution is to divide the linguistic labor between euphemisms, which refer to an entity without evoking the unwanted emotions, and dysphemisms, including taboo words, for those rhetorical occasions on which we want to rub in how truly awful the entity is.

Euphemisms and dysphemisms for taboo concepts materialize and turn over rapidly. Allan and Burridge estimate that English has accumulated more than eight hundred expressions for copulation, a thousand for a penis, twelve hundred for a vagina, and two thousand for a wanton woman (making you wonder why people make so much of a fuss about the number of Eskimo words for snow).[66] In contemporary English we find several dozen specialized terms for feces, presumably because it is both so disgusting and so unavoidable:

> taboo: *shit*
> mildly dysphemistic: *crap, turd*
> mildly euphemistic: *waste, fecal matter, filth, muck*
> formal: *feces, excrement, excreta, defecation, ordure*
> with children: *poop, poo, poo-poo, doo-doo, doody, ka-ka, job, business, Number 2, BM*
> of diapers: *soil, dirt, load*
> medical: *stool, bowel movement*
> animal, large units: *pats, chips, pies*
> animal, small units: *droppings*
> animal, scientific: *scat, coprolites, dung*
> animal, agricultural: *manure, guano*
> human, agricultural: *night soil, humanure, biosolids*

Most of the polite terms are specific to a context in which feces must be discussed and to the actions that are appropriate in that context (spreading it as fertilizer, changing a diaper, analyzing it for medical or scientific purposes, and so on). Using the euphemism thus leaves no doubt as to why a conversant is bringing up the subject.

When it comes to the referents of taboo terms, the English language has gone overboard with specialization, and fails to provide us with neutral terms for casual conversation. Even people who swear only in moments of extreme and justifiable excitement would sound rather stuffy if in conversation with a friend they used *feces, flatulence,* or *anus* rather than their taboo alternatives. And the words *penis* and *vagina* force us to speak in Latin, whereas our other body parts have concise Anglo-Saxon roots, in keeping with the rest of our casual vocabulary. As C. S. Lewis put it, "As soon as you deal with [sex] explicitly, you are forced to choose between the language of the nursery, the gutter, and the anatomy class."[67]

There are times, of course, when we want to remind our listeners of the disagreeable aspects of something, and that is when we turn to the language of the gutter. Sometimes for the sake of narrative vividness, sometimes out of anger, we use taboo words to convey just how vile something is:

So the plumber wanted to talk to me while he was working
  under the sink, and I had to look at the crack in his ass the
  whole time.
His motto in life is: If it moves, fuck it; if it doesn't, stab it.
Will you pick up your dog's shit, and stop him from pissing on
  my roses!
Then John showed me the album, and I'm supposed to say,
  "Oh, that's nice," as if his dick weren't hanging out there
  [Ringo Starr, describing his reaction to being shown the
  Two Virgins album cover, on which Lennon and Yoko
  Ono had posed nude].

Try replacing the taboo terms in these sentences with their polite synonyms (*buttocks, having sex,* and so on). They lack a certain something, because the emotional force of the speaker's reaction is no longer being conveyed. And because taboo words evoke carnal details in the minds of listeners and readers, they are often put to use in pornography and in the formula for sexual arousal requested by many consenting adults: "Talk dirty to me."

Needless to say, not everyone reserves taboo words for special rhetorical effect. The expressions "to swear like a sailor," "to cuss like a stevedore," and "locker-room language" point to the fact that swearing is the language of choice in many male-dominated and working-class circles. One reason is that swearing, which forces a listener to think about disagreeable things, is mildly aggressive, so it fits with the other trappings that men in rough-and-tumble settings brandish to advertise that they can inflict and endure pain (heavy boots, metal studs, exposed muscles, and so on). The other reason is that a conspicuous willingness to break taboos conveys an atmosphere of informality, a freedom from having to watch what you say. Of course swearing has expanded in recent decades to women and the middle class. (When I was a teenager during the heyday of the "generation gap," the father of one of my friends used to say to her, "Nancy, your mouth is like a toilet.") The trend was part of a larger development in the twentieth century toward informality, egalitarianism, and the spread of macho and cool-pose styles.

The ability of taboo words to evoke an emotional reaction is useful not just when speakers wish to convey their own distress to a listener but also when they want to create that distress in a listener from scratch. Hence we have the use of profanity in insults, execrations, and other forms of verbal abuse.

There are moments in everyone's life when one feels the urge to intimidate, punish, or downgrade the reputational stock of some other person. The crafting of maledicta has probably exercised people's language instinct more vigorously than all the other kinds of speech acts put together, and in many cultures it has been raised to a high art, sometimes called flyting. For example, there are Shakespearean insults:

> PRINCE HENRY: . . . this sanguine coward, this bed-presser, this horseback-breaker, this huge hill of flesh,—
>
> FALSTAFF: 'Sblood, you starveling, you elf-skin, you dried neat's tongue, you bull's pizzle, you stock-fish! O for breath to utter what is like thee! you tailor's-yard, you sheath, you bowcase; you vile standing-tuck,—

And Yiddish curses:

> She should have stones and not children.
> May all your teeth fall out but one, so you can have a toothache.
> He should give it all away to doctors.

And the African American tradition known as sounding, snapping, signifying, ranking, and the dozens:

> You're so ugly, when you were born the doctor looked at your face and looked at your ass and said, "It's twins."
> Your mama's like a bowling ball—she gets picked up, fingered, thrown in the gutter, and then comes back for more.
> Your mama's so dumb she thinks Moby Dick is a venereal disease.

When crafting a curse, the availability of words that trigger unpleasant thoughts in a listener or bystander is a weapon that is too handy to forbear, and so taboo words figure prominently in imprecations. People or their parts may be likened to effluvia and their associated organs and accessories (*piece of shit, asshole, cunt, twat, prick, schmuck, putz, old fart, shithead, dickhead, asswipe, scumbag, douchebag*). They can be accused of engaging in undignified sexual activities such as incest (*motherfucker*), sodomy (*bugger, sod*), fellatio (*cocksucker; You suck! You bite! You blow!*), and masturbation (*wanker, jerk*). They can be advised to undertake degrading activities (*Kiss my ass, Eat shit, Fuck yourself, Shove it up your ass,* and—my favorite—*Kiss the cunt of a cow,* from 1585).[68] They can be threatened with violence accompanied by degradation, as in *I'll stick a pig's leg up your cunt until your back teeth rattle* (from Japan), and *I'll rip your head off and shit down your windpipe,* which I overheard at a Boston bus stop. Surveys of maledicta in other languages uncover similar themes.[69] And then there is the most common obscene curse in English, *Fuck you,* but to understand it we must take a closer look at taboo terms for sex.

Verbs for sex show a curious pattern. The anthropologist Ashley Montagu referred to *fuck* as "a transitive verb for the most transitive of human actions," and therein lies a tale.[70] Think of the transitive verbs for sex—the ones that fit in the slot *John verbed Mary:*

> fuck, screw, hump, ball, dick, bonk, bang, shag, pork, shtup

They're not very nice, are they? The verbs are jocular or disrespectful at best and offensive at worst. So what are the verbs that we do use in polite company when referring to the act of love?

> have sex, make love, sleep together, go to bed, have relations,
> have intercourse, be intimate, mate, copulate

They are *intransitive,* every one of them. The word for the sexual partner is always introduced by a preposition: have sex *with,* make love *to,* and so on. Indeed, most of them aren't even verbs of their own, but idioms that join a noun or an adjective to an insubstantial "light verb" like *have, be,* or *make.* (In *Crazy English,* Richard Lederer asks: "*To sleep with someone.* So who's sleeping? *A one-night stand.* Who's standing?") In the last section we saw many cases where a sense of decorum mandates the choice of a word. But why should it mandate something as abstruse as a grammatical construction?

Here is another payoff of the analysis of verb constructions in chapter 2. Remember that every construction chooses its verbs from a set of microclasses, each with a meaning that is conceptually compatible with the construction, if only metaphorically. Using this principle, can we discover anything about human sexuality from the syntax of the verbs for sex—the "copulative verbs," in a sense very different from the one in traditional grammar?

The polite idioms have a number of telltale grammatical traits. By lacking a distinctive verb root, they fail to specify an action with a characteristic manner of motion or kind of effect. By lacking a direct object, they specify no entity that is impinged upon or caused to change. Moreover, they are semantically symmetrical: if John had sex with Mary, it implies that Mary had sex with John, and vice versa. And all of them alternate with another intransitive construction in which the partner is not mentioned in a prepositional object but rather is part of a plural subject: *John and Mary had sex, John and Mary made love, John and Mary were intimate,* and so on. The semantics of the non-sexual verbs that behave in this way implies joint voluntary action, like *dance, talk, trade,* and *work: John danced with Mary, John and Mary danced,* and so on. So in the mental model presupposed by the polite verbs for sex, sex is an activity, manner unspecified, that two people jointly engage in.

Compare the ruder, transitive verbs for sex. Recall from chapter 2 that transitive verbs describe an agent that deliberately carries out an action that impinges on an entity, or affects the entity, or both. Though *fuck* doesn't fall perfectly into any of the five classes of transitive verbs we visited in chapter 3, it does have affinities with the microclass of verbs of motion-contact-effect.[71] It can be accepted into the conative, possessor-raising, and middle constructions, but not into the contact-locative or anticausative construction. (A sense of propriety, or what's left of it, compels me to put the examples in an endnote.)[72] This is consistent with the verb's etymology from an Old Norse word for beating, striking, or thrusting, and with the fact that its transitive synonyms include *bang* and *bonk.* (The Yinglish *shtup* comes from a different metaphor: the verb in Yiddish means "to stuff.")

In a well-known paper, the linguist Quang Fuc Dong notes that *fuck* occurs far more often with a male subject than a female one, and that some speakers use the verb only with a male subject.[73] To be more exact, its semantic requirement is that the subject be the active party. In sexual encounters between two men, he notes, *Boris fucked Lionel* is grammatical if

Boris is on top, and in encounters between two women *Cynthia fucked Gwendolyn* is grammatical if Cynthia used a dildo. The object of the verb, on the other hand, does not have to be female, or human, or even animate. In a memorable passage in Philip Roth's *Portnoy's Complaint,* the narrator confesses to an event that took place when he was a teenager and had discovered a piece of raw liver in the refrigerator: "Now you know the worst thing I have ever done. I fucked my own family's dinner."

If the transitive verbs for sex imply that the direct object is affected, exactly how must it be affected? The answer may be found in a Lakoffian analysis of the way that the verbs for sex take part in conceptual metaphors. Many of the transitive verbs for sex can be used metaphorically to refer to exploitation, as in the joke we used to tell about why the Québec government planned to change the provincial symbol from the fleur-de-lys to a condom: It prevents conception, allows inflation, protects a bunch of pricks, and gives you a false sense of security when you're being screwed. Metaphors in this family include *I was screwed, They fucked me over, We got shafted, I was reamed,* and *Stop dicking me around.*

The other metaphorical topic of the transitive verbs for sex is grievous damage, as in *fucked up, screwed up, buggered up,* and the British *bollixed* and *cockup.* World War II army slang included the acronyms *snafu* (Situation Normal, All Fucked Up), *tarfu* (Things Are Really Fucked Up), and *fubar* (Fucked Up Beyond All Recognition). The terms were absorbed into the argot of engineers, and today when computer programmers create a temporary file or teach a novice how to name one, they use the filename foo.bar—a bit of nerd humor. The metaphors underlying the transitive verbs for sex, then, are TO HAVE SEX IS TO EXPLOIT SOMEONE and TO HAVE SEX IS TO DAMAGE SOMEONE.

These conceptual metaphors are found in many other languages, too. In Brazilian Portuguese, the vulgar equivalent of *fuck* is *comer,* "to eat," with the man (or active homosexual partner) as the subject. This would be mysterious if the verb were a metaphor based on the mechanics of copulation, because it should be the woman's body that metaphorically eats the man's. But it fits the understanding of sex in which a woman is enjoyed and exploited by the man.

So the syntax of the verbs of sex uncovers two very different mental models of sexuality. The first is reminiscent of sex-education curricula, marriage manuals, and other sanctioned views: Sex is a joint activity, details unspecified, which is mutually engaged in by two equal partners. The second is a darker view, somewhere between mammalian sociobiology and Dworkin-style feminism: Sex is a forceful act, instigated by an active male and impinging on a passive female, exploiting her or damaging her. Both models capture human sexuality in its full range of manifestations, and if language is our guide, the first model is approved for public discourse, while the second is taboo, though widely recognized in private.

As I've mentioned, the dividing line between terms that are merely dysphemistic and those that cross over to taboo is mysterious. For many people, *excrement* has a far more unpleasant connotation than *shit,* because *excrement* is reserved for descriptions of filth and squalor whereas *shit* is used in a wider range of idioms and casual contexts. Nonetheless, *shit* is less acceptable than *excrement.* Similarly, the behavior labeled by *fuck* is nowhere near as upsetting as the behavior labeled by *rape,* yet *rape* is not a taboo word. People treat an unpleasant word as taboo to the extent that everyone else treats it as taboo, so the status of the words may be at the mercy of the boom-and-bust epidemiology that sets the fate of words and names in general.

What this all entails is that taboo words, though evocative of the nastier aspects of their referents, don't get their punch from those connotations alone. Taboo status *itself* gives a word an emotional zing, regardless of its actual referent. This gives rise to the countless idioms that incorporate taboo terms. Some of them, like *bullshit, They fucked me over, He pissed on my proposal,* and *She pissed away her inheritance,* are clearly metaphorical, projecting one of the unpleasant aspects of their vehicle onto an aspect of their topic. But a much larger number show no discernible analogy to their subject matter, and incorporate the taboo word only for its ability to pique the listener's interest:

> He went through a lot of shit. Tough shit! We're up shit's creek.
> We're shit out of luck. A shitload of money. Shit oh dear!
> [New Zealand]. Shit, eh? [New Zealand]. Let's shoot the
> shit. Let's smoke some shit. Put your shit over there. A lot
> of fancy shit. He doesn't know shit. He can't write for
> shit. Get your shit together. Are you shitting me? He

thinks he's hot shit. No shit! All that shit. A shit-eating
grin. Shitfaced [drunk]. Apeshit. Diddly-shit. Sure as shit.
It's piss-poor. Piss off! I'm pissed at him. He's pissed off. He's
pissed [drunk]. Full of piss and vinegar. They took the
piss out of him [British].
My ass! Get your ass in gear. Ass-backwards. Dumb-ass. Your
ass is grass. Kiss your ass goodbye. Get your ass over here.
That's one big-ass car! Ass-out [broke]. You bet your ass!
A pain in the ass.
Don't get your tits in a tangle [New Zealand]. My supervisor
has been getting on my tits [British].
Fuckin-A! Aw, fuck it! Sweet fuck-all. He's a dumb fuck. Stop
fucking around. He's such a fuckwit [New Zealand]. This
place is a real clusterfuck ["disorganized situation,"
army]. Fuck a duck! That's a real mindfucker. Fuck this
shit.

More than 250 entries of this sort may be found in a specialized dictionary
by the lexicographer Jesse Sheidlower called *The F-Word*.[74] As we saw in
chapter 5, metaphors and idioms can congeal into formulas that people no
longer analyze. This seems to have happened, at least in part, with the vulgar idioms, and together with expletives like *fucking amazing* they are the
least offensive way to use taboo words.

The affective clout of taboo words can make them into strange synonyms:
they substitute for one another in idioms even when they have no affinity in
syntax or meaning. Many bafflingly ungrammatical profanities must have
originated in more intelligible *religious* profanities during the transition
from religious to sexual and scatological swearing in English-speaking
countries:

Who (in) the hell are you? → Who the fuck are you? (Also:
Where the fuck are you going? What the fuck are you
doing? Get the fuck out of here, etc.)
I don't give a damn → I don't give a fuck; I don't give a shit; I
don't give a sod.
Holy Mary! → Holy shit! Holy fuck!
For God's sake → For fuck's sake; For shit's sake.

When it comes to the family ties among taboo words, then, connotation is a stronger filament than meaning or syntax. That helps explain the two great mysteries in the syntax of English profanity: what the word *fuck* is doing in *Close the fucking door,* and what it is doing in *Fuck you!*

The mysteries were first explored in what must be the strangest festschrift in the history of academia, *Studies Out in Left Field: Defamatory Essays Presented to James D. McCawley on the Occasion of His 33rd or 34th Birthday.*[75] The late linguist Jim McCawley was one of the founders, together with George Lakoff and Haj Ross, of the school of linguistics called Generative Semantics. His contributions included a guide to the fractious field called *Thirty Million Theories of Grammar,* a primer called *Everything That Linguists Have Always Wanted to Know about Logic (But Were Ashamed to Ask),* and *The Eater's Guide to Chinese Characters,* a tutorial that empowers readers to order from the Chinese side of the menu and get the really good dishes that the Chinese patrons are always eating. Among the many unconventional aspects of this 1971 festschrift is that several of the contributions were penned by McCawley himself under the pseudonyms Quang Fuc Dong and Yuck Foo, both of the "South Hanoi Institute of Technology" (get it?). Despite the sometimes sophomoric humor and tasteless examples, the papers are sophisticated analyses of the grammar of English taboo expressions, and are still cited in scholarly work today (sometimes as "Quang (1971)" or "Dong, Q. F.").

Expletives like *bloody* and *fucking* are probably the most commonly used taboo words in casual speech, despite their nonsensical semantics and syntax. A century-old British slang dictionary includes the following in its entry for *bloody:* "Most frequently . . . as it falls with wearisome reiteration every two or three syllables from the mouths of London roughs of the lowest type; no special meaning, much less a sanguinary one, can be attached to its use."[76] Similar observations have been made about the dialect called Fuck Patois, like the story about the soldier who said, "I come home to my fucking house after three fucking years in the fucking war, and what do I fucking-well find? My wife in bed, engaging in illicit sexual relations with a male!"

The grammar of *fucking* in its expletive role made the news in 2003 when NBC broadcasted the Golden Globe Awards and Bono said, "This is really, really, fucking brilliant" on the air. The FCC originally chose not to sanction the network because their guidelines define "indecency" as "material that describes or depicts sexual or excretory organs or activities," and Bono had used the word as "an adjective or expletive to emphasize an exclamation." Cultural conservatives were outraged, and California Representative Doug Ose tried to close the loophole with the filthiest piece of legislation ever considered by Congress, the Clean Airwaves Act:

A BILL

To amend section 1464 of title 18, United States Code, to provide for the punishment of certain profane broadcasts, and for other purposes.

Be it enacted by the Senate and House of Representatives of the United States of America in Congress assembled, That section 1464 of title 18, United States Code, is amended—

(1) by inserting '(a)' before 'Whoever'; and

(2) by adding at the end the following: '(b) As used in this section, the term 'profane', used with respect to language, includes the words 'shit', 'piss', 'fuck', 'cunt', 'asshole', and the phrases 'cock sucker', 'mother fucker', and 'ass hole', compound use (including hyphenated compounds) of such words and phrases with each other or with other words or phrases, and other grammatical forms of such words and phrases (including verb, adjective, gerund, participle, and infinitive forms).

Unfortunately for Rep. Ose, the bill would not have closed the loophole after all, because it fails to specify the syntax of Bono's expletive properly (to say nothing of its misspelling of *cocksucker, motherfucker,* and *asshole,* or its misidentifying them as "phrases").

The Clean Airwaves Act assumes that *fucking* is a participial adjective. But this is not correct. As Quang notes, with a true adjective like *lazy,* you can alternate between *Drown the lazy cat* and *Drown the cat which is lazy.*[77] But *Drown the fucking cat* is certainly not interchangeable with *Drown the cat which is fucking.* (Likewise, *Drown the bloody cat* does not mean the same thing as *Drown the cat which is bloody.*) Nor can you say *The cat seemed fucking, How fucking was the cat?,* or *the very fucking cat,* three more tests for adjectivehood.[78]

Some critics have poked fun at the Clean Airwaves Act for another bit of grammatical illiteracy. If anything, the *fucking* in *fucking brilliant* should be an adverb, because it modifies an adjective, and only adverbs can do that, as in *truly bad, very nice,* and *really big.* Yet "adverb" is the one grammatical category that Ose forgot to include in his list! As it happens, taboo expletives aren't genuine adverbs, either. Another "study out in left field" notes that while you can say *That's too fucking bad* and *That's no bloody good,* you can't say *That's too very bad* or *That's no really good.*[79] Also, as the linguist Geoffrey Nunberg has pointed out, while you can imagine the dialogue *How brilliant was it? Very,* you would never hear the dialogue *How brilliant was it? Fucking.*[80]

Most anarchically of all, expletives can appear in the middle of a word or compound, as in *in-fucking-credible, hot fucking dog, Rip van Fucking Winkel, cappu-fucking-ccino,* and *Christ al-fucking-mighty*—the only known case in English of the morphological process known as infixation. *Bloody* also may be infixed, as in *abso-bloody-lutely* and *fan-bloody-tastic.* In his memoir *Portrait of the Artist as a Young Dog,* Dylan Thomas writes, "You can always tell a cuckoo from Bridge End . . . it goes cuck-BLOODY-oo, cuck-BLOODY-oo, cuck-BLOODY-oo."

The semantics of expletives are as strange as their syntax. *Bloody* and *fucking* generally express disapproval, yet the disapproval is not necessarily directed at the modified noun:

> INTERVIEWER: Why is British food so bad?
> JOHN CLEESE: Because we had a bloody empire to run, you see?[81]

Cleese was not casting aspersions on the empire on which the sun never set; he was expressing mock exasperation with the interviewer's question. Likewise, if I say, "They stole my fucking laptop!" the laptop need not have been execrable; it could have been a sleek titanium PowerBook with a 17-inch screen and a 1.67-gigahertz processor.[82] Expletives indicate that something is lamentable about an entire state of affairs, not the entity named by the noun, though properties of that entity may have something to do with why the state of affairs is lamentable. Just as important, the situation has to be lamentable from the point of view of the speaker, not of any of the characters mentioned in the sentence or discourse. If someone reports to you that

John says his landlord is a fucking scoutmaster, you attribute the disrespect for scoutmasters to the speaker, not John, despite the fact that the *fucking* is inside the clause that conveys what John said.[83]

Part of the solution to this linguistic puzzle is that expletives like *bloody* and *fucking* arose from the process by which one taboo word substitutes for another despite their having nothing else in common (the process that allowed *Where in hell* to beget *Where the fuck,* and *Holy Mary* to inspire *Holy shit*). With the expletives in *fucking scoutmaster* or *bloody empire,* the historical source is *damned* or *God-damned,* which persist today in expletives like *Damn Yankees, They stole my goddam laptop,* and *abso-goddam-lutely.* (*Damned* became *damn* when the insubstantial *-ed* got swallowed in pronunciation and overlooked in perception, as it did in *ice cream, mincemeat,* and *box set,* formerly *iced cream, minced meat,* and *boxed set.*) If something has been damned, it is condemnable, pitiable, and no longer of earthly use. One can imagine this connotation summoning words with similar emotional overtones like *fucking, bloody, dirty, lousy,* and *stupid.* They probably took their place alongside *damned* once religious expletives started to lose their sting in the history of English.

The other part of the solution is that affect-laden words can sometimes escape the usual grammatical machinery that computes who did what to whom from the arrangement of words in a parse tree. Linguists such as Christopher Potts argue that the grammar of English not only allows speakers to make an assertion in a sentence—what is "at issue"—but also provides them with ways to editorialize and comment on the assertion.[84] Sometimes called conventional implicatures, these devices allow a speaker to convey his attitude about what is being talked about, such as his opinion of the outcome or his degree of respect for one of the participants. One such device allows an attitude-laden word to break free of the actors in the drama being described and gravitate to the worldview of the speaker. For instance, if I say *Sue believes that that jerk Dave got promoted,* it's quite possible that Sue has a high opinion of Dave, but I am implying that I do not. This is just the scheme of interpretation that governs taboo expletives like *fucking* and *bloody.*

The swappability of taboo terms also explains the mystery of *Fuck you.* Woody Allen's joke about telling a driver to be fruitful and multiply but not in those words assumes that *Fuck you* is a second-person imperative, like *Get fucked* or *Fuck yourself.* Lenny Bruce made the same assumption, as did

Bill Bryson in his delightful book *The Mother Tongue: English and How It Got That Way:*

> English is unusual in including the impossible and the pleasurable. It is a strange and little-noted idiosyncrasy of our tongue that when we wish to express extreme fury we entreat the object of our rage to undertake an anatomical impossibility, or, stranger still, to engage in the one activity that is bound to give him more pleasure than almost anything else. Can there be, when you think about it, a more improbable sentiment than "Get fucked"? We might as well snarl, "Make a lot of money" or "Have a nice day."[85]

But Quang makes short work of this theory.[86] For one thing, in a second-person imperative the pronoun has to be *yourself,* not *you*—Madonna's hit song was titled "Express Yourself," not "Express You." For another, true imperatives like *Close the door* can be embedded in a number of other constructions:

> I said to close the door.
> Don't close the door.
> Go close the door.
> Close the door or I'll take away your cookies.
> Close the door and turn off the light.
> Close the door when you leave tonight.

But *Fuck you* cannot:

> *I said to fuck you.
> *Don't fuck you.
> *Go fuck you.
> *Fuck you or I'll take away your cookies.
> *Fuck you and turn off the light.
> *Fuck you when you leave tonight.

The difference can be seen with third-person objects as well, as in *Fuck communism!* Though you can conjoin two imperatives sharing an object, as in *Clean and press these pants,* you can't conjoin the imprecation with a true imperative, as in **Describe and fuck communism.*

Quang does not evaluate another folk etymology of *Fuck you*, namely that it's short for *I fuck you* (as in the story in chapter 1 about the impatient customer and the airline clerk). This is certainly compatible with the conceptual metaphor in which sex is a kind of exploitation or damage, but it makes no sense on linguistic grounds. The tense is wrong, the missing subject is unexplained, and there are no parallel constructions. Nor is there any historical evidence that *I fuck you* was ever a common curse in English.

The simplest explanation is that the *fuck* in *Fuck you* is like the *fuck* in *Where the fuck* and *a fucking scoutmaster*: a substitution for an older religious epithet with similar emotional force. In this case, the obvious source is *Damn you* (perhaps shortened from *God damn you* and *May God damn you*). The original semantics would have been a kind of third-person imperative meaning "May it be so," which is common in blessings ("May you be forever young") and curses (as in "May you live like a chandelier: hang by day and burn by night"). But the curse melted into a holistic pronouncement of disapproval. As Quang notes, *Fuck you* resembles not just *Damn you* but other constructions that express nothing but a strong attitude of the speaker toward the object: *To hell with you!*, *Shit on you!*, *Bless you!*, *Hooray for you!*, and the sarcastic *Bully for you!*

The remaining use of taboo language is cathartic—the blurting out of *damn, hell, shit, fuck*, or *bugger* in moments of sudden pain, frustration, or regret. If you ask people why they do it, they'll say that it "releases tension" or helps them "let off steam." This is the hydraulic metaphor for emotion, also seen in *venting one's feelings, finding an outlet, bottling up rage*, and *exploding with anger*. Though it undoubtedly captures what it feels like to express frustration, the hydraulic metaphor doesn't *explain* the feeling. Neuroscientists have not found vessels or pipes in the brain carrying heated fluid, just networks of neurons that fire in complex patterns. Nor is there a law of thermodynamics that could explain why the articulation of *Oh, fuck* would enable a dissipation of energy any more effectively than *Oh, my* or *Fiddle-dee-dee*.

The brain has other mechanisms, though, that may play a role in cathartic swearing. One of them is an electrophysiological response that kicks in when people notice they have just made an error.[87] It emanates from the anterior cingulate cortex, a part of the limbic system involved in the monitoring of cognitive conflict. In public, cognitive neuroscientists call this response the Error-Related Negativity; in private they call it the Oh-Shit Wave.

Also relevant are the limbic circuits in mammals that underlie the experience of anger. One of them, called the Rage circuit, runs from a part of the amygdala down through the hypothalamus (the tiny brain cluster that regulates motivation) and then into the gray matter of the midbrain.[88] The Rage circuit originally housed a reflex in which a suddenly wounded or confined animal would erupt in a furious struggle to startle, injure, and escape from a predator, often accompanied by a bloodcurdling yowl. Anyone who has accidentally sat on a cat or stepped on the tail of a dog may discover a new sound in their pet's repertoire, sometimes followed by fresh clawmarks or teethmarks on their leg. The reaction has been studied in a long line of research in experimental psychology on an idea called the Frustration-Aggression hypothesis.[89] When two rats are put together in a chamber and given an electric shock, for example, they will start to fight. A rat will also attack another rat if a reward such as food is suddenly withdrawn, presumably an adaptation to the sudden theft of food, space, or other resources by a fellow animal. The underlying brain circuits have been conserved in human evolution. When this pathway is electrically stimulated in neurological patients during brain surgery, they experience a sudden intense rage.[90]

So here is a hypothesis about cathartic swearing. A sudden pain or frustration engages the Rage circuit, which activates parts of the limbic brain connected with negative emotion. Among them are representations of concepts with a strong emotional charge and the words connected to them, particularly the versions in the right hemisphere, with its heavier involvement in unpleasant emotions. The surge of an impulse for defensive violence may also remove the safety catches on aggressive acts ordinarily held in place by the basal ganglia, since discretion is not the better part of valor during what could be the last five seconds of your life. In humans, these inhibited responses may include the uttering of taboo words. Recall that the Rage response in animals also includes a fearsome yelp. Perhaps the combination of a firing up of negative concepts and words, a release of inhibition on antisocial acts, and the urge to make a sudden sharp noise culminates in an obscenity rather than the traditional mammalian shriek. (Of course, when people experience severe pain, they show that our species has also retained the ability to holler and howl.) Cathartic swearing, then, would come from a cross-wiring of the mammalian Rage circuit with human concepts and vocal routines.

The problem with the cross-wiring theory is that angry expletives are *conventional*. Like our other words and formulas, they depend on a memorized pairing between a sound and a meaning which is shared throughout a language community. When we bump our heads, we don't shout *Cunt!* or *Whore!* or *Prick!*, though these words are just as taboo as *shit, fuck,* and *damn* (and indeed are the translations of toe-stubbing cries in other languages). Also, the expletive is keyed to the cause of the misfortune. People shout *Asshole!* when they suffer a sudden affront from a human perpetrator, but not when they pick up a hot casserole or have a mousetrap snap on their finger. So the cathartic swear words are specific to the occasion and specific to the language. As Mrs. Pearce said of Eliza's use of the *b*-word, we learn them at our mother's knee, or perhaps more often at our father's. When I was four years old and sitting in the front seat of the car next to my father, the door swung open as we rounded a curve, and I said, "Oh, shit!," proud of knowing what an adult would say in those circumstances. I was quickly reprimanded in a show of hypocrisy that is one of the perquisites of parenthood.

Why should we go to the trouble of learning specific words for cathartic swearing rather than just letting our rage fire up any old taboo word? Cathartic swearing is part of a larger linguistic phenomenon called ejaculations or response cries.[91] Consider the following list:

> aha, ah, aw, bah, bleh, boy, brrr, eek, eeuw, eh, goody, ha, hey, hmm, hmph, huh, mmm, my, oh, ohgod, omigod, ooh, oops, ouch, ow, oy, phew, pooh, shh, shoo, ugh, uh, uh-oh, um, whee, whoa, whoops, wow, yay, yes!, yikes, yipe, yuck

At first glance these seem less like real words than transliterations of the noises that escape from our mouths when we are in the throes of a strong feeling. They can't be used in a grammatical sentence (**I like goody; *I hate ouch*), and many of them violate the sound pattern of English (like *eeuw, hmph,* and *shh*). Nor are they spoken on cue when a speaker takes the floor in the give-and-take of a conversation.

But they really are words, with conventional sounds and meanings. The sounds are standardized, not just emitted as the feeling arises, to the point that many people articulate the spelling that cartoonists use to render people's exclamations onomatopoeically, like "Gulp!" "Tisk, tisk!" and "Phew!" And one of the most obvious giveaways that a speaker is foreign is the use of the wrong exclamation, as when an American interrupts his fluent French with the unmistakably Anglophone *um* or *ouch*. According to the joke, a Jewish woman trying to pass as a WASP at an exclusive country club wades into an ice-cold swimming pool. She shouts, "Oy vey! . . . whatever that means."

What *oy vey* and other response cries mean are as conventional as other words in a language. What do you say when you see an adorable baby? When you're cold? When you discover a worm in your apple? When you drop a napkin? When you discover the open window that's been letting in a draft? When you warm yourself with a spoonful of hot soup? Every English speaker knows which word to choose from the list.

The sociologist Erving Goffman was a theater critic of everyday life, analyzing the staging and dialogue that we use to manage the impression of a real or imagined audience.[92] One goal in this performance, he suggested, is to reassure onlookers that we are sane, competent, reasonable human beings, with transparent goals and intelligible responses to the current situation. Ordinarily this requires that we not talk to ourselves in public, but we make an exception when a sudden turn of events puts our rationality or effectiveness to the test. My favorite example is when we do an about-face in a hallway and mutter a soliloquy explaining to no one in particular that we forgot something in our office, as if to reassure any onlookers that we are not a lunatic who lurches around at random.

Goffman argues that there is a good reason we utter response cries: to signal our competence and shared understanding of the situation to a generic audience.[93] A person who knocks over a glass might be a klutz, but if he says *whoops,* then at least we know that he didn't intend the outcome and regrets that it happened. A person who says *yuck* after dripping pizza sauce on his shirt or stepping in dog feces is someone we understand better than someone who would seem not to care. And so it is with cathartic swearing. Faced with a sudden challenge to our goals or well-being, we inform the world that the setback matters to us, indeed, that it matters at an emotional level that calls up our worst thoughts and is at the boundaries of voluntary control. Like other response cries, taboo outbursts are calibrated to the severity of the setback, *shoot* indicating a minor annoyance and *fuck* a more serious one. And depending on the choice of word and the tone in which it is uttered, an outburst can summon help, intimidate an adversary, or warn a careless actor of the harm he is inadvertently causing. Goffman sums up his theory: "Response cries, then, do not mark a flooding of emotion outward, but a flooding of relevance in."[94]

The Rage-circuit theory, which views cathartic swearing as a by-product, and the response-cry theory, which views it as an adaptation, are not mutually exclusive. Many ordinary response cries must have arisen as conventional renderings of vocal sounds, like *brr* for chattering teeth or *yuck* for expelling something from the back of the mouth. This ritualization may have shaped cathartic swearing as well. The epithets may have originated as Tourette-like outbursts of taboo words released by the Rage circuit and then were conventionalized into standard response cries for that kind of trespass or misfortune. Some cognitive neuroscientists have revived Darwin's suggestion that verbalized outbursts were the evolutionary missing link between primate calls and human languages.[95] If so, swearing would have played a more important role in the human career than most people would acknowledge.

SWEARING, CON AND PRO

So what should we do about profanity? Does the science of swearing cast any light on the controversies over shock jocks, clean airwaves, and broadcast decency? As far as policy is concerned, my remarks will be few and banal. It seems to me that free speech is the bedrock of democracy and that it is not among the legitimate functions of government to punish people who use certain words or allow others to use them. On the other hand, private media have the prerogative of enforcing a house style, driven by standards of taste and the demands of the market, that excludes words their audience doesn't enjoy hearing. In other words, if an entertainer says *fucking brilliant,* it's none of the government's business, but if some people would rather not explain to their young children what a blow job is, there should be television channels that don't force them to. Rather than review policy issues any more deeply than this, I hope to say a few words about how the psycholinguistics of taboo language might inform our judgment about when to discourage, when to tolerate, and even when to welcome profanity.

Language has often been called a weapon, and people should be mindful about where to aim it and when to fire. The common denominator of taboo words is the act of forcing a disagreeable thought on someone, and it's worth considering how often one really wants one's audience to be reminded of excrement, urine, and exploitative sex. Even in its mildest form, intended only to keep the listener's attention, the lazy use of profanity can feel like a series of jabs in the ribs. They are annoying to the listener, and a confession by the speaker that he can think of no other way to make his words worth attending to. It's all the more damning for writers, who have the luxury of choosing their words off-line from the half-million-word phantasmagoria of the English lexicon. A journalist who, in writing about the cruelty of an East German Stasi guard, can do no better than to call him a *fucker* needs to get a good thesaurus.[96]

Also calling for reflection is whether a linguistic taboo is always a bad thing. Why are we offended—why *should* we be offended—when an outsider refers to an African American as a *nigger,* or a woman as a *cunt,* or a Jewish person as *a fucking Jew*? The terms have no real meaning, so the offense cannot come from their perpetuating a stereotype or endorsing oppression. Nor is it a reaction to learning that the speaker harbors an abominable attitude. These days someone who displayed the same attitude by simply saying "I hate African Americans, women, and Jews" would be stigmatizing himself far more than his targets, and would quickly be written off as a loathsome kook. I suspect that our sense of offense comes from the nature of speech recognition and from what it means to understand the connotation of a word. If you're an English speaker, you can't hear the words *nigger* or *cunt* or *fucking* without calling to mind what they mean to an implicit community of speakers, including the emotions that cling to them. To hear *nigger* is to try on, however briefly, the thought that there is something contemptible about African Americans, and thus to be complicit in a community that standardized that judgment by putting it into a word. The same thing happens with other taboo imprecations: just hearing the words feels morally corrosive, so we consider them not just unpleasant to think but not to be thought at all—that is, taboo. None of this means that the words should be banned, only that their effects on listeners should be understood and anticipated.

Also deserving of reflection is why previous generations of speakers bequeathed us a language that treats certain topics with circumspection and restraint. Recall that the lexical libertines of the 1960s believed that taboos on sexual language were pointless and even harmful. They argued that removing the stigma from sexuality would eliminate shame and ignorance and thereby reduce venereal disease, illegitimate births, and other hazards of sex. But on this matter Saint Lenny turned out to be mistaken. Sexual language has become far more common since the early 1960s, but so have illegitimacy, sexually transmitted infections, rape, and the fallout of sexual competition, like anorexia in girls and swagger culture in boys. Though no one can pin down cause and effect, the changes are of a piece with the weakening of the fear and awe that used to surround thoughts about sex and that charged sexual language with taboo.

Those are some of the reasons to think twice about giving carte blanche to swearing. But there is another reason. If an overuse of taboo words, whether by design or laziness, blunts their emotional edge, it will have deprived us of a linguistic instrument that we sometimes sorely need. And this brings me to the arguments on the pro-swearing side.

To begin with, it's a fact of life that people swear. The responsibility of writers is to give a "just and lively image of human nature," and that includes portraying a character's language realistically when their art calls for it. When Norman Mailer wrote his true-to-life novel about World War II, *The Naked and the Dead,* in 1948, he knew it would be a betrayal of his depiction of the soldiers to have them speak without swearing. His compromise with the sensibilities of the day was to have them use the pseudo-epithet *fug.* (When Dorothy Parker met him she said, "So you're the man who doesn't know how to spell *fuck.*") Sadly, this prissiness is not a thing of the past. Some public television stations today are afraid to broadcast Martin Scorsese's documentary on the history of the blues and Ken Burns's documentary on World War II because of the salty language in their interviews with musicians and soldiers. The prohibition against swearing in broadcast media makes artists and historians into liars, and subverts the responsibility of grown-ups to learn how life is lived in worlds distant from their own.

Even when their characters are not soldiers, writers must sometimes let them swear in order to render human passion compellingly. In the film adaptation of Isaac Bashevis Singer's *Enemies: A Love Story,* a sweet Polish peasant girl has hidden a Jewish man in a hayloft during the Nazi occupation and becomes his doting wife when the war is over. When she confronts him over an affair he has been having, he loses control and slaps her in the face. Fighting back tears of rage, she looks him in the eye and says slowly, "I saved your life. I took the last bite of food out of my mouth and gave it to you in the hayloft. I carried out your *shit!*" No other word could convey the depth of her fury at his ingratitude.

For language lovers, the joys of swearing are not confined to the works of famous writers. Every idiom must have been the brainchild of some creative speaker lost in the mists of time, and many of the profane ones deserve our admiration. We should pause to applaud the poetic genius who gave us the soldiers' term for chipped beef on toast, *shit on a shingle,* and the male-to-male advisory for discretion in sexual matters, *Keep your pecker in your pocket.* Hats off, too, to the wordsmiths who thought up the indispensable *pissing contest, crock of shit, pussy-whipped,* and *horse's ass,* not to mention that fine descriptor of the clueless, *He doesn't know shit from Shinola.* Among those in the historical record, Lyndon Johnson had a certain way with words when it came to summing up the people he distrusted, including a Kennedy aide ("He wouldn't know how to pour piss out of a boot if the instructions were printed on the heel"), Gerald Ford ("He can't fart and chew gum at the same time"), and J. Edgar Hoover ("I'd rather have him inside the tent pissing out than outside pissing in").

Profanity can be used effectively in poetry, such as in Philip Larkin's "This Be the Verse," his 1974 poem on how "man hands on misery to man":

> They fuck you up, your mum and dad.
> They may not mean to, but they do.
> They fill you with the faults they had
> And add some extra, just for you.[97]

And it can be used in scientific argumentation, as in Judith Rich Harris's case against the belief that parents shape their children's characters:

> Poor old Mum and Dad: publicly accused by their son, the poet,
> and never given a chance to reply to his charges. They shall have
> one now, if I may take the liberty of speaking for them:

How sharper than a serpent's tooth
To hear your child make such a fuss.
It isn't fair—it's not the truth—
He's fucked up, yes, but not by us.[98]

It can be even used in a protest against government sanctions on profanity, as in "The FCC Song" by Monty Python's Eric Idle:

Fuck you very much, the FCC.
Fuck you very much for fining me.
Five thousand bucks a fuck,
So I'm really out of luck.
That's more than Heidi Fleiss was charging me.

Which is also the clearest illustration I know of the logician's distinction between the "mention" and "use" of words.

When used judiciously, swearing can be hilarious, poignant, and uncannily descriptive. More than any other form of language, it recruits our expressive faculties to the fullest: the combinatorial power of syntax; the evocativeness of metaphor; the pleasure of alliteration, meter, and rhyme; and the emotional charge of our attitudes, both thinkable and unthinkable. It engages the full expanse of the brain: left and right, high and low, ancient and modern. Shakespeare, no stranger to earthy imprecations himself, had Caliban speak for the entire human race when he said, "You taught me language, and my profit on't is, I know how to curse."

8

GAMES PEOPLE PLAY

 Mistaken identity is a plot device so revealing of human foibles that Shakespeare used it in no fewer than eight of his comedies.[1] Deception and self-deception, illusion and reality, and the presentation of self in everyday life are just some of the psychological themes that may be illuminated as we watch characters react to people who aren't who they seem to be. Our words are as much a part of our social selves as our appearance and demeanor, and so a mistaken-identity plot can lay bare some of the ways we cloak our intentions in language.

In the 1982 comedy *Tootsie*, Michael Dorsey (played by Dustin Hoffman) is an unemployed actor who disguises himself as a middle-aged actress, Dorothy Michaels, and improbably wins a major role in a soap opera. In a plot twist worthy of *Twelfth Night*, Dorothy befriends an attractive young actress, Julie Nichols (played by Jessica Lange), while the disguised Michael falls in love with her. During a session of late-night girl talk, Julie commiserates with Dorothy on the difficulties of being a single woman in modern times:

> You know what I wish? That a guy could be honest enough just to walk right up to me and say, "Listen. You know, I'm confused about this, too. I could lay a big line on you, do a lot of role-playing. But the simple truth is: I find you very interesting. And I'd really like to make love with you." Simple as that. Wouldn't that be a relief?

Later in the film, a twist of fate throws Julie together with an unrecognized Michael at a cocktail party in New York. Michael approaches her on the balcony:

> Hi. Mike Dorsey. Great view, huh? You know, I could lay a big line on you. And we could do a lot of role-playing. But the simple truth is that I find you very interesting. And I'd really like to make love to you. It's—

And before he can say "It's as simple as that," she has thrown a glass of wine in his face and stormed away.

When people talk, they lay lines on each other, do a lot of role-playing, sidestep, shilly-shally, and engage in other forms of vagueness and innuendo. We all do this, and we expect others to do it, yet at the same time we profess to long for plain speaking, for people to get to the point and say what they mean, simple as that. Such hypocrisy is a human universal. Even in the bluntest societies, people don't just blurt out what they mean but cloak their intentions in various forms of politeness, evasion, and euphemism.[2]

In the first chapter I offered some examples of indirect speech—snatches of dialogue in which a speaker says something that he doesn't literally mean, knowing that the hearer will interpret it as he intended. At the same time, the hearer knows that the speaker intended it to be interpreted that way, the speaker knows that the hearer knows that the speaker intended for the hearer to it interpret it that way, and so on.

Sexual come-ons are a classic example. "Would you like to come up and see my etchings?" has been recognized as a double entendre for so long that by 1939 James Thurber could draw a cartoon with a hapless man in an apartment lobby saying to his date, "You wait here and I'll bring the etchings down."[3] The concept of a veiled threat also has a stereotype: the Mafia wise guy offering protection with the soft sell, "Nice store you got there. Would be a real shame if something happened to it." The veiled bribe is another recognizable plot device, as when the kidnapper in *Fargo* shows a police officer his driver's license in a wallet with a fifty-dollar bill protruding from it and suggests, "So maybe the best thing would be to take care of that here in Brainerd." Anyone who has sat through a fund-raising dinner is familiar with euphemistic schnorring such as "We're counting on you to show leadership" or "We hope you will help to shape the future of Huxley College." And polite requests are made in countless mealymouthed forms: "I was wondering if you could pass the guacamole," "Do you think you could pass the guacamole," "If you could pass the guacamole, that would be awesome," and so on, anything to avoid the straightforward "Pass the guacamole."

This chapter is about indirect speech—why we so often don't just come out and say what we mean. Indirect speech figures in many arenas in which our choice of words can make a difference, including rhetoric and persuasion, negotiation and diplomacy, intimacy and seduction, and the prosecution of extortion, bribery, and sexual harassment. But they also pose questions about our nature as social beings. As with so many aspects of the mind, a danger we face is the temptation to explain a puzzle by appealing to intuitions that feel thoroughly natural but that themselves need an explanation. The theory that we swear to let off steam is an example of this fallacy, and the theory that we use weasel words to escape embarrassment, avoid awkwardness, save face, or reduce social tension is another. These rationales are true enough, but they are scientifically unsatisfying. We need to know what "face" is, and why we have emotions like embarrassment, tension, and shame that trade in it. Ideally, those enigmas will be explained in terms of the inherent problems faced by social agents who exchange information. We have no guarantee, of course, that indirect speech has a hidden rationale. But as we shall see, its details are so precisely tuned to the particularities of the speaker, the hearer, and the situation that it is almost certain to embody a hidden logic rather than being an arbitrary ritual.

We'll begin with a famous theory from the philosophy of language that tries to ground indirect speech in pure rationality—the demands of efficient communication between two cooperating agents. This Spock-like theory will then be enhanced by a dose of social psychology, which reminds us that people don't just exchange data like modems but try to save face, both the speaker's and the hearer's. Even this theory will turn out to be too simple, because it assumes that people in conversation always cooperate. A reflection on how a pair of talkers may have goals that *conflict* as well as coincide will bring us to the tricky logic of plausible denial, both in legal contexts, where people's words may be held against them, and in everyday life, where the sanctions are social rather than judicial. This will require nothing less than a theory of the distinct kinds of relationships that make up human social life, a consideration of the dizzying machinations in which A may know that B knows that A knows that B knows that A knows that *x,* and the paradox of rational ignorance, where we choose not to know something relevant to our interests.

TÊTE À TÊTE

Suppose a reviewer said of a musical performance, "Ms. Winterbottom produced a series of sounds that corresponded closely with the score of 'Candle in the Wind.' " You would probably interpret it as a snide comment on the quality of her singing. Why? Because the reviewer chose a verbose circumlocution rather than the concise term *sang,* and we assume he must have done so for a reason, namely, to comment that her performance fell short of the standards to which the word *singing* usually applies.

This is an example of the way we read between the lines when understanding language, a part of our intelligence introduced by the philosopher Paul Grice in one of the most important papers in the history of linguistics, "Logic and Conversation."[4] Grice began with the well-known fact that logical terms like *and, not, or, all,* and *some* have different meanings in everyday language from the ones they have in formal logic. In ordinary conversation, *He sat down AND told me he was a Republican* implies that he did the acts in that order, not merely that he did both of them (the logical meaning of *and*). *Your money OR your life* implies that you can keep your money or you can keep your life but not both, whereas *or,* strictly speaking, is compatible with both disjuncts being true. And *A horse is a horse* is logically circular and hence should be meaningless, yet people use these tautologies for a definite purpose, such as to point out that most horses have stereotypical horselike properties.

Grice had no desire to accuse ordinary speakers of being sloppy or illogical. On the contrary, he proposed that the use of language in conversation has a rationality of its own, rooted in the needs of conversational partners to cooperate to get their messages across. Speakers tacitly adhere to a "cooperative principle," he said: they tailor their utterances to the momentary purpose and direction of the conversation. That requires monitoring the knowledge and expectations of one's interlocutor and anticipating her reaction to one's words. (By the way, in this chapter I will refer to the generic speaker as a "he" and the generic hearer as a "she," just to help you keep track of who's who; this is a common convention in the linguistics literature.) Grice fleshed out the cooperative principle in four conversational "maxims," which are commandments that people tacitly follow (or should follow) to further the conversation efficiently:

Quantity:
- Say no less than the conversation requires.
- Say no more than the conversation requires.

Quality:
- Don't say what you believe to be false.
- Don't say things for which you lack evidence.

Manner:
- Don't be obscure.
- Don't be ambiguous.
- Be brief.
- Be orderly.

Relevance:
- Be relevant.

At first glance the maxims may strike you as ludicrous. If people were required to say only those things for which they had evidence, to avoid verbosity and obscurity, to stick to the topic, and to expound their ideas in an orderly fashion, an eerie hush would descend over academia, government, and the nation's barrooms. But the maxims are not all that preposterous, for two reasons.

People undoubtedly can be tight-lipped, long-winded, mendacious, cavalier, obscure, ambiguous, verbose, rambling, or off-topic. But on closer examination they are far less so than they *could* be, given the possibilities. If I ask a friend how to buy a movie ticket online, he will neither begin his answer with lessons on how to type, nor leave it at "Go to a Web site where they sell movie tickets." He won't direct me to a porn site, nor will he make up the address www.buymovieticketsonline.com just because it sounds plausible. He won't bury his answer in a half-hour disquisition on how the Internet is changing our lives, or give me tips on how to cook scrod. None of these feats can be taken for granted. Computer and voice-menu systems can be infuriatingly obtuse, and legalese, which is directed to an adversarial reader rather than a cooperative one, is a thicket of prolixity. Because human hearers can count on *some* degree of adherence to the maxims, they can read between the lines, weed out unintended ambiguities, and connect the dots when they listen and read. In a joke that was voted the world's funniest in a multinational Web-based experiment in 2002, two hunters are out in the woods when one of them collapses and stops breathing. His companion calls an emergency number on his cell phone and cries, "My friend is dead! What should I do?" The operator says, "Calm down; I can help. First, let's make sure he's dead." There is a silence, then a gunshot, and the hunter says, "OK, now what?"[5] Two million people laughed because the hunter failed to apply Grice's maxims in interpreting the ambiguous phrase *make sure*.

But the more interesting way in which the maxims apply to our conversation is when they are observed in the breach. Speakers often flout the maxims, counting on their listeners to interpret their intent in a way that would make it consistent with the cooperative principle after all. That's why we interpret the review about a performer "producing a series of notes" as a snark. The author intentionally violated the maxim of Manner (he was not succinct); readers assume he was providing the kind of information they seek in a review; the readers conclude that the reviewer was implicating that the performance was substandard. Grice called this line of reasoning a conversational implicature. It is not logically necessary; the reviewer could have canceled the inference, without contradicting himself, by continuing, "and they were the most glorious sounds this reviewer has heard in years." Yet in the absence of such a proviso, the speaker counts on the listener to get his drift.

Conversational implicatures power the comprehension of many kinds of nonliteral speech. In the film *Shakespeare in Love,* Will excitedly recounts the plot of his newest play:

> For killing Juliet's kinsman Tybalt, the one who killed Romeo's friend Mercutio, Romeo is banished. But the Friar who married them gives Juliet a potion to drink. It is a secret potion. It makes her seeming dead. She is placed in the tomb of the Capulets. She will awake to life and love when Romeo comes to her side again. By malign fate, the message goes astray which would tell Romeo of the Friar's plan. He hears only that Juliet is dead. And thus he goes to the apothecary. And buys a deadly poison. He

enters the tomb to say farewell to Juliet who lies there cold as death. He drinks the poison. He dies by her side. And then she wakes and sees him dead. And so Juliet takes his dagger and kills herself.

One of the producers says, "Well, that will have them rolling in the aisles." He has flouted the maxim of Quality, saying something that is patently not true. But thanks to an implicature, his audience interprets him as meaning that the play is overly depressing. This is the logic behind irony and sarcasm, including overstatements and understatements.

A violation of the remaining maxims can explain other rhetorical devices (some of them discussed in chapter 3). When the writer of a letter of recommendation says of a student that he has excellent hair, she is implicating by her violation of the maxim of Relevance that he is a mediocre scholar—the logic of the backhanded compliment. And when a student's T-shirt reads "I went to Harvard University for four years. It was OK," the violation of the maxim of Quantity tips us off that he thinks the university is overrated, that he is a cool detached dude, or both.

Grice came to conversation from the bloodless world of logic and said little about *why* people bother to implicate their meanings rather than just blurting them out. We discover the answer when we remember that people are not just in the business of downloading information into each other's heads but are social animals concerned with the impressions they make. An implicature involves two meanings: the literal content (sometimes called the sentence meaning) and the intended message (sometimes called the speaker meaning). The literal sentence meaning must be doing some work or the speaker would not bother to use it in the first place. In the implicatures we have just seen, the intended messages have been negative but the literal content has been positive or neutral. Perhaps speakers are trying to eat their cake and have it too—they want to impugn something they dislike while staving off the impression that they are whiners or malcontents. The psychologist Ellen Winner and her colleagues have shown that people have a better impression of speakers who express a criticism with sarcasm ("What a great game you just played!") than with direct language ("What a lousy game you just played!"). The sarcastic speakers, compared with the blunt ones, are seen as less angry, less critical, and more in control.[6] This may be cold comfort to the target of the sarcasm, of course, since criticism is more damaging when it is seen to come from a judicious critic than from a dyspeptic one.

TOUCHY, TOUCHY: THE LOGIC OF POLITENESS

The double message conveyed with an implicature is nowhere put to greater use than in the commonest kind of indirect speech of all, politeness. Politeness in linguistics does not refer to social etiquette, like eating your peas without using your knife, but to the countless adjustments that speakers make to avoid the equally countless ways that their listeners might be put off. People are very, very touchy, and speakers go to great lengths not to step on their toes. In their magisterial work *Politeness: Some Universals in Language Use,* the anthropologists Penelope Brown and Stephen Levinson (the same Levinson we met in chapters 3 and 4) extended Grice's theory by showing how people all over the world use politeness to lubricate their social interactions.[7]

Politeness Theory begins with Erving Goffman's observation that when people interact they constantly worry about maintaining a nebulous yet vital commodity called "face" (from the idiom "to save face").[8] Goffman defined face as a positive social value that a person claims for himself. Brown and Levinson divide it into positive face, the desire to be approved (specifically, that other people want for you what you want for yourself), and negative face, the desire to be unimpeded or autonomous. The terminology, though clumsy, points to a fundamental duality in social life, which has been discovered in many guises and goes by many names: solidarity and status, connection and autonomy, communion and agency, intimacy and power, communal sharing and authority ranking.[9] Later we will see how these wants come from two of the three major social relations in human life.

Brown and Levinson argue that Grice's Cooperative Principle applies to the maintenance of face as well as to the communication of data. Conversational partners work together, each trying to maintain his own face and the face of his counterpart. The challenge is that most kinds of speech pose at least some threat to the face of the hearer. The mere act of initiating a conversation imposes a demand on the hearer's time and attention. Issuing an imperative challenges her status and autonomy, as if the speaker feels entitled to boss her around. Making a request puts her in the position where she might have to refuse, earning her a reputation as stingy or selfish. Telling something to someone implies that she was ignorant of the fact in the first place. And then there are criticisms, boasts, interruptions, outbursts, the telling of bad news, and the broaching of divisive topics, all of which can injure the hearer's face directly. It's no surprise that the first thing out of our mouths when we address a stranger is a request for forgiveness: *Excuse me.*

But despite the many ways in which a speaker can touch a nerve, he can't be constantly walking on eggshells. People have to get on with the business of life, and in doing so they have to convey requests and news and complaints. The solution is to make amends with politeness: the speaker sugarcoats his utterances with niceties that reaffirm his concern for the hearer or that acknowledge her autonomy. Brown and Levinson call the stratagems positive and negative politeness, but better terms are sympathy and deference.

The essence of politeness-as-sympathy is to simulate a degree of closeness by pretending to want what the hearer wants for herself. Two witty lexicographers commented on this stratagem when defining the word *politeness* in their dictionaries. Samuel Johnson explained it as "fictitious benevolence"; Ambrose Bierce called it "the most acceptable hypocrisy." Two familiar examples of politeness are the impotent bidding of good fortune (*Be well, Have a nice day*) and the feigned inquiry into the person's well-being (*How are you?, How's it going?*). There are also lame compliments (*you look marvelous*), presumptions about the hearer's needs (*You must be hungry*), sound but useless advice (*Take care*), and the broaching of topics where agreement is inevitable, prompting Charles Dudley Warner's complaint that everyone talks about the weather but no one does anything about it.[10]

One step beyond fictitious benevolence is fictitious solidarity.[11] Speakers may address their hearers with bogus terms of endearment like *my friend, mate, buddy, pal, honey, dear, luv, brother, guys,* and *fellas.* They may use slang connected to an in-group they are presumed to belong to, as in *Lend me two bucks* or *two quid.* They may include the listener in their plans, as in *Let's have another beer.* They may combine a number of devices, such as soliciting the listener's agreement (*you know?*), hedging their opinions (*like, sort of*), acknowledging the listener's familiarity with the situation (*you know*), and confirming their listener's attention and approval by proffering statements as if they were questions. These, of course, are the signatures of the dialect that is commonly attributed to adolescents and

Californians (especially Valley Girls), but which has been rapidly spreading to other demographics. In 1993 the journalist James Gorman wrote:

> I used to speak in a regular voice. I was able to assert, demand, question. Then I started teaching. At a university? And my students had this rising intonation thing? It was particularly noticeable on telephone messages. "Hello? Professor Gorman? This is Albert? From feature writing?"
>
> I had no idea that a change in the "intonation contour" of a sentence, as linguists put it, could be as contagious as the common cold. But before long I noticed a Jekyll-and-Hyde transformation in my own speech. I first heard it when I myself was leaving a message. "This is Jim Gorman? I'm doing an article on Klingon? The language? From 'Star Trek'?" I realized then that I was unwittingly, unwillingly speaking uptalk. I was, like, appalled?[12]

Though uptalk probably began as a reflex of politeness (part of the twentieth-century trend toward egalitarianism and social closeness), it is becoming a neutral feature of standard American English, as it has been for centuries in some Irish, English, and Southern American dialects. The spread of uptalk is a rare case in which we can feel what it's like to be part of a historical change in the language, watching a construction as it tips from having a transparent rationale to being just a convention.

The politeness story gets more interesting, and gets us back on track toward indirect speech, when we turn to the gestures of verbal deference (what Brown and Levinson call negative politeness). Commands and requests are among the most face-threatening speech acts, because they challenge the hearer's autonomy by assuming her readiness to comply. The speaker is ordering the hearer around, or at least putting her out, something you don't do to a stranger or a superior and might even think twice about doing with an intimate. So requests are often accompanied by various forms of groveling:

> *Questioning rather than commanding:* Will you lend me your car?
> *Expressing pessimism:* I don't suppose you might close the window.

Hedging the request: Close the door, if you can.

Minimizing the imposition: I just want to borrow a little bit of paper.

Hesitating: Can I, uh, borrow your bicycle?

Acknowledging the impingement: I'm sure you're busy, but . . .

Indicating reluctance: I normally wouldn't ask, but . . .

Apologizing: I'm sorry to bother you, but . . .

Impersonalizing: Smoking is not permitted.

Acknowledging a debt: I'd be eternally grateful if you would . . .

Politeness is calibrated to the level of the threat to the hearer's face. The threat level in turn depends on the size of the imposition, the social distance from the hearer (the lack of intimacy or solidarity), and the power gap between them. People kiss up more obsequiously when they are asking for a bigger favor, when the hearer is a stranger, and when the hearer has more status or power. A fully loaded request like "I'm terribly sorry to trouble you, and I wouldn't ask unless I were desperate, but I'd be eternally grateful if you think you could possibly . . ." would sound smarmy if it were used to ask a stranger for a small favor like the time, or if it were used to ask a bigger favor (like the use of a computer) of a spouse or an assistant.

The two social dimensions that go into face threats, solidarity and power, are nowhere more present than in forms of address, in which a speaker has grabbed the hearer's attention and now must be careful what he does with her, like having a tiger by the tail. Many languages have two forms of the second-person pronoun, such as the French *tu* and *vous,* the Spanish *tu* and *usted,* and the German *du* and *Sie.* English used to have this distinction in *thou* and *ye* (now *you*), but *thou* is now restricted to prayers and other old-fashioned speech styles. The distinction between the T and the V pronouns, as linguists call them, hinges on both solidarity and power, with T being familiar (addressed to intimates or subordinates) and V being respectful (addressed to strangers or superiors).[13] Intimates typically address each other with T, while strangers address each other with V. Subordinates address superiors with V; superiors address subordinates with T. The details vary with the language, the dyad (children and parents, waiters and customers, teachers and students), and the historical period. V pronouns are often banned in self-consciously egalitarian contexts, such as the aftermath of the French Revolution, the meetings of socialist parties, and religious communities like the traditional Quakers, who preserved the archaic *thou* and *thee.* The psycholinguist Roger Brown and the English scholar Albert Gilman showed that most Western languages have hosted an inexorable trend. Power has been losing ground to solidarity, so that strangers of all ranks are addressed with V (such as a customer addressing a salesclerk), and intimates of all ranks are addressed with T (such as an adult addressing a parent). With these conventions as a backdrop, people can deliberately wield pronouns as a face-challenging weapon. To address someone with T (in French, *tutoyer*) who ordinarily expects V can convey disrespect; to address someone with V (*vouvoyer*) who ordinarily expects T can convey chilliness. In *Twelfth Night,* Sir Toby Belch tries to egg on Andrew Aguecheek to challenge the disguised Viola as follows: "Taunt him with the license of ink, if thou thou'st him some thrice, it shall not be amiss."

The respectful V pronoun often evolves from a generic second-person *plural* pronoun, and may continue to do double duty in the language. That is, speakers use the V-form when addressing more than one hearer, regardless of power or solidarity. The historical source of these mergers is that a language without a respectful pronoun tends to co-opt a plural pronoun for that purpose because its speakers are repeatedly tempted to use the plural pronoun in addressing a superior or a stranger.[14] One reason is to bow to a hearer's power: rather than calling attention to the fact that the hearer is the sole target of one's communiqué, one speaks as if she had a formidable entourage behind her. The other reason is that a plural pronoun seems to give the hearer the option of ignoring you, as if you were addressing a diffuse crowd rather than drawing a bead on her alone. In many societies the trepidation about using the second-person singular pronoun extends to a taboo on addressing people by their names at all; people are addressed only by honorifics or euphemisms.[15] In our own society, certain demigods cannot abide having the riffraff single them out as if they were mere individuals; a supplicant has to address himself to one of their exalted qualities by using a term such as *Your Highness, Your Worship, Your Eminence, Your Excellency, Your Honor,* or *Your Grace.* Even with commoners, to finger a hearer with the singular *you* can be patently disrespectful, as when the time for politeness has passed and one yells at a child, *You get that dog out of here right now.* And what is the quintessentially rude way to claim someone's attention? *Hey, you!*

Though English speakers have lost the distinction between *thou* and *you,* the delicacy of referring to people in their presence plays out in other forms of address. There is still an etiquette to the use of *Professor Pinker* versus *Steve,* the former typically used by undergraduates in large courses and strangers writing for a favor, the latter by undergraduates working in my lab and graduate students and colleagues in my department. When it comes to addressing untitled adults in a respectful manner, the default honorifics *Mr., Mrs., Miss,* and *Ms.* are available, though people don't use them with children, subordinates, or intimates—hence the expression *to be on a first-name basis.* In *The Complete Upmanship,* Stephen Potter shows how the two dimensions of politeness define an arc of formality in the ways that a British company president addresses his subordinates:[16]

The Guv'nor addresses:

Co-director Michael Yates as	MIKE
Assistant director Michael Yates as	MICHAEL
Sectional manager Michael Yates as	MR. YATES
Sectional assistant Michael Yates as	YATES
Indispensable secretary Michael Yates as	MR. YATES
Apprentice Michael Yates as	MICHAEL
Night-watchman Michael Yates as	MIKE

As with T and V pronouns, terms of address can change over time. Graduate students used to address their professors as *Professor,* and my grandparents addressed even their closest friends as *Mr. and Mrs. So-and-So.* A common complaint to hospitals from many older patients is that they feel disrespected when younger doctors address them by their first names. Oddly enough, though people are less likely these days to be addressed with fancy titles in real life, they are more likely to be granted them in officialdom. In universities nowadays, almost every administrator is a Dean, and in corporations every executive is a Vice President. Title inflation is most keenly felt at the top, where potentates have to bestow themselves with ever more lofty titles. Haile Selassie, the emperor of Ethiopia from 1934 to 1974, contented himself with "King of Kings, Elect of God, Conquering Lion of Judah." King Juan Carlos of Spain has thirty-eight titles, including "Duke of Athens" and "Sovereign Grand Master of the Order of the Golden

Fleece." Not to be outdone, North Korea's Kim Jong Il claims the titles "Guardian of Our Planet," "Lodestar of the Twenty-first Century," and a thousand others of his own design.[17]

Politeness, in the linguists' sense, appears to be a human universal. Forms of address that distinguish power and distance have been documented in more than twenty-five languages in which linguists have looked for them. Brown and Levinson meticulously documented a full range of polite forms in two languages that are geographically and culturally remote from Europe: Tzeltal, the Mayan language spoken in Mexico and discussed in chapter 3, and Tamil, a non-Indo-European language spoken in south India and Sri Lanka. They found counterparts to each of the forms of deference and sympathy they had documented in English, differing only in the details. Pessimistic requests, for example, complete with a negation, a subjunctive, and a postscript, can be heard in Tamil ("Here you wouldn't have brought money, would you?") and in Tzeltal ("You wouldn't perhaps sell your chicken, it was said"). The asking for reasons not to do something, as in the English *Why don't we go to the seashore?*, finds counterparts in the Tzeltal "Why don't you lend us-all your record-player?" and the Tamil "Why shouldn't we go to Kangayam?" Brown and Levinson bolster their compendium of politeness strategies with examples from many other languages as well.[18]

Needless to say, cultures also differ in their expected use of politeness. That's why the verbal gaffe that threatens to cause an international incident is a staple of lighthearted travel writing. Cultures differ not only in the exact wording of their politeness strategies but in their sensibilities as to when each strategy should be used. They differ in how readily people perceive everyday acts to be face-threatening and hence in need of mitigating politeness. They differ in whether the denizens are more obsessed with power or with distance and hence feel more entitled to deference or to sympathy. And they differ in who merits which kind of politeness from which kinds of speakers (women, children, teachers, waitstaff, strangers, and so on). Differences in the settings of these knobs can make a culture feel distant and suspicious to outsiders, or boastful and self-important, or warm and friendly, or stiff and formal. In *Dave Barry Does Japan,* the humorist tries to capture the difference between a culture that calls for a great deal of deferential politeness and a culture that calls for a modicum of sympathetic politeness:

Typical Japanese business meeting:
First businessman: Hello, sir.
Second businessman: Hello, sir.
First businessman: I am sorry.
Second businessman: I am extremely sorry.
First businessman: I cannot stand myself.
Second businessman: I am swamp scum.
First businessman: I am toenail dirt.
Second businessman: I should be put to death.

Typical American business meeting:
First businessman: Bob!
Second businessman: Ed!
First businessman: How they hangin'?
Second businessman: One lower than the other!
First businessman: Har!
Second businessman: Listen, about those R-243-J's, the best we
    can do for you is $3.80 a unit.
First businessman: My ass, Bob.
Second businessman: Har!

Some cultures are famous for how little call they have for deferential politeness. One of them is Israel, whose native-born citizens are called sabras, the Hebrew word for an edible cactus that is said to be prickly on the outside and sweet on the inside. Another is identified in the joke in which four people are walking down the street: a Saudi Arabian, a Russian, a North Korean, and a New Yorker. A reporter rushes up to them and says, "Excuse me, can I get your opinion of the meat shortage?" The Saudi Arabian says, "Shortage—what's a shortage?" The Russian says, "Meat—what's meat?" The North Korean says, "Opinion—what's an opinion?" The New Yorker says, "Excuse me—what's excuse me?"[19]

Why do cultures differ in their degrees and kinds of politeness? Hierarchical societies are more obsessed with reinforcing power than are egalitarian societies, and expect more deferential politeness. Elites, whose status rests on social conventions and who resent impositions, tend to keep more social distance around them, and use and expect more politeness. The masses, who are powerless and more interdependent, use less politeness all around, and what they do use tends to be more sympathetic than deferential. Other aspects of history, class, ideology, and ecology may go into this tuning as well. The social psychologist Richard Nisbett has noted that societies with a "culture of honor," where affronts must be quickly and sometimes violently redressed, are often extremely polite, because an unintended insult can escalate into a duel or a feud.[20] The American South was historically a culture of honor, and Nisbett, a native Texan, recalls that his first impression on arriving at Yale as a freshman was that everyone seemed to be really rude.

Now that we have explored the mentality of politeness, let's get back to our topic, indirect speech. Beneath all the layers of honorifics and apologies and hedges that go into deferential politeness, there has to be a request in there somewhere, and it, too, must be couched with delicacy. Often the request is not stated baldly but conveyed indirectly with the help of an implicature. The result is a whimperative like *Can you pass the salt?* or *If you could pass the salt, that would be terrific.* Taken literally, the first example violates the maxim of Relevance, because the answer to the question is already known. The second one violates the maxim of Quality, because the consequent of the conditional is an overstatement. So the hearer interprets them as requests (a reasonable conclusion if she's sitting at the end of the table with the salt shaker and the speaker is sitting at the end without it), while noting from the literal wording that the speaker was seeking to avoid the appearance of treating her like a flunky.

The literal wording of a polite request can't be just any non sequitur. Typically it brings up one of the prerequisites to making a sensible request, which linguists call felicity conditions. It makes no sense to ask someone to pass the salt if you already have the salt, if you don't like salt, if you don't want the hearer to pass you the salt, if the hearer is incapable of passing the salt, if the hearer is unwilling to pass the salt, or if you're certain that the hearer will not pass the salt. Every one of these prerequisites can be asserted, questioned, or doubted as a roundabout way of asking for the salt:[21]

> There isn't any salt down here.
> I could use some salt.

> I'd appreciate it if you could pass the salt.
> Can you pass the salt?
> Could you possibly pass the salt?
> Would you pass the salt?
> Is there any salt down there?
> I was wondering if you might pass the salt.

The underlying rationale is that the hearer is not given a command or request but is asked or advised about one of the necessary conditions for her to pass the salt. Should she be disinclined to comply, she could exercise her prerogative not to do so without having to issue a face-threatening refusal. For example, if the indirect request was made as a comment on the state of affairs, all she would have to do is to ignore it, since a comment does not demand a response. If it was framed as a question about one of the prerequisites for passing the salt, she could, in theory, deny that they are in place. In all these cases, she is off the hook.

Now, no one believes that the salt-seeker and the salt-passer literally work through these deductions in their heads as they negotiate over the salt. At this stage in the history of the language, the implicatures have been frozen into conventions.[22] Formulas like *Can you pass the salt?* are mostly opaque, like idioms and dead metaphors, and are used as straightforward requests. We see this in the way that whimperatives use pro forma openings like *Can you* rather than other wordings with the same meaning, such as *Are you capable of passing the salt?* And often we ignore the literal meaning of an indirect request until it is called to our attention, as when a snarky thirteen-year-old who is asked if he can pass the salt mischievously nods and then just sits there. In a joke told by a certain ethnic group, an old couple is lying in bed when the wife says, "Irving, it's cold outside." Irving gets up, closes the window, and says to her, "So now it's *warm* outside?"

Nonetheless, the wording of indirect requests embalms a line of reasoning that must have been at work in the minds of earlier speakers. We don't, after all, request salt with just any old reference to the stuff, such as *Salt was first harvested from seawater by the Phoenicians* or *Salt is an ionic compound of sodium and chlorine.* Also, the literal content of indirect requests is similar in many languages, too many to be a coincidence. Indirect requests have fossilized a certain line of reasoning: the logic of plausible deniability, of giving the hearer an "out." Indeed, some aspects of the literal meaning still have to be kept in mind. The psychologist Herbert Clark points out that in couching a request as a question, the speaker must single out an obstacle to a reply that is at least minimally plausible. You can request the time by asking *Do you know the time?* but you would never request someone's middle name by asking *Do you know your middle name?* We all know our middle name, so even within the make-believe world of politeness, a reluctant hearer could never use such ignorance as an out.

Because a clichéd indirect request is recognized *as* a request by any competent English speaker, it is effectively "on the record." A speaker who says *Can you pass the salt?* in ordinary dinnertime circumstances cannot plausibly deny that he has asked for something. But according to Brown and Levinson, if an indirect speech act is freshly minted rather than pulled off the shelf, its effect on the hearer is different. The request is now "off the record." When a speaker thinks up a novel indirect request, like *The chowder is pretty bland* or *They never seem to have enough salt shakers at this restaurant,* the hearer can ignore the comment without publicly and indubitably rebuffing the request. For this reason, Brown and Levinson argue that off-record indirect speech acts coined for the occasion—hints, understatements, idle generalizations, and rhetorical questions—are the politest forms of all. A speaker can say *It's too dark to read* as a way to ask a hearer to turn on the lights, or *The lawn has got to be mowed* instead of "Mow the lawn," or *It looks like someone may have had too much to drink* instead of "You're drunk." Since polite indirect speech can use any hint that cannot be pinned down as a request by its literal content, but that can lead an intelligent hearer to infer its intended meaning, there is no limit to the number of forms it can take.

And now we can reconnect with the problem that kicked off this chapter, indirect speech. Politeness theory gives us insight into one variety, the off-record request. The hearer is implicitly given the opportunity to ignore the request without a public refusal, which also means that if she *complies* with the request, it's not because she's taking orders. And according to Brown and Levinson, this saves face for both of them, especially the hearer with her desire for autonomy.

So does this explain the puzzle? Not yet. First we need to see how good the evidence is. Politeness Theory has been tested in many experiments in which psychologists ask people what they would say in certain circumstances, or ask them to rate the politeness of things that other people might say.[23] Many parts of the theory have been confirmed. Not surprisingly, the use of the proposed politeness strategies indeed makes a request sound more polite. Indirect requests do sound more polite than direct ones. And the degree of imposition matters, as does the relative power of the speaker and the hearer.

But one hypothesis didn't come out as well. Brown and Levinson claimed that face threat was a single scale, the result of adding up the power disparity, the social distance, and the degree of imposition. They claimed that the three kinds of politeness were arranged along a scale, too. Sympathy expresses a little bit of politeness, and is suitable for smaller face threats. Deference expresses more, and is suitable for bigger ones. And off-record indirect speech acts (ones coined for the occasion) express the most politeness, and are suitable for the biggest threats.

But in both cases Brown and Levinson may have jammed qualitatively different dimensions onto a single scale. Rather than having a single face-threat meter in their heads, and a single politeness meter that tracks it, people tend to target certain *kinds* of face threat with certain *kinds* of politeness. For instance, to criticize a friend (which threatens solidarity), people tend to emphasize sympathetic politeness ("Hey, buddy, let's go over this paper and see if we can find ways to bring it up to your high standards"). But to ask a big favor (which threatens power), people tend to emphasize deferential politeness, as in the cringing request to borrow someone's computer ("I'm terribly sorry to trouble you . . .") that we saw on page 383.[24]

Also, off-record indirect speech—the obsession of this chapter—didn't fit into the scale at all. Politeness Theory deemed it the politest strategy of all, but people said it was far *less* polite than deferential politeness.[25] In fact, in some cases it can be downright rude, like *Didn't I tell you yesterday to pick up your room?* or *Shouldn't you tell me who is coming to the party?* One reason is that if the hearer's competence and willingness are questioned too blatantly, it suggests that she is inept or uncooperative. Another is that an indirect request can make the speaker sound devious and manipulative, and force the listener to do a lot of mental spadework to figure out what he was trying to say. This was the crux of Julie's complaint in the girl-talk scene from *Tootsie*.

The fact that indirect speech acts are not so considerate to the hearer after all brings up another problem. The examples with which I began the chapter—veiled threats, oblique bribes, sexual come-ons—are hardly examples of a speaker being *polite*! I'm sure a merchant listening to an advisory from the local racketeer on the many accidents that can befall a store doesn't see it that way. And the cop with his ticket book, or the woman at the elevator door, sensing the indecent proposal in the innocent question, could be forgiven for thinking that the propositioner was looking out for his interests, not theirs (though as we shall see, there can be complicity in those cases as well).

A final problem for Politeness Theory is the built-in dilemma in its treatment of off-record requests. If an implicature is too much of a treasure hunt, the speaker will have missed an opportunity. The hearer might have been perfectly happy to comply with his request, if only she knew he was making one! (Remember George on *Seinfeld*, who failed to realize until too late that *coffee* didn't mean coffee.) On the other hand, if the implicature is such a doddle that the hearer can figure it out without fail, then it should be obvious enough for any other intelligent person to figure out, too, so it's not clear why the request should be perceived as being "off the record." Who could claim to be fooled by the line about the etchings, or about settling the ticket here in Brainerd?

Politeness Theory is a good start, but it's not enough. Like many good-of-the-group theories in social science, it assumes that the speaker and the hearer are working in perfect harmony, each trying to save the other's face.[26] (Grice himself was guilty of this in trying to deduce the laws of conversation from a "Cooperative Principle.")[27] We need to understand what happens when the interests of a speaker and a hearer are partly in conflict, as they so often are in real life. And we need to distinguish the *kinds* of relationships people have, and how each is negotiated and maintained, rather than stringing all forms of face threat into a single scale, and doing the same with all forms of face saving. Finally, we need a deeper analysis of the enigmatic commodity called "face," and how it depends on the equally elusive "record" such that requests can be "on" it or "off" it.

MUDDY IT UP: VAGUENESS, DENIABILITY, AND OTHER STRATEGIES OF CONFLICT

To get some purchase on nebulous concepts like "providing an out," "plausible deniability," and "on the record," let's begin with a scenario in which their meanings are clear-cut. Consider a perfect Gricean speaker who says exactly what he means when he says anything at all. Maxim Man is pulled over for running a red light and is pondering whether to bribe the officer. Since he obeys the maxims of conversation more assiduously than he obeys the laws of traffic or the laws of bribery, the only way he can bribe the officer is by saying, "If you let me go without a ticket, I'll pay you fifty dollars."

Unfortunately, he doesn't know whether the officer is dishonest and will accept the bribe or is honest and will arrest him for attempting to bribe an officer. Any scenario like this in which the best course of action depends on the choices of another actor is in the province of game theory. In game theory, the conundrum where one actor does not know the values of the other has been explored by Thomas Schelling, who calls it the Identification Problem.[28] The payoffs can be summarized like this, where the rows represent the driver's choices, the columns represent the different kinds of officer he might be facing, and the contents of the squares represent what will happen to the driver:

	Dishonest officer	Honest officer
Don't bribe	Traffic ticket	Traffic ticket
Bribe	Go free	Arrest for bribery

The allure of each choice (row) is determined by the sum of the payoffs of the two cells in that row weighted by their probabilities. Let's consider them in turn. If the driver doesn't try to bribe the officer (first row), then it doesn't matter how honest the officer is; either way the driver gets a ticket. Nothing ventured, nothing gained. But if he does offer the bribe (second row), the stakes are much higher either way. If Maxim Man is lucky and is facing a dishonest cop, the cop will accept the bribe and send him on his way without a ticket. But if he is unlucky and is facing an honest cop, he will be handcuffed, read his rights, and arrested for bribery. The rational choice between bribing and not bribing will depend on the size of the traffic fine, the proportion of bad and good cops on the roads, and the penalties for bribery, but neither choice is appealing. Maxim Man is caught between the devil and the deep blue sea.

But now consider a different driver, Implicature Man, who knows how to implicate an ambiguous bribe, as in "So maybe the best thing would be to take care of it here." Suppose he knows that the officer can work through the implicature and recognize it as an intended bribe, and he also knows that the officer knows that he couldn't make a bribery charge stick in court because the ambiguous wording would prevent a prosecutor from proving his guilt beyond a reasonable doubt. Implicature Man now has a third option:

	Dishonest officer	Honest officer
Don't bribe	Traffic ticket	Traffic ticket
Bribe	Go free	Arrest for bribery
Implicated bribe	Go free	Traffic ticket

The payoffs in this new, third row combine the very big advantage of bribing a dishonest cop with the relatively small penalty of failing to bribe an honest one. It's an easy choice. We have explained the evolution of Implicature Man.

Well, almost. We also have to take the point of view of an honest officer and the legal system he serves. Why wouldn't an honest officer arrest anyone who offered a veiled bribe? If it's obvious to him, it might be obvious to a jury, so he has a chance of putting a bad guy behind bars. To explain why the officer wouldn't arrest people at the hint of a bribe, making implicature as dangerous as naked bribery, we must assume two things, both reasonable. One is that even if all dishonest drivers offer remarks that can be interpreted (correctly) as implicated bribes, some honest drivers make those remarks, too, as innocent observations. So any arrest might be a false arrest. The second assumption is that an unsuccessful arrest is costly, exposing the officer to a charge of false arrest and the police department to punitive damages. Then the *officer's* decision matrix would look like this:

	Dishonest driver	Honest driver
Don't arrest	Traffic ticket	Traffic ticket
Arrest	Successful conviction	False arrest

(Of course from his point of view a traffic ticket is a good thing, not a bad thing.) The appeal of arresting the driver will depend on the values of the outcomes in the four cells and on their probabilities. And those probabilities will depend on the proportion of dishonest and honest drivers who utter the ambiguous remark, that is, on the ratio of the numbers of events in the left and right columns. If the remark sounds close enough to an innocuous remark that plenty of honest drivers might make it (or, at least, enough of them so that a jury could not convict the speaker for those words beyond a reasonable doubt), then the odds of a successful conviction go down, the odds of a false arrest go up, and the appeal of the "Arrest" row would be lowered. And that is how Implicature Man can force the officer's hand. He can craft his remark so that a dishonest officer will detect it as an implicated bribe, but an honest officer can't be sure (or at least can't take the chance) that it is one. (Note, by the way, that contra most linguists, this analysis shows that indirect speech is not an example of pure cooperation. Implicature Man is manipulating an honest officer's choices to his own advantage and to the officer's disadvantage. It is consistent with the theory by the biologists Richard Dawkins and John Krebs that communication in the animal kingdom can often be a form of manipulation, not just information-sharing.)[29]

In real life, veiled bribes are a conundrum for law enforcement and the legal system. In most cases, American courts go with common sense and treat a veiled bribe like a naked one. If a defendant were to say, "I didn't try to bribe him; I was just asking (out of sheer curiosity) if there was some way to pay the ticket there," it would not, as they say, pass the giggle test. So real-life bribes have to be veiled more discreetly, with some risk that they might go over the head of a bribable officer. But there are other real-life circumstances in which even an obvious veiled bribe can be effective. These are cases in which freedom of speech is an issue. Bribery, by definition, must be treated as an exception to the right of free speech, so courts are mindful that it be defined narrowly when the political process is involved, such as in speaking with a political representative. In that case, the distinction between direct and indirect speech can be decisive.[30]

This lesson in linguistics came too late to a lobbyist for the National Organization for Women named Wanda Brandstetter, who in 1980 was convicted of bribery for her effort to get an Illinois state representative to vote for ratification of the Equal Rights Amendment. Brandstetter had handed him a business card on which she had written, "Mr. Swanstrom the offer for help in your election plus $1000 for your campaign for the pro-E.R.A. vote." The prosecutor called it "a contract for bribery," and the jury agreed. The verdict may sound incredible to most Americans, who are thoroughly jaded about the influence of money in the political process. What are all those lobbyists in Gucci Gulch doing if not bribing legislators? The answer is that they do it with implicatures. If Brandstetter had said, "As you know, Mr. Swanstrom, NOW has a history of contributing to political campaigns. And it has tended to contribute more to candidates with a voting record that is compatible with our goals. These days one of our goals is the ratification of the ERA," she would not have had to pay five hundred dollars, put in 150 hours of community service, and spend a year on probation.[31]

The strategic use of vagueness also has a long history in diplomacy. According to a joke from another era, here is the difference between a lady and a diplomat:

> When a lady says "no," she means "maybe."
> When she says "maybe," she means "yes."
> If she says "yes," she's no lady.
> When a diplomat says "yes," he means "maybe."
> When he says "maybe," he means "no."
> If he says "no," he's no diplomat.

This prompted a feminist revision:

> When a woman says "yes," she means "yes."
> When a woman says "maybe," she means "maybe."
> When a woman says "no," she means "no."
> If the man persists, he's a rapist.

The revision may be a good policy for relations between the sexes, but when it comes to diplomacy, the original stands. In an op-ed entitled "The Language of Diplomacy," Michael Langan, a former American treasury official, recounts:

> At one point in my federal government career, I wrote up an explanation of a complicated matter in what I considered an extremely clear, cogent manner. The senior government official to whom I reported read it carefully, ruminating and adjusting his glasses as he read. Then he looked up at me and said, "This isn't any good. I understand it completely. Take it back and muddy it up. I want the statement to be able to be interpreted two or three ways." The resulting ambiguity enabled some compromise between competing governmental interests.[32]

In the wake of the Six-Day War in 1967, the United Nations Security Council passed its famous Resolution 242, which, among other things, called for "withdrawal of Israeli armed forces from territories occupied in the recent conflict." The phrase *territories occupied in the recent conflict,* a noun phrase lacking an overt quantifier, is ambiguous between "some of the territories" and "all of the territories." The resolution was palatable to Israel and its allies under the former interpretation, and to the concerned Arab states and their allies under the latter. Any unambiguous version would have been rejected by one side or the other.

But as the feminist response to the joke about the lady and the diplomat suggests, creative vagueness can also be dangerous. For forty years partisans have been arguing over the semantics of Resolution 242 like graduate students in a linguistics seminar (another illustration of the high stakes of "mere semantics").[33] And of course the larger dispute between Israel and the Arab countries remains unresolved, to put it mildly. Strategic vagueness can also backfire in business agreements. In one famous case, Johnson & Johnson invested ten million dollars in Amgen at an early point in its history, and obtained the right to a compound for "secondary uses." The meaning of *secondary* was left ambiguous, and the ambiguity may have helped them to "get to yes," as negotiators like to put it. But since then the two companies have spent $350 million in legal fees to resolve the ambiguity, and each one hates the other's guts.[34]

So is creative vagueness a smart tactic? If all it does is postpone the day of reckoning, why don't negotiators either hammer out a clear-cut agreement or acknowledge the irreconcilable differences and walk away? In the case of international diplomacy, one reason is that the language of an agreement has to be palatable not just to the leaders but to their citizenries. Reasonable leaders might come to an understanding between themselves while each exploits the ambiguity of the agreement to sell it to his more bellicose constituency. Another is that merely having an agreement, even one that is tricky to implement, can yoke two enemies in a common purpose, possibly easing their hatred by pure symbolism. Finally, diplomats can gamble that times will change and circumstances bring the two sides together, at which point they can resolve the vagueness amicably. According to an old story, a man sentenced to be hanged for offending the sultan offered a deal to the court: If they would give him a year, he would teach the sultan's horse to sing, earning his freedom; if he failed, he would go to the gallows willingly. When he returned to the dock, a fellow prisoner said, "Are you crazy?" The man replied, "I figure, over the course of a year, a lot can happen. Maybe the sultan will die, and the new sultan will pardon me. Maybe I'll die; in that case I wouldn't have lost a thing. Maybe the horse will die; then I'll be off the hook. And who knows? Maybe I'll teach the horse to sing!"

These examples of creative vagueness come from arenas in which a person's words are on the record and the stakes are tangible: traffic tickets, arrests for bribery, war and peace. What about everyday life, where offers and requests can be tendered without fear of legal penalties? After all, no one expects the Spanish Inquisition. In the give-and-take of ordinary conversation we should be free to speak our minds, without worrying that the way a hearer parses our words could land us in jail. But in fact it doesn't work that way. When it comes to everyday bribes, threats, and offers, our own emotions make us watch our words as carefully as if we were in legal jeopardy, and we all turn into Implicature Man.

"Everyday bribes"? When would a law-abiding citizen be tempted to offer a bribe? How about this: You want to go to the hottest restaurant in town. You have no reservation. Why not offer fifty dollars to the maitre d' if he will seat you immediately? This was the assignment given to the writer Bruce Feiler by *Gourmet* magazine in 2000.[35] The results are eye-opening.

The first result is predictable to most people who imagine themselves in Feiler's shoes: the assignment is terrifying. Though no one, to my knowledge, has ever been arrested for bribing a maitre d', Feiler felt like the kidnapper in *Fargo:*

> I am nervous, truly nervous. As the taxi bounces southward through the trendier neighborhoods of Manhattan—Flatiron, the Village, SoHo—I keep imagining the possible retorts of some incensed maitre d'.
> "What kind of establishment do you think this is?"
> "How dare you insult me?"
> "You think you can get in with *that*?"

The second result is that when Feiler did screw up the courage to bribe a maitre d', he thought up an indirect speech act on the spot. He showed up at Balthazar, a popular restaurant in Manhattan, and with sweaty skin and a racing heart he looked the maitre d' in the eye, handed him a folded twenty-dollar bill, and mumbled, "I hope you can fit us in." Two minutes later they were seated, to the astonishment of his girlfriend. On subsequent assignments he implicated the bribes with similar indirectness:

> I was wondering if you might have a cancellation.
> Is there any way you could speed up my wait?
> We were wondering if you had a table for two.
> This is a really important night for me.

The payoff matrix is identical in structure to the one for bribing a police officer:

	Dishonest maitre d'	Honest maitre d'
Don't bribe	Long wait	Long wait
Bribe	Instant seating	Public humiliation
Implicated bribe	Instant seating	Long wait

The third lesson—as intriguing to a restaurant-goer as it is to a psycholinguist—is that *it worked every time.* Feiler was invariably seated within two to four minutes. On a dare from his editor, he then tried to get into Alain

Ducasse, a new French restaurant where a meal cost $375, the seats were booked six months in advance, and "if you order verbena tea they bring the *plant* to your table and a white-gloved waiter snips the leaves with silver shears." Feiler showed up, asked the maitre d' whether he might have a cancellation, and slid a hundred-dollar bill his way. The maitre d' responded with a look of "complete and utter horror." He said, "No, no, monsieur. You don't understand! We only have sixteen tables. There is absolutely no way!" Feiler left his business card anyway, discreetly covering the money. Two days later his phone rang and he was offered a table for four. Thanks to a C-note and a Gricean implicature, Feiler had jumped a 2,700-person queue.

When a maitre d' says "no," he means "maybe." Like Julie in *Tootsie*—like all of us—the restaurants were systematically hypocritical. When Feiler called them up and asked about their policy on accepting money for a table, the responses ranged from "It's disgusting" to "The maitre d' will be fired if he is caught doing that." Yet in every case their palms could be greased. The logic of plausible deniability is part of the answer. The indirectness of the speech act solved the game-theoretic Identification Problem by allowing Feiler to tender a bribe without risking a social penalty. But it must have done something more, since *all* the restaurants were in fact bribable. Somehow the implicated nature of the bribe allowed *both* sides to pretend that they could deny that they had transacted a bribe, as if they thought a hidden tape recorder might be running and they could be indicted by a prosecutor in a court of fancy-restaurant etiquette.

The plot thickens. Why would the prospect of being turned down for a bribe—or some other offer, like a sexual advance or a request for a donation—be so terrifying? And if the transaction does occur, why should keeping it "off the record" (what record?) make it easier on both parties? To answer these questions we need to leave linguistics and game theory for evolutionary social psychology, where we can seek out the roots of embarrassment and taboo.

SHARING, RANKING, TRADING: THOUGHTS ABOUT RELATIONSHIPS

What's so scary about bribing a maitre d'? The worst he can do is say no. It seems somehow unethical, but why? We pay for expedited service in parcel delivery, first-class travel, and other commercial transactions, and we tip all kinds of personnel, such as cabdrivers and tour guides, for better service after the fact. But somehow we feel that a maitre d' claims a different *kind* of relationship with us, one that excludes the quid pro quo exchanges that are unexceptionable elsewhere. To attempt to transgress that boundary feels embarrassing, even immoral.

The anthropologist Alan Fiske has developed a sweeping theory of human sociality which lays out the major kinds of relationships we have with one another, and the thoughts, emotions, and social practices that maintain them.[36] As in Brown and Levinson's Politeness Theory, one of the relationship types centers on solidarity and the other on power. But Fiske argues that the two have very different logics, rather than being two sides of a coin called "face." And Fiske adds a third relationship type, centered on social exchange. The three relationship types are rooted in our evolutionary history, and each applies instinctively to certain kinds of human dyads. But with the use of certain channels of communication, including language, we can try to force the mindset of a given relationship type onto other dyads. These negotiations drive many of the cultural practices documented by anthropology, and as we shall see, they seem to motivate the "acceptable hypocrisy" of indirect speech.

The first relationship type is called Communal Sharing, communality for short. Its underlying logic is "What's mine is thine; what's thine is mine." This is the relationship measured out as "social distance" in Politeness Theory, and it is the one safeguarded by the emotion called "positive face," the desire that other people want for you what you want for yourself. Communality arises naturally among blood relatives, for reasons that are obvious to an evolutionary biologist. In our evolutionary history, any gene that predisposed a person to be nice to a relative would have had some chance of helping a copy of *itself* inside that relative (since relatives share genes), and it and its copies would have been favored by natural selection and entrenched in the genome.[37] A shared genetic inheritance is not the only tie that binds. A lifelong monogamous couple have their genetic fates merged in a single package, their children, so what is good for one is good for the other (at least if they downplay the competing ties of their blood relatives). Also, shared tastes or shared enemies can bind friends in a compact of common interest. If two roommates have similar tastes in music, one roommate will benefit the other every time she brings home a CD, so each ought to value the other's well-being, a social analogue to the relationship that ecologists call mutualism and economists call positive externalities.[38]

Many readers misunderstand evolutionary arguments to mean that organisms actually calculate their genetic relatedness to one another and calibrate their behavior accordingly, a preposterous idea. Even most sociobiologists give little thought to how common interests in genes or resources translate into altruistic behavior, treating organisms as if they were zombies programmed to carry out the dictates of their genes. In fact, communality is implemented in the mind as an *emotion* and as a set of *ideas*. The emotion is the warm and fuzzy feeling we get when we commune with our friends and relations. The ideas consist of conceptual metaphors. One is that SOLIDARITY IS PHYSICAL CLOSENESS, the source of many idioms for solidarity, including the technical term *social distance* itself. Another is that SOLIDARITY IS BEING CONNECTED, which we met at the outset of chapter 5 in the Declaration of Independence (*bonding, ties,* and so on). And a particularly potent metaphor is SOLIDARITY IS BEING MADE OF THE SAME FLESH. These intuitions go into the primal acts that bond us to those we love: the cuddling of an infant by a mother, the embrace of two lovers, the handshaking and hugging among friends, the exchange of bodily fluids in breastfeeding and sex, the sharing of food within a family and among close friends.

Communal sharing is never perfect, even among close kin and friends, and it is even harder to maintain among casual acquaintances and loose affiliates. The reason is that when there is an ethos for everyone to take what they want without encumbrance or accounting, people with looser genetic or mutualist ties to their fellows are tempted to take more than their fair share of the goodies. Communality is vulnerable to greed. This is a tragedy for the group and especially for its leaders, since a one-for-all-and-all-for-one mentality would lead to a more powerful and prosperous community, if only it could be made instinctive. So people have devised ingenious mind-control techniques to implant and nurture communal thoughts in others.

One technique is the use of kinship metaphors, which permeate cults, religions, clubs, political parties, and social movements: *brethren, brotherhood, fraternity, sisterhood, sorority, the fatherland, the mother country, the family of man.*[39] Recall that one of the tools of sympathetic politeness is the use of terms of endearment, as in *Brother, can you spare a dime?* Now we see that sympathetic politeness is not so much a cooperative effort to save face as a ploy to trigger communal feelings in strangers or to reinforce them in friends and allies.

The other technique is to reinforce the folk-biological intuition that people are made of the same flesh and hence are parts of one big superorganism. Communal meals are one of the commonest bonding rituals the world over, as if people believed that you are what you eat and so if you eat the same stuff you are the same stuff. A corollary is that food taboos protect the boundaries of the group by preventing its members from sharing food with neighboring groups. Many tribes and coalitions (such as the Mafia) cut their fingers and rub them together to allow their blood to mingle, hence the expression *blood brothers*. People also disfigure their bodies—by scarring, tattooing, piercing, stretching, filing, hairstyling, circumcision and other forms of genital mutilation—as if to make the group look like a separate species, biologically distinct from other human groups.

People in groups also engage in synchronized movements, like dancing, bowing, standing, sitting, marching, drilling, and exercising. The impression from the outside is of a single communal body rather than many individual ones, because of a law of perception called "common fate": things that move together are seen as attached together. The impression from within is even more insidious. People ordinarily sense the boundaries of their own bodies by observing what moves when they will a part of their body to move. That's why people feel that tools, bicycles, and cars are extensions of their own bodies, and why a psychologist using mirrors or video displays can fool people into thinking that they are voluntarily controlling someone else's hand, or even a phantom hand of their own which had been amputated years before.[40] Combine this bit of our sensorimotor psychology with a groupwide rhythmic movement and a person can literally feel as if he is a part of one communal body. Personal boundaries can also be eroded when people undergo an intense emotional experience together, like an ordeal of hunger, terror, or pain, or a drug-induced altered state of consciousness. These cheap tricks are common in initiation rites all over the world.

A sense of communality via folk biology can also be reinforced by myths and ideologies. People are told that they are descended from a patriarch or a primeval couple, or that they are connected to a natal land, or came into being in the same act of creation, or are related to the same totemic animal. Here is a rule of thumb in anthropology: whenever a society (including ours) has a cultural practice that seems bizarre, its members may be manipulating their intuitive biology to enhance feelings of communality.

Conspicuous by its absence is the one mechanism that social and political theorists treat as the foundation of society: a social contract. Friends, families, couples, and clans don't sit down and verbally articulate the rights and responsibilities that bind them together. If they use language at all, it's to avow their solidarity in unison or close succession, as in *I love you, I pledge allegiance,* and *I believe with a perfect faith.* What they don't like to do is negotiate the terms of their communality.[41] The very act of delineating perquisites and obligations in words undermines the nature of the emotional (and in their minds physical) fusion that allows them to share instinctively, without concern for who takes what and who gets what. Of course, when people get into conflicts they often do resort to verbal negotiation, from couples therapy to courts of law. But it isn't the core of a communal relationship and often feels awkward and out of place. A proposed 1996 date-rape policy at Antioch College was widely ridiculed for requiring students in flagrante delicto to obtain explicit verbal consent for every escalation. A song lyric from *My Fair Lady* (which I used in chapter 2 to illustrate the dative construction) can also be used to illustrate the preferred medium by which we cement our intimate relationships: "Sing me no song, read me no rhyme, don't waste my time; show me!"

Not only do people not like to talk about their closest bonds, they don't like to think about them either. A person known to ruminate over the terms of his marriage, his parenthood, his friendship, or his loyalty to the group is considered a cad, a bad parent, a fair-weather friend, a traitor, an infidel, someone who "just doesn't get it." Once again we see the fingerprints of the psychology of taboo.[42] A clear example is the Prenup Paradox. In one conception of rationality, every affianced couple should sign a prenuptial contract that sets out the terms of property division should the marriage end in divorce, as half of all marriages do. Yet many couples resist the suggestion, and they are not completely irrational. The very act of negotiating a prenuptial contract makes it more likely that it will be needed, because the couple is being forced to think just the thoughts that they should not be thinking if their marriage is to be based on the right communal emotions.

The second relationship type is called Authority Ranking, also known as power, status, autonomy, and dominance. The logic of Authority is "Don't mess with me." Its biological roots are in the dominance hierarchies that are widespread in the animal kingdom. One organism claims the right to a contested resource based on its size, strength, seniority, or allies, and another animal cedes it to him when the outcome of a fight can be predicted in advance and both parties have a stake in not getting bloodied in a battle whose outcome would be a foregone conclusion.[43] In this way they sort themselves into a linear hierarchy.

Authority Ranking, like Communal Sharing, is signaled in humans not primarily through words but by co-opting perceptual faculties tailored to another domain of life. In the case of Communality it was intuitive biology; in the case of Authority it is intuitive physics, namely, the Kantian categories of space, time, substance, and force explored in chapter 4. The ranking of people in a dominance hierarchy is usually symbolized as an ordering in time, space, size, or strength. Dominant individuals (chiefs, presidents, priests, shamans, generals) strut ahead of their subordinates, enter and exit first, stand taller (often on platforms and balconies), look bigger (with the help of hats, helmets, and headdresses), *are* bigger (leaders, including American presidents, tend to be taller than the runners-up), are depicted as bigger (in outsize images and statues), and have bigger offices, palaces, and monuments. Hundreds of metaphors express this equivalence, like *first among equals* (time), *strongman* (force), *big shot* (size), and *top dog* (position in space).

Though visible signs of the ability to prevail in a fistfight are the most salient advertisements of authority, they are not necessarily the qualifications that earned the authority in the first place. Dominance in humans is tied up with status: the possession of assets like talent, beauty, intelligence, skill, and wisdom. And in the end, dominance and status are social constructions that depend crucially on the perception of others and of oneself. How much authority you possess depends on how much authority you are prepared to claim, and on how much authority others are willing to cede to you. This, I believe, is the real nature of the concept of "face" that was bandied around by Brown and Levinson without a satisfying theory (though it was broached now and again by Goffman). Their "negative face," the desire not to be impeded, is a claim of dominance; their "positive face," which they sometimes defined as a desire for approval or esteem, is a claim of status. (At other times they define it as a desire for sympathy, but I suspect that's a different emotion, more closely tied to communality.)

When thought of in terms of Authority, face is not just the warm bath of self-esteem but a kind of social currency with real value. In many arenas of life, what we get depends on what we feel entitled to demand. When a buyer and a seller engage in bargaining, there is always a range of prices in which both would prefer to consummate the transaction than to walk away. For instance, a car that costs a dealer $20,000 may be worth up to $30,000 to a customer, and any price within that range leaves them both better off than if the deal falls through. Which price they settle on within that range depends on each one's resolve. The dealer will relent with a lower price if the buyer convinces him that he won't budge from it; the buyer will muster a higher price if he is convinced the seller won't budge from it.[44] Similarly, when two people stand off over a taxi or a parking space, the victor will be the one who appears most prepared to stand his ground, verbally or physically. In both cases, appearances matter. Each claimant will back off to the extent that he thinks the other will stand his ground, and will stand his ground to the extent that he thinks the other will back off. Of course, either can test the other's mettle through brinkmanship, but the costs to both sides—walking away from the deal, coming to blows—can be high. Bluster and self-confidence, backed up by the deference and esteem of third parties, can be a decisive weapon. This respect can be won by possessing assets that others value or by having prevailed in previous battles of will or of force. To be disarmed of these weapons through a public defeat or disrespect that all can see—to "lose face"—is painful.

People naturally protect their face, and those with no desire to challenge it—such as tablemates who would like them to pass the salt but don't want to cause a scene—will use techniques of deferential politeness, including indirect speech. Implicatures can be used to protect one's own face, as well as to defer to another's, and sometimes the implicated message is insinuated at dog-whistle pitch. In the movie *Crimson Tide,* an authoritarian navy captain in command of a submarine (Gene Hackman) is joined by an intelligent lieutenant (Denzel Washington); their relationship is respectful but chilly. In the course of the film's outlandish plot, they get a garbled communiqué ordering a launch of nuclear missiles, which would surely begin World War III. The lieutenant defies the commander, and after much shouting and fighting and destruction of property, he succeeds in preventing the launch, which is a good thing because the order was an error. In the film's denouement, after Washington has been cleared of a mutiny charge, Hackman walks up to him and says, "You were right and I was wrong." Washington raises his eyebrows. "About the horses—the Lipizzaners. They *are* from

Spain, not Portugal." The linguist Deborah Tannen, in an article about why men don't apologize, recounts the scene and writes in exasperation: "Why couldn't he just say it? 'I made a mistake. You were right. I was wrong about starting that nuclear war.' " Even as astute an observer as Tannen seems not to have noticed at that moment that he *did* apologize, but in an ultrasonic message that perhaps only men can hear, one whose literal form allowed him not to relinquish his dominance. All this would support Tannen's larger theory (made famous in her book *You Just Don't Understand*) that men and women often communicate differently, not in their words or syntax but in their implicatures.

Implicature can be used to challenge authority as well as to maintain it, and when it does we call the discourse "humor." In 2001 Dick Cheney was hospitalized for a cardiac arrhythmia, and a comedian said that the situation was serious because George W. Bush was now a heartbeat away from the presidency. This violates the Maxim of Quality, because the comedian got the cliché about vice presidents and the presidency backwards. But the audience could infer that he was really saying that Bush was out of his depth as president and Cheney was running the show. There is almost always a butt to a joke, someone you are laughing at rather than laughing with. The butt is depicted as inept or foolish or undignified and thus loses authority in the eyes of onlookers. As George Santayana said, "To knock a thing down, especially if it is cocked at an arrogant angle, is a deep delight of the blood." By hiding the insult in an implicature the challenger makes it all the more compelling, because any listener who can retrace the implicature successfully—who "gets the joke"—realizes that she knew about the target's infirmity all along, and that others who are laughing with her knew about it, too.

Much humor, though, consists of friendly teasing and self-deprecation, where the butt of the joke is the speaker or a chum rather than an authority in need of being taken down a few pegs. Here are some laugh lines from real conversations among university students, surreptitiously recorded by the psychologist Robert Provine:[45]

> Is that considered clothing or shelter?
> I'd pay a hundred dollars to wade through her shit [an
>   expression of endearment].
> Do you date within your species?
> I try to lead a normal life!

The humor is convivial, not aggressive, yet a key component is still a reduction in dignity. I think it is being used as a signal that the basis of a relationship is communality rather than authority. Whenever people interact, the seeds of an authority relationship are present, because, as Samuel Johnson noted, "no two men can be half an hour together but one shall acquire an evident superiority over the other." At first one might think that the superior party would always relish the chance at dominance, but that isn't always true: uneasy lies the head that wears a crown. Dominance is useful for as long as you have it, but it can disappear with age and circumstances. Friends are there through thick and thin, and all of us need them. One way to signal to a companion that the basis for a relationship is friendship rather than dominance is to call attention to an undignified trait in yourself or in the companion, disavowing the possibility that one of you has something to lord over the other.[46]

The third relationship type is called Equality Matching, though more familiar terms are reciprocity, exchange, and fairness. Its logic is "If you scratch my back, I'll scratch yours," and its evolutionary basis is reciprocal altruism.[47] In an Exchange relationship, people divvy up resources equally, or take turns, or barter goods and services for equivalent goods and services, or trade favors in a tit-for-tat exchange. When it is used to divide up a resource among people who can't share it communally but who each have a claim on it, Equality Matching can avert a contest of authority and the costs to everyone of a violent scrum. It also allows people to enjoy gains in trade, in which two parties with more of some commodity than they can enjoy at one time can barter their surpluses to each other, benefiting them both.

Fiske suggests that the psychological implementation of Exchange is a concrete operation: a behavioral algorithm that ensures that the players put in an equal effort and take out an equal benefit. They flip a coin, draw straws, count out rhymes like "eeny-meeny-miney-moe," line things up in rows, or weigh them in balances. But Exchange is a realm in which literal language comes into its own. "If you do this, then I'll do that" is a convenient way of trading intangible goods and services, or tangible ones that are given and taken at different times. Language is also the channel with which we spread information about a person's trustworthiness, the phenomenon called gossip. It may not be a coincidence that we use "face" as a metaphor for reputation in matters of authority (possibly going back to the stare of dominance in primates), but we use "good name" as the metaphor for reputation in matters of fairness.

Fiske's taxonomy also accommodates a fourth relationship type, which he calls Market Pricing. It embraces the entire apparatus of modern market economies: currency, prices, salaries, benefits, rents, interest, credit, options, derivatives, and so on. The medium of communication is symbolic numerals, mathematical operations, digital accounting and transfers, and the language of formal contracts. Unlike the other three relationship types, Market Pricing is nowhere near universal. A culture with no written language and with a number system that peters out at "3" cannot handle even the rudiments of Market Pricing. And the logic of the market remains cognitively unnatural as well. People all over the world think that every object has an intrinsic fair price (as opposed to being worth whatever people are willing to pay for it at the time), that middlemen are parasites (despite the service they render in gathering goods from distant places and making them conveniently available to buyers), and that charging interest is immoral (despite the fact that money is more valuable to people at some times than at others).[48] These fallacies come naturally to an Exchange mindset in which distributions are fair only when equivalent quantities of stuff change hands. The mental model of face-to-face, tit-for-tat exchanges is ill-equipped to handle the abstruse apparatus of a market economy, which makes diverse goods and services fungible among a vast number of people over great distances of time and space.

As far as I can see, this takes Market Pricing out of the realm of human nature, and there seem to be no naturally developing thoughts or emotions tailored to it. In this regard Market Pricing can be lumped with other examples of formal social organization that have been honed over the centuries as a good way to organize millions of people in a technologically developed society, but that don't occur spontaneously to untutored minds. A parallel example may be found in the political institutions of a democracy, where power is assigned not to a strongman (Authority) but to representatives who are selected by a formal voting procedure and whose prerogatives are delineated by a complex system of checks and balances. Yet another is a large institution like a corporation, university, or nonprofit organization. The people who work in them aren't free to hire their friends and relations (Communality) or to dole out spoils as favors (Exchange), though these are universal human temptations. Instead, their behavior is hemmed in by fiduciary duties and regulations.

Though Communality, Authority, and Exchange are universal modes in which people conceive of their relationships, cultures differ in which relationship type may be applied to which resource for which kind of dyad in which context. In Western cultures we buy, sell, and trade our land (Exchange), but don't buy, sell, or trade women betrothed as brides; in other cultures it's the other way around. A boss in a corporation can control an employee's salary and office space (Authority), but may not help himself to his possessions or his wife, though these *droits de seigneur* were the perquisites of many kings and despots in other times and places. A guest at an American dinner party (Communality) should not pull out his wallet at the end of the evening to pay his hosts for the meal, nor should he reciprocate with a conspicuous invitation for dinner the very next night. But in many cultures such reciprocity is calculated openly, a bit like the way people in our culture privately reckon the annual exchange of Christmas cards.

When a person in a particular culture misjudges which relationship type applies to a given situation, emotions can run high. We are, after all, dealing with the culture's approved way of distributing resources and power. Helping yourself to a goody can be a prerogative in the context of one relationship type but grand larceny in another. Ordering someone around can be a requirement of your job in one setting but a case of extortion in another.

Sometimes the mismatch is a one-time event, the result of a misunderstanding, the testing of a new relationship, or a unique exigency. This triggers the emotion we call "awkwardness" and the events called "gaffes" or "faux pas." The awkward person is now self-conscious, acutely attentive to the details of the situation (especially other people's reactions to his demeanor and actions), and paralyzed in word and deed until a repair strategy is hatched. Just about any relationship mismatch can trigger a sense of awkwardness. It's widely appreciated that good friends (Communality) should not undertake a large financial transaction between themselves (Exchange), like selling a car or a house; it can endanger the friendship. There can be touchy moments when a supervisor (Authority) eases into friendship with an employee or a student (Communality), a transition which may be signaled by a change in the form of address or a shift in pronouns from

V to T. When the Authority relationship threatens to morph into a sexual one, the result can be not just awkwardness but a sexual harassment suit. Sexuality, for its part, is a special kind of communal relationship, which may clash with other kinds of communality such as friendship, creating another trigger for social tension. The entertainment industry exploits this awkwardness as a plot device in comedies of sexual manners like *Seinfeld* and *Friends*.

If a mismatch is not a one-shot but is deliberate and protracted, the emotion may escalate from awkwardness to moral condemnation. A mother who sells a baby, a teacher who demands sex from a student, a friend who uses another friend to get ahead or make a profit, are all considered beneath contempt. As with the prenup paradox, the psychology of taboo kicks in: even *thinking* about crossing the wires of one relationship type with another can be incriminating. The psychologist Philip Tetlock, a sometime collaborator of Fiske's, has shown that people are outraged even to be asked about applying an exchange or market mentality to communal or authority relationships. For example, they take umbrage at being asked for their opinions on whether there should be a market in adoption rights, whether people should have the right to sell their vote or their organs, and whether people should be able to buy their way out of jury duty or military service.[49]

At the comfortable distance of fiction, we can be riveted by characters who are forced to think the unthinkable about their intimate relationships, as in *Sophie's Choice* and *Indecent Proposal*. And at the even more comfortable distance of humor, we can laugh at them. In a *New Yorker* cartoon, a gentleman seated in an armchair says to a young man, "Son, you're all grown up now. You owe me two hundred and fourteen thousand dollars." And in another archaic joke about the sexes that is too apposite to omit, a man asks a woman, "Would you sleep with me for a million dollars?" She answers, "Hmm . . . I suppose I might." He then asks, "Would you sleep with me for a hundred dollars?" She replies, "What kind of woman do you think I am?" He says, "We've already established that; we're just haggling over price."

And now we can return to off-record indirect speech acts in everyday life. Take our food writer, sweating and trembling as he proffers the maitre d' a bribe. If he was facing an honest maitre d', he'd have been creating a mismatch between Authority (the usual relationship over customers claimed by a maitre d') and Exchange (the terms Feiler was trying to offer). No wonder Feiler felt awkward, if not immoral; his trepidation was right out of Relational Models theory. Feiler was saved by an implicature. The literal content (*I was wondering if you had a cancellation*) was consistent with the Authority relationship, but the implicated content ("I'll give you a hundred dollars for a quick table") conveyed the desired exchange. An honest maitre d' could not be offended, and a dishonest one could accept the bribe. (One also senses that the maitre d' would feel that he had maintained his authority, though why this lame pretense should work is another puzzle, the last to be addressed in the chapter.)

The logic of seduction is mostly the same, though with a twist. The game-theoretic matrix is familiar:

	Willing partner	**Unwilling partner**
Say nothing	Handshake	Handshake
"I'd really like to make love to you"	Sex	Wine thrown in face
"Would you like to come up and see my etchings?"	Sex	Handshake

But the wine-in-the-face cell deserves a comment. In the scene from *Tootsie,* why should Michael's transgression result not just in awkwardness—the usual penalty for a relationship mismatch—but the taking of offense? Though a committed romantic relationship is a communal one, a nascent sexual relationship has an element of exchange in it because of the greater demand for casual sex by men than by women.[50] A desirable woman may expect special attention and generosity from a prospective partner, and may have high standards for his minimum desirability, especially his status. Julie was a gorgeous starlet; Michael was a nobody who barely offered her a pleasantry. His come-on was not just any relationship mismatch but one that threatened her face—the bargaining position she felt entitled to in the sexual marketplace—and she was compelled to defend it.

Soliciting a benefactor for a large donation has much in common with sexual courtship. A sumptuous meal is mandatory, establishing an atmosphere of warm conviviality. Throughout the proceedings an aura of friendship is maintained, and much enjoyed by the target of the seduction.

Sometimes entertainment is supplied (that would be me, and other presentable professors). For most of the evening the business at hand is never mentioned, though it is very much on people's minds. The seducer has to be careful not to let the evening slip away without making his move, but not to make it too early, before the mood is right. One difference is that at the moment of truth a dean can't very well sidle up to the donor and nonchalantly slip his hand into the donor's checkbook. But "the ask," as they call it in the trade, has to be couched with delicacy, with the donor called a "leader" and a "friend" and the altruistic nature of the "gift" repeatedly pointed out.[51] A cynical analysis of the transaction—that the university is selling the donor naming rights, prestige, and a simulacrum of friendship with interesting people—is taboo. The exchange is only implicated, preserving the aura of communality; the actual terms are deferred to the my-people-will-call-your-people stage. And as with courtship, the parties sometimes really do end up as good friends.

Threats, for their part, tend to be veiled in implicatures for two reasons. One is familiar: a plain-speaking extortionist would be exposing himself and would thus risk legal penalties, just like a briber. But he also faces the risk that the target will call his bluff by defying the threat. In order to maintain the reputation on which his livelihood depends, the extortionist would have to carry out the threat, which can be risky and expensive and yet is pointless after it has failed in its purpose of coercing the target. An implicated threat solves both problems. If the threat is off the record, the extortionist is harder to convict, and if defied he can choose not to carry it out without going back on his word and undermining his credibility. The English language provides him with a particularly handy option: the inherent ambiguity of the future tense between futurity and volition (explored in chapter 4). "If you don't contract with our trucking company, your union will go on strike" can always be defended as a farsighted prediction rather than an intentional policy. Another creative example comes from *The Godfather Part II*. Frank Pentangeli is about to testify at a congressional hearing against Michael Corleone and is sequestered by the FBI so he can't receive a threat. Michael appears in the spectator gallery accompanied by Pentangeli's brother, who has just flown in from Sicily, in full view of Pentangeli as he is about to testify. (Tom Hagen explains: "He came, at his own expense, to aid his brother in his trouble.") Frankie quickly retracts his story.

Corleone's veiling of a threat in the mere glimpse of a brother is one of many ways in which people have solved the problem of conveying a message when words would not just be awkward but dangerous. In his autobiography, Roger Brown explains the payoff matrix faced by gay men in the early 1950s when he was a university student, and how they solved the Identification Problem in the men's rooms in which they cruised:

> You must bring a book and, sitting on the pot, door closed, read with as much concentration as is possible in a duck blind. When your duck arrives and settles his feathers in an adjoining booth, you abandon your text and concentrate on the foot and ankle visible to you. A certain amount of shifting or foot tapping is meaningless neurological overflow. What you want to be on the lookout for is PATTERN. There are several kinds of interpretable patterns, but the easiest to read is the foot that moves consistently in your direction; first the toe, then the heel, always by very small increments, never enough to justify a challenge or even to be seen by any but the devoted cryptanalyst. The nightmare possibility is a deep voice saying: "Hey bo, what's up." . . .
>
> Back in the stalls, one's own foot should move in the manner of the chap next door, edging in his direction. There is in the end an inescapable Rubicon. One or the other must risk . . . CONTACT. Actually, it is not a great risk; unless there is a clearly voluntary companionate pressure, anything less can be erased with an apology. However, once pressure is assayed and returned, the scene becomes electric with erotic potential.[52]

The ingenuity of these subterfuges shows that implicatures recruit the entirety of social intelligence and are not restricted to interpreting language itself.

PASSING THE GIGGLE TEST:
THE LOGIC OF NOT-SO-PLAUSIBLE DENIAL

One problem remains unsolved: the psychological import of whether an overture is "on" or "off" the record in everyday conversation. The puzzle arises in cases in which two things are true. First, the Identification Problem has been solved and each party knows the other's intentions. Second, the implicature is so obvious as to leave no doubt in the hearer's mind as to what was intended. The etchings, the restaurant cancellation, the leadership, the possibility of accidents, are transparent ruses, so any "plausible deniability" is not, in reality, plausible: it would not pass the giggle test. In a courtroom, the standard of proving guilt beyond a reasonable doubt, especially in cases where free speech is an issue, can explain why even a scintilla of deniability can get someone off the hook. But why should we act like defense lawyers in everyday life? Why would it have been worse for the maitre d' who has accepted a bribe if the customer had stated the bargain in so many words? Why would a donor be put off by a direct solicitation couched as a quid pro quo when deep down he must have some awareness that that is what it is? And why is a rebuffed sexual overture more uncomfortable when it was put forth as a bald proposition than if it had been conveyed by unmistakable innuendo or body language?

As with many human interactions, the dynamics of seduction are too subtle and touchy to be reproduced in a lab. In these cases our best methodology may be the kind of thought experiment called the "comedy of manners," where plausible characters play out the unstated rules of social engagement before our eyes. Harry, having met Sally just hours before, has miscalibrated the optimal level of indirectness in a remark about her looks, and Sally accuses him of coming on to her.

> HARRY: What? Can't a man say a woman is attractive without it being a come-on? All right, all right. Let's just say, just for the sake of argument, that it was a come-on. What do you want me to do about it? I take it back, OK? I take it back.
> SALLY: You can't take it back.
> HARRY: Why not?
> SALLY: Because it's already out there.
> HARRY: Oh jeez. What are we supposed to do? Call the cops? It's already out there!

What exactly is this concept of a proposition being "out there," such that you "can't take it back"? As Harry points out, it's not as if you can call the cops.

Though many features of implicature may recruit general processes of rational inference, at the end of the day we are faced with something in the minds of people that is specific to language itself. Expressing a sentiment in a sentence—baldly, on record, in so many words—makes a difference. Some things once said can never be unsaid.

In the cases we have examined, one or both parties want to preserve the relationship type that is consistent with the literal wording of a sentence (the man and woman are friends or colleagues, the maitre d' is an authority figure, the donor and dean are pals) while transacting the business at hand (the sexual advance, the bribe, the donation), which presupposes a different relationship type and is implicated between the lines. Why do people feel that indirect speech lets them get away with this hypocrisy in a way that plain speech would not? I don't know the answer, but here are some ideas.

The token bow.[53] By couching a proposition as an indirect speech act, the speaker signals to the hearer that he is making an effort to spare her dignity, feelings, or face. The mere perception that the speaker has gone to this effort makes the hearer appreciative of his considerateness, and feelings are eased all around. A bald proposition, by its very efficiency, signals that the speaker has put no effort into soothing the hearer's feelings.

Don't talk at all; show me. A relationship of communality is not negotiated with language but is consecrated by physical signs of communion such as rituals, feasts, and bodily contact. The very act of trying to articulate a relationship in words is a signal that it cannot be a communal one, because communal relationships are felt in the marrow, not decided upon rationally. The same is true for authority relationships, which are projected nonverbally by signs of size, strength, and priority.

The virtual audience. The speaker and the hearer may have no doubt about the intent of an indirect speech act because they know the backstory and can witness each other's bearing and mannerisms. But an eavesdropper or a third party, learning about the event from a distance, lacks this information, and has only the actual words to go on. Of course, overhearers are also capable of reading implicatures, but their level of certainty is far less than that of the speaker and the hearer, and the deniability may be plausible to them, even if it isn't to the participants. Compare the effects of a bald proposition. Not only is it more transparent to an earwitness, but it can be conveyed more accurately in a chain of gossip. That is because language is a digital medium and digital messages can be transmitted without loss. Of course language, even at its most precise, is rife with vagueness, and people's memory for the wording of a sentence is very far from perfect. Still, the content of a sentence is more reproducible than the analogue information in a speaker's tone of voice or how close two people were sitting. According to this line of thought, we always play to an imagined audience (as Goffman so often insisted), if only to manage any information that might leak out to an eavesdropper or a gossip.

Preserving the spell. The public relationship shared by the speaker and the hearer is a pleasant illusion, like watching a play, sitting in a planetarium, or viewing a vase of silk flowers. A maitre d' can reign over his opulent fiefdom. A woman can enjoy the attention and kindness lavished by a man who is interested in her mind. A donor can enjoy the conviviality of a warm dinner with prestigious people. The illusion can survive the implicated message of an indirect speech act, but would be shattered by a bald proposition, just as an actor who flubs his lines breaks the spell of a play or the sight of a MADE IN TAIWAN tag spoils the illusion of a silk rose. Once the spell is broken, the participants can enjoy the pretense only at the cost of being liars and dupes. In this theory, the self is divided for the purpose of self-deception, and the deniability is plausible to one part of the self even while it is disbelieved by another part.

Certainty as a focal point. The legal policy of convicting a person only if the evidence establishes his guilt beyond a reasonable doubt may have a counterpart in everyday life. Relationship types are discrete and very different modes of interaction, and for a dyad to switch from one to another is no small matter. Since it takes two to tango, people must jointly recognize a policy about when to switch. The threshold can't be openly negotiated any more than the relationship itself can be negotiated, and so it has to evolve as an unstated compact. Exactly how close can a man sit to a woman, how lavishly can he compliment her, how slim a pretext for inviting her to his apartment can he offer, before she concludes that his intentions are sexual? Her private wariness can track the cues in an analogue fashion, but her relationship with him must be fish or fowl. She might have to tolerate a considerable amount of suggestiveness before calling things to a halt, because there are costs to switching the relationship and it's hard to know where to draw the line. A bald proposition certainly falls on the other side, and the difference between it and the continuum of innuendo may be the only clear place to draw that line. The plausibility of the denial may be very small—one percent, or one tenth of one percent—but as long as it isn't absolutely zero (as it would be for a bald proposition), she may not be able to call him on it.

This is an example of the Coordination Game, another scenario explored by Schelling.[54] A couple gets separated in a department store, and each has to guess where to meet the other. Or two paratroopers are dropped into a foreign territory, equipped only with maps, and have to rendezvous without communicating. Each has to anticipate where the other will anticipate that he will anticipate that the other will anticipate that he will turn up, ad infinitum. In a Coordination Game any focal point that stands out to the two players can emerge as a solution, even if nothing makes it intrinsically suitable for the job other than standing out to them. The paratroopers might meet at the only tree in a desert, or the highest hill in the territory, or the intersection between two rivers, even if that focal point was a long hike from where either one was dropped, simply because it is the only place that can be singled out in the featureless landscape. Schelling notes that this is why two negotiators often split the difference between their opening positions, or settle on a round number: "The salesman who works out the arithmetic for his 'rock-bottom' price on the automobile at $2,507.63 is fairly pleading to be relieved of $7.63."[55] Similarly, "if one has been demanding 60 percent and recedes to 50 percent, he can get his heels in; if he recedes to 49 percent, the other will assume that he has hit the skids and will keep sliding."[56]

Mutual knowledge. Say a woman has just declined a man's invitation to see his etchings. She knows—or at least is highly confident—that she has turned down an invitation for sex. And he knows that she has turned down the invitation. But does he know that she knows that he knows? And does she know that he knows that she knows? A small uncertainty within one's own mind can translate into a much bigger uncertainty when someone else is trying to read it. After all, the woman may privately base her confidence that he made a sexual overture on her social astuteness, her extensive knowledge of the opposite sex, and her due diligence on this man's behavior from gossip with women who have dated him in the past. But all *he* has to go on is what a generic person might infer in the circumstances he has created. Similarly, while he is wise enough to know that her *no* means "no," she can't be sure that he isn't a naïf who is hoping that maybe she didn't get the point. A denial of the sexual intent may not be plausible, but a denial that the *other party* knew about the sexual intent may be. Compare this with what happens when the man makes an overt proposition and the woman rebuffs it. The lid is blown off this higher-order uncertainty. Not only does each party know that she has turned him down, but each one knows that the other knows.

This is the state of affairs that scholars call mutual knowledge, joint knowledge, common knowledge, and common ground.[57] Beginning with Grice, many theorists have assumed that mutual knowledge of the rules of a language, of the background beliefs in a culture, and of human rationality is necessary for successful communication to take place, especially via implicature. But mutual knowledge may play another role in language as well. It's possible that mutual knowledge of a specific request or offer is a prerequisite to two people being forced to change their relationship type, and mere individual knowledge (two people knowing the same thing, but neither knowing whether the other knows it) is not. If you know that I've asked you for sex and have been turned down, and I know that I've asked you for sex and have been turned down, we can pretend that it never happened and continue to be (or at least pretend to be) friends. But if I know that you know, and you know that I know that you know, and so on, then the charade can no longer be maintained.

The ability of language to explode individual knowledge into mutual knowledge is the basis for a number of fables and brainteasers. The most famous is the story of the emperor's new clothes. Every onlooker knew that the king was naked, but none of them could be sure that the others knew, and so they were all intimidated into silence. All it took was for one boy to say "The emperor has no clothes!" and the crowd could burst into laughter. Crucially, the boy was not telling a single person anything he or she didn't already know. But his words still conveyed information—the information that all the *other* people now knew the same thing that each one of them did.

A more surprising example has popped up in various isomorphs and may be called the Barbecue Sauce Problem.[58] Twenty logicians are at a picnic where they are served spare ribs in barbecue sauce. Three of the logicians get barbecue sauce on their faces, but there aren't any mirrors around, so they don't know it, nor do the others know that their own faces are clean. Of course each logician can see that some of the other logicians have messy faces, but no one wants to embarrass the others by telling them that, nor does anyone want to look foolish by wiping his or her own face if it's clean. The chef returns with a plate of watermelon, surveys the scene, and announces, "At least one of you has barbecue sauce on your face. I'll ring the dinner bell to give you a chance to wipe it off. Then I'll ring the bell again, and again, and so on. When everyone's face is clean I'll serve the water-melon." He rings the bell, and no one moves. He rings it a second time, and still no one moves. He rings it a third time, and this time the three logicians with barbecue sauce on their faces wipe it off. Watermelon is served.

Before the chef's announcement, everyone knew that at least one person had a messy face, so the chef wasn't telling anyone anything they didn't already know. But the fact that he uttered those words in the presence of all the picnickers changed their knowledge nonetheless. It informed each of them that *everyone else* knew what they knew. And that was information they could put to use. Here's how.

Imagine a simpler version of the story, in which only one logician has a messy face. When the chef says that at least one guest has a messy face, she looks around, sees that everyone else's face is clean, and concludes that she must be the messy one. When the bell rings, she cleans her face. Easy. Now suppose that there are two messy eaters. When the chef makes his announcement, the first logician sees that another logician's face is messy, but she has no idea about her own face, so she does nothing. The second messy logician thinks the same thing. But after the bell has rung and no one has moved, the first logician realizes that her face must be messy, because if the second logician were the only one, he would have known to wipe his face, as we saw in the one-messy-face scenario. The fact that he didn't means that he must have seen another messy face, and since all the other faces visible to her are clean, she concludes that it must be hers. When the bell rings the second time, she wipes her face. So does the second logician, because he has cranked through the identical deduction. The same logic applies on the third ring when there are three messy eaters: each one deduces from the lack of activity after two rings, together with their seeing two messy faces, that they must be messy themselves. The logic can be extended to any number of messy eaters, who will simultaneously clean themselves after the corresponding number of rings (ten messy eaters would clean themselves on the tenth ring, eleven messy eaters on the eleventh ring, and so on).

Though it's easiest to explain mutual knowledge by saying that A knows *x,* and B knows *x,* and A knows that B knows *x,* and B knows that A knows *x,* ad infinitum, clearly no finite head can hold an infinite set of propositions. And aside from the barbecue-sauce problem, people generally don't *need* to think themselves dizzy with layers upon layers of "A-knows-B-knows" propositions. As in other cases in linguistics in which a person is said to "know" an infinite set of expressions (words, sentences, propositions), the knowledge in mutual knowledge is *implicit*. All the person really needs to have in her head is a formula that is recursive, that is, a formula that contains an example of itself. What people would share in their minds is the following statement, which we can call *y:* "Everyone knows *x,* and everyone knows *y.*"[59] If necessary, people can reel out however many levels of propositions they need for a given problem, as long as they can keep track of them in memory. But they can grasp that they *have* mutual knowledge simply by noticing the recursive nature of that piece of information in their minds. Even more easily, they can infer the commonality of their knowledge by noticing the public circumstances in which they and other people gained it, such as from the ringing of an audible bell or the shouting of an audible boy.

Mutual knowledge may explain much of the face-saving and face-losing we see in everyday life, because "face" is inherently a phenomenon of mutual knowledge. You feel emboldened to press for a favorable bargaining position because you know that others know that you know that others know you know that you are esteemed or powerful enough to stand your ground. Expressions of disrespect are damaging when they are public because they can nip this cycle in the bud. Every mature person knows that other people, even their close friends, may gossip about them behind their back. You might even overhear an unflattering remark about you in a phone conversation, or catch a glimpse of one in an e-mail message. Yet as long as no one knows that you know, the barb can pass without incident. If, in contrast, an unkind remark gets back to you from a third party, or you overhear it as you join a cluster of people who suddenly discover that you were in earshot the whole time, or if it is inadvertently disseminated in an e-mail by someone who doesn't know the difference between "Reply" and "Reply to All," the hurt is deeper and the desire to confront it greater. The difference is that now everyone knows that you know that they know and so on, threatening your face if you accept it without seeking redress.

The skill called "tact" consists of preventing touchy individual knowledge from becoming mutual knowledge. At a dinner party, everyone may know that one of the guests is overweight, or that another has a speech impediment. Yet to mention it out loud, creating mutual knowledge, would be deeply embarrassing. Or take a more subtle example that I have experienced in real life and have explored in an experiment. A student complains to me about his grade from a graduate-student teaching assistant (TA). I read the paper and agree that the grade is too harsh. I tell the student I will speak to the TA about changing his grade, and I tell the TA that I have spoken to the student about changing his grade. Yet I'm careful not to tell them in each other's presence, or even to cc an e-mail addressed to one of them to the other, because I sense that doing so would undermine the TA's authority. The individual messages prevent mutual knowledge. The TA may know that his judgment has been overruled, and the student may know it. But the student doesn't know that the TA knows that he knows it; for all the student knows, the TA might think that the student just wore me down. And the TA doesn't know that the student knows *he* knows it; for all the TA knows, the student might think that I asked the TA to play along just to increase the course's popularity. A joint message would have become mutual knowledge, and mutual knowledge closes off these avenues for saving face.

Though I think that mutual knowledge is the deepest explanation for why people play along with indirect speech even when they can see through it, the other five explanations are not incompatible with it. Perhaps there is a conspiracy of reasons why a bald proposition is so much more damaging to a relationship than a veiled one. Not only can the bald proposition not be ignored if it's mutual knowledge, but it's the only clear line in the sand, and it would be too easily recognized by a virtual audience. This shatters the illusion of the relationship, spoiling any pleasure people get from it. All the more so with communal relationships, which are corroded by the very act of openly negotiating their terms. And perhaps a considerate speaker takes steps to avoid this hazard and implicates his intention in indirect speech instead, earning points for his bow to the speaker's face. For all these reasons we have the feeling that a bald proposition is "out there," and that the speaker cannot "take it back."

CHOOSING NOT TO KNOW:
THE PARADOX OF RATIONAL IGNORANCE

The games people play as they use language are anything but frivolous. They come from the fact that conversation is a quintessentially social activity. People do things with words—they offer, they command, they threaten, they proposition—and the things they do necessarily affect the relationships they have. We choose our words carefully because they have to accomplish two tasks at once: convey our intentions, and maintain or renegotiate our ties with our fellows.

Is the avoidance of plain speaking a bug in our mind design, or might it have a deeper rationale—a rationale that would predict that *any* social communicator would engage in indirect speech? At first glance, a rationale might seem unlikely. The whole reason to *have* a language is to convey information, and since knowledge is power, it stands to reason that the more information it conveys, the better. One might naively think that it's always better to know something than not to know it for the same reason that it's better to be rich than to be poor: if you're rich, you can always give your money away and be poor. And if you know something, you can always decide to ignore it.

Of course, it is a cliché of our times that we suffer from information overload because of the ubiquity of electronic media. And for fifty years, cognitive scientists have been harping on the limitations of the brain in processing information. Some have argued that Grice's cooperative maxims are a way to manage the flow of information in a conversation, maximizing the rate of transmission of usable knowledge.[60]

But the ultimate reason our speech is so indirect may lie in a different danger of information—not that we might be overwhelmed by how much there is, but that we might be poisoned by what it says. The paradox of rational ignorance is that even if we could accommodate as much information as we wanted, and could always separate the wheat from the chaff, there are certain messages a rational mind may not want to receive.[61]

Sometimes we choose not to know things because we can anticipate that they would have an uncontrollable effect on our emotions. In support of his "Law of Indispensable Ignorance," the psychologist Gerd Gigerenzer lists some examples. People who haven't seen a movie or read a book will shun a review that gives away the ending. A basketball fan who videotapes a game will sequester himself from media outlets so as not to learn the outcome before he watches it. Many expectant parents choose not to learn the sex of their unborn child, and in countries ravaged by the selective abortion of girls, divulging the information can be a crime. The considerable number of families in which a child is unrelated to the nominal father would probably be happier if none of them took a DNA test. The children of parents with Huntington's disease usually refuse to take the test that would tell them whether they carry the gene for it. And most of us would rather not know the day on which we will die.

Another reason a rational system might choose to be ignorant is that if it is designed to come to an unbiased decision, the slightest bit of extraneous information can tip it one way or another. So juries are prevented from knowing the criminal record of the accused, or information that the police obtained by illegal means. Scientists test drugs in double-blind studies, where they keep themselves from knowing who got the drug and who got the placebo. Scholarly manuscripts are refereed anonymously, with the identities of the author and the reviewers hidden. And government contracts are awarded through sealed bids.[62]

But the kind of rational ignorance that can best explain why we veil our speech comes from the Strangelovian dilemmas in which our own rationality can be turned against us and a unilateral disarmament of knowledge is the only countermeasure (another set of paradoxes first explored by Schelling).[63] People are better off if they can't receive a threat. Hence misbehaving children avoid their parents' glances, state's witnesses may be held incommunicado, and I know a colleague who kept a nice jacket and perhaps his life because he couldn't understand some muggers who were threatening him in a heavy accent. Being in possession of a secret makes one vulnerable to extortion by those who want to know it and to silencing by those who don't want it known. Hence kidnap victims are better off if they don't see the kidnapper's face, envoys are kept ignorant of sensitive information for their own safety, and we have the spy-movie cliché, "I could tell you, but then I would have to kill you." In a Coordination Game, the person with the least information is in the better position: if two friends are negotiating over where to have dinner, the one who suggests a restaurant convenient to her just before her cell phone goes dead will have the shorter walk.

Merely being asked certain questions can put a person at a disadvantage, since one answer might be damaging, the other would be a lie, and a refusal to answer would be a de facto confession that those are the respondent's two options. Witnesses who exercise their Fifth Amendment right against self-incrimination by refusing to answer a question often do incriminate themselves in the court of public opinion. When a prestigious position is open and the headhunting begins, candidates can't admit to wanting it, because if it goes to someone else they would be humbled; nor can they say they don't want it, because that might take them out of the running. They can't even say "No comment," for why would they have to if they had no interest in the position? During the recent search for a Harvard dean, the newspapers found that the plausible candidates mysteriously (and incredibly) could not be reached by their assistants. And of course we have seen many examples in which mutual knowledge can transform negative information into a damaging loss of face. Many authors refuse to read their unfavorable reviews so they can honestly say they have no reply to them. Some authors won't read *any* of their reviews, lest acquaintances conclude the worst about the ones they avoided.

Knowledge, then, can be dangerous because a rational mind may be compelled to use it in rational ways, allowing malevolent or careless speakers to commandeer our faculties against us. This makes the expressive power of language a mixed blessing: it lets us learn what we want to know, but it also lets us learn what we don't want to know. Language is not just a window into human nature but a fistula: an open wound through which our innards are exposed to an infectious world. It's not surprising that we expect people to sheathe their words in politeness and innuendo and other forms of doublespeak.

9

ESCAPING THE CAVE

Like the pachyderm palpated by the visually impaired, human nature can be explored in many ways. Anthropology can catalogue the ways in which people in all cultures are alike and the ways in which they differ. Biology can map out the systems of the brain, or the developmental program of the genes, or the adaptive problems that must be solved in the human niche. Psychology can trick people into disclosing their foibles in the lab, or can document how they vary within the normal range and cross the line into pathology. Literature can explore the themes that eternally obsess people in the world's myths and stories, or even in the works of Shakespeare alone.

In this book I have given you the view from language—what we can learn about human nature from the meanings of words and constructions and how they are used. Like all vantage points, it brings some things into visibility while leaving others out of view. Language is a medium that is public and digital, and so must hide the aspects of our experience that are private and smoothly blended: our sensations, our emotions, our inklings and intuitions, and the choreography of our bodies. Still, we are gregarious animals who like to teach and gossip and boss one another around, and few aspects of our lives are unaffected by our dealings with other people. As the channel in which much of this information is trafficked, language is adapted to every feature of our experience that is shareable with others, and a large part of the human condition falls into its purview.

How might the proverbial Martian scientist—in this case a Martian linguist—characterize our species, knowing only the semantics and pragmatics of our language? In this chapter I will provide a word's-eye view of human nature, one that emerges from the phenomena of the chapters that precede it. Though most of the examples have come from English, I have stuck to phenomena that have the best chance of saying something about people in general because they are found in historically unrelated languages all over the world. The phenomena may not be literally universal, since the words and constructions in a given language depend not just on the psychology of its speakers but on its history of fads, conquests, and neighbors. Nor are the phenomena necessarily direct reflections of the genetic patterning of our brains; some may emerge from brains and bodies interacting in human ecologies over the course of human history. Even with these caveats, the view from language reveals a species with distinctive ways of thinking, feeling, and interacting.

Humans construct an understanding of the world that is very different from the analogue flow of sensation the world presents to them. They package their experience into objects and events. They assemble these objects and events into propositions, which they take to be characterizations of real and possible worlds. The characterizations are highly schematic: they pick out some aspects of a situation and ignore others, allowing the same situation to be construed in multiple ways. People thereby can disagree about what a given situation really is even when they agree on how matter has moved through space.

Human characterizations of reality are built out of a recognizable inventory of thoughts. The inventory begins with some basic units, like events, states, things, substances, places, and goals. It specifies the basic ways in which these units can do things: acting, going, changing, being, having. One event may be seen as impinging on another, by causing or enabling or preventing it. An action can be initiated with a goal in mind, in particular, the destination of a motion (as in loading hay) or the state resulting from a change (as in loading a wagon). Objects are differentiated by whether they are human or nonhuman, animate or inanimate, solid or aggregate, and how they are laid out along the three dimensions of space. Events are conceived as taking up stretches of time and as being ordered with respect to one another.

Each of these ideas has a distinctive anatomy. Humans recognize unique individuals, and also pigeonhole them into categories. They distinguish stable categories that capture an individual's essence from transitory and superficial properties they may happen to possess. They have a mental zoom lens that can home in on the substance an entity is made of (plastic) or pan back to see its boundaries (a cup). A substance can be seen as a continuous medium (like applesauce) or as an aggregate of parts (like pebbles).

Humans have a primitive concept of number, which distinguishes only one, two, and many, though they can also estimate larger quantities approximately. They use this coarse way of quantifying not just when tallying objects (as in singular, dual, and plural) but also when locating things in space (as in *at, near,* and *far*) and when locating things in time (as in the present, the recent past, and the remote past).

When humans think about where an entity is, or what it is, or how it changes and moves, they tend to conceive of it holistically, as a blob or point without internal parts. The entire object is thought to be located in a spot, or to move as a whole, or to have a trait that suffuses it, or to change from one state to another in its entirety (as in a wagon loaded with hay, or a garden swarming with bees). But humans are also capable of articulating an object into its parts and registering how they are related to one another (as in *the bottom of the wagon* or *the edge of the garden*). When the object is a human body, another entity comes into play: the person, who is thought both to *be* his body parts and to *have* his body parts. Among people's possessions are not just their body parts and their chattels but also their ideas (which they can send to one another) and their good fortune.

When humans see the world or visualize it in a mental image, they situate objects and events in a continuous medium of space. But that is not the only way they grasp the physical world. In another mental system, humans don't reckon space in smooth coordinates like those revealed by rulers, protractors, and surveyors' levels; instead, they impose a coordinate frame onto a reference object and locate a figure relative to the frame, using qualitative spatial relations like *in, on,* and *above.* Humans can align a reference frame with the earth, their bodies, or a prominent object, and can mentally flip among these frames, allowing them to reason about a figure's location in multiple ways. The reference frames distinguish up from down and front from back, but are unreliable in distinguishing left from right. Humans readily notice topological relationships between a figure and a reference object, such as whether one is touching, attached to, or inside the other, and whether the figure is on or over its reference object, and whether it is near or far from it. A mental zoom lens allows these spatial concepts to be applied at any scale, from the subatomic to the intergalactic.

The human mind melts objects down into schematic models, which are composed of generic matter stretched into some number of dimensions (zero, one, two, or three). The matter aligned with one or more of these dimensions can be lopped off by a boundary, can be reeled out indefinitely, or can have a bit of adjacent matter stuck to it (as in ribbons, beams, and slabs). This geometry can also be applied to the boundary separating a hunk of matter from the space around it (as in ends, crusts, and borders), or to the nothingness left behind when a piece of matter is scooped out of the whole. The contours, angles, and lengths making up a shape fade into the background when people think spatially (so *across* can apply to a hand or to a country), though they can be brought back into awareness when people put the object into a category labeled by a noun. Spatial thinking is tailored to the demands of manipulating things, so it is not defined by geometry alone, but to an intuitive physics of fitting, supporting, containing, covering, and other ways that humans put objects to use.

Though a continuous flow of time is the medium of our consciousness, that is not how time is treated within the division of thought connected to language. In that division, time is treated like a dimension of space, and humans think of events as material laid along it. Time may be thought of as a road on which we march, or as a parade that marches past us. It is not measured with a stopwatch or calendar but divided into discrete regions. Humans tend to trichotomize time into a psychological present (a moment of awareness about three seconds long), an indefinite past (sometimes split into recent and distant), and an indefinite future (also splittable into impending and distant). The past and the future are often not pure temporal concepts but are infected with metaphysics: the past is merged with the actual, the nonpast with the hypothetical, the future with the willable. The events populating the mental time line are conceived as extrusions of time-stuff: like objects, they can be punctate or extended, can have crisp boundaries or fade out indefinitely, and can be composed of a single happening or an aggregate of repetitions. A mental zoom lens can zero in on the microscopic nature of the activity (*crossing the street*) or can pan back to take in the event in its entirety (*crossed the street*). And like the mental tape measure, the mental stopwatch is calibrated to human purposes. It runs differently when an act is seen as voluntarily enacted and when it seen as just occurring, and it starts when actors exert their will and stops when they consummate their goals.

Humans see some things as just happening and others as being caused. Causality is assessed not just by correlating things in time or by pondering what would have happened if things were otherwise, but by sensing an impetus that is transferred from a potent agent with a tendency toward motion to a weaker entity that would rather stay put. Variations on this mental cartoon of pushing and resisting give rise to intuitions of helping, hindering, preventing, and allowing.

The first link in a chain of causes is generally construed as an action initiated by an agent, typically a person. Humans demarcate actions by their manner, the change they bring about, or both. Humans care a lot about whether a change is brought about intentionally or accidentally, directly or with an intervening agent, and as a means or an end. They moralize these distinctions, holding agents blameworthy for the events they voluntarily, intentionally, and directly cause to happen.

No human is an island. Humans stock their minds with mental artifacts, such as names and other kinds of words, which are products of the minds of other humans. Some of these artifacts are ubiquitous in a given society at a given time, and collectively make up what we call a culture, one part of which is its language. Though a mental artifact such as a word may be omnipresent among the people in a society, it must have originated in the mind of an inventor who cast her bread upon the waters, and its fate depends both on its appeal to other minds and on the networks of influence that connect them. Every human is both a producer and a consumer of these artifacts, most notably when naming a child, and humans are ambivalent about their roles in this network of influence, torn between desires to fit in and to stand out.

Humans don't just entertain ideas but steep them with emotion. They stand in awe of deities, their parts and possessions, and the supernatural realms they control. They are terrified by disease, death, and infirmity. They are revolted by bodily secretions. They take a prurient interest in sexuality in all its variations. They loathe enemies, traitors, and subordinate peoples. As unpleasant as these thoughts are, people willingly inflict them on one another, sometimes to intimidate or denigrate them, sometimes to get their attention, sometimes to show that they can willingly endure the thoughts. As humans make it through the day, they react emotionally to its ups and downs, especially its frustrations and setbacks, and sometimes advertise these reactions to others.

Humans are touchy about their relationships. They maintain a "face" which emboldens them to stake out claims in negotiation and conflict. They are sensitive to their social rank, and also to their solidarity and empathy with others. With some of their fellows—typically kin, lovers, and friends—humans share resources, freely extend favors, and feel ties of empathy and closeness, which they blur with an intuition of being one flesh. With other people, they jockey for dominance, or show off their status, entitling them to the exercise of power or influence. With still others, they trade goods and services on a tit-for-tat basis, or divide things and responsibilities into even portions.

People invest their relationships with a moral coloring. They feel embarrassment when they breach the logic of a relationship with an inadvertent act, and feel contempt for others who breach it deliberately. Human relationships are ratified by mutual knowledge, in which people know that others know that they know that the relationship is of a certain kind. This makes humans especially sensitive to public acknowledgment of an act that violates the logic of a relationship, such as a threat, an offer, a request, or an insult. Nonetheless, humans often risk these breaches, sometimes to get on with the business of life, sometimes to renegotiate a relationship. As a result they engage in hypocrisy and taboo, designed to preserve the mutual knowledge that maintains a relationship even as they transact business inconsistent with it.

Any inventory of human nature is bound to cause some apprehension in hopeful people, because it would seem to set limits on the ways we can think, feel, and interact. "Is that all there is?" one is tempted to ask. "Are we doomed to picking our thinkable thoughts, our feelable feelings, our possible moves in the game of life, from a short menu of options?"

It is an anxiety that goes back to Plato's famous allegory of the prisoners in the cave. Captives are shackled in a grotto, their heads and bodies chained so that they can look only at the rear wall. The cave is a kind of movie theater out of *The Flintstones*, with a fire behind a balcony on which projectionists hold up cutouts and puppets, which cast moving shadows onto the wall. This movie is all that the prisoners know of the world. What they think of as objects are mere likenesses, and if they ever managed to leave the cave, the sight of the objects in the light of day would dazzle their dark-adapted eyes. In one interpretation of the allegory, the cave is our skull, and our acquaintance with the world consists only of the shadowy representations our minds make available to us.[1]

In these pages I have tried to lay out the major kinds of thoughts, feelings, and social relationships that go into the meaning and use of language. Are they the shadows on the wall of a cave in which our minds are forever trapped? Many of the book's discussions raise this fear, because they suggest that the machinery of conceptual semantics makes us permanently vulnerable to fallacies in reasoning and to corruption in our institutions.

One apparent limitation on our rationality is that our ability to frame an event in different ways makes us capable of flip-flopping on a course of action depending only on how the action is described (such as ensuring a gain versus preventing a loss). Another is that our concepts themselves, though useful enough in everyday toolmaking and cooperation, are hard-pressed to deal with the new conceptual worlds that science and society have made for us to live in. People are apt to think of an entity holistically, making them confuse statistical differences between groups with absolute superiority of one over the other. They think of possessions as physical commodities that can be in only one place at a time, rendering them ill-equipped to regulate a new marketplace of digital media. They conceive of motion as being powered by an impetus transferred from an antagonist to an agonist, leading them to misunderstand elementary physics (thinking, for example, that a ball in flight has a force pushing it along). Their grasp of evolution is just as tenuous; even people who profess to believe in Darwin's theory think that organisms change their traits in response to a need (as if they were agents pursuing a goal) and that all the members of a population evolve in lockstep (as if the species had an essence that changed holistically).[2] They think that human actors are uncaused causes, and bring about effects with the immediacy of billiard balls, with the result that a better understanding of causes inside the brain, and a closer examination of the circuitous causal structure of the world, can leave judges and juries befuddled.

Like natural concepts of the physical world, natural concepts of the social world may delude the minds that house them. Our instinctive guidelines as to how to deal with our fellows (communality, dominance, and reciprocity) may have served us well in face-to-face village life, but they can lead us astray in the formal arenas of a modern society. Nepotism and cronyism are permanent dangers to institutions. Arrogations of authority constantly threaten a democracy. And an expectation of tit-for-tat reciprocity poisons an understanding of the role of intermediaries in a complex economy. The economist Thomas Sowell has documented that middlemen—retailers and moneylenders—have been despised throughout history because they don't cause things to come into being, or barter equals for equals, but take a profit as goods or money pass through their hands. When ethnic minorities specialize in the middleman niche, they are seen as a race of bloodsuckers and become a target of persecution or genocide, despite their indispensable role in the local economy.[3] In *World on Fire*, the legal scholar Amy Chua argues that exporting free-market democracy may breed ethnic hatred and global instability, because a free market opens up middleman niches in which certain minorities prosper, inflaming the uncomprehending majority.[4]

We can also be diverted from the brightly lit world of reality by the emotions infusing our language. The automatic punch of emotionally laced words can fool us into thinking that the words have magical powers rather than being arbitrary conventions. And the taboos on thinking and speaking that shield our personal relationships from the mutual knowledge that might break their spell can leave us incapacitated as we try to deal with problems at the unprecedented scale of a modern society. Scientific findings that seem to challenge authority or threaten social solidarity, from Copernican astronomy to evolutionary biology, have been shushed as if they were social faux pas, or condemned as if they were personal betrayals.[5] And problems screaming for technical fixes, such as the American Social Security system, remain third rails that would electrocute any politician who touched them. Opponents can frame any solution as "putting a price on the welfare of our elderly citizens" (or our children, or our veterans), activating a taboo mentality that has a place in our dealings with family and friends but not in making policy for a nation of three hundred million people.[6]

Though language exposes the walls of our cave, it also shows us how we venture out of it, at least partway. People do, after all, catch glimpses of the sunlit world of reality. Even with our infirmities, we have managed to achieve the freedom of a liberal democracy, the wealth of a technological economy, and the truths of modern science. Though I doubt we will ever reach a cognitive utopia in which all the problems we dream up for ourselves are solvable, the human mind does have the means to go beyond a few reruns perpetually showing on the wall of the cave. Indeed, language offers the clearest window on how we can transcend our cognitive and emotional limitations.

The first way out is conceptual metaphor. Humans take their concepts of space, time, causality, and substance, etch away the leaden physical contents they were designed for, and apply the residual framework to airier subject matters. People co-opt the concept of an object at a location and use it for an entity in a circumstance (*going from Detroit to Chicago* → *going from bad to worse*). They co-opt the concept of an antagonist exerting force and use it for other kinds of causation, such as social pressure or internal conflict (*force the drawer to close* → *force Anne to leave*). Together, these abstractions provide the means to express a variable with a value and a cause and its effect—enough conceptual machinery to frame the basic laws of science. And humans have at their disposal more articulated metaphors which they exploit for more complex collections of thoughts—journeys for love, wars for argument, cords for political affiliation. The metaphors are not just literary tropes but can capture deep equivalences in causal networks, and people use them not just to talk but to reason.

The second way out is the combinatorial power of language—the "infinite use of finite means" by which words are composed into phrases and sentences whose meanings can be deduced from the meanings of the words and the way they are arranged. The combinatorial apparatus of grammar mirrors the combinatorial apparatus of thought, each phrase expressing a complex idea. Because my previous books have celebrated the infinite compositionality of language, I have not belabored it in the preceding chapters. But now is the time to be reminded that the combinatorial nature of language and thought allows us to entertain an explosion of ideas even though we are equipped with a finite inventory of concepts and relations. Just as a customer at Starbucks can order a cup of coffee in almost a hundred thousand varieties (multiplying out the sizes, roasts, caffeination, syrups, brewing styles, and kinds of milk), the owner of a human mind can brew up a staggering variety of thoughts by multiplying out the way that objects, events, causes, and goals can be combined.

Purveyors of linguistics have thought up many ways to show off the fecundity of language and the thoughts it expresses. My new favorite is the family of Web sites in which people post the strange things they overhear in offices, subways, and other walks of life. These snippets of real conversation from real people (no one could make them up) are the best advertisement for the combinatorial exuberance of the human mind, for its eternal power to flabbergast us with thoughts that have never before been thunk:[7]

> CONDUCTOR: Passengers, please do not use your valuables, or
> your child, to stop the train doors from closing!

> GIRL #1: As Shakespeare once said, "Thou shalt not kill."
> GIRL #2: No, that would be God.

> CO-WORKER IN OFFICE: I think from now on, I'm going to
> speak in the third person about myself, and I'll call myself
> "Angry Chinese clam." Angry Chinese clam is most displeased with your actions.

> MAN ON CELL PHONE: I tried to call you yesterday, but you
> weren't home. Where were you? What? Colonoscopy? Did
> he at least buy you flowers and talk dirty to you? Sorry,

yeah, that was out of line. I'll cut the crap now. Oh, ha, ha, I just made an unintended pun. No . . . no, sorry. . . . Hello? Hello?

When you combine these two aptitudes—metaphor and compositionality—the language of thought can be pressed into service to conceive and express a ceaseless geyser of ideas. People can discover new metaphors in their efforts to understand something, and can combine them to form still newer and more complex metaphors and analogies.

Of course these abilities can also feed a ceaseless geyser of *bad* ideas. But other mental talents provide a means to distinguish better from worse. People are not handcuffed to a single metaphor when thinking about something but can switch among them, sifting them for the best match between the relations among the concepts in the metaphor and the relations among the things they are trying to understand. And this sifting can be driven by a core intuition. People sense that their words are *about* things in the world, and are not just definitions trapped in a self-referential circle of terms (as we see in intuitions about the semantics of names). In a similar way, people can think of propositions as being *objectively* true or false, not just as things they *suppose* to be true or false (as we see in intuitions about the semantics of factive verbs like *learn* and *know*). The intuition that ideas can point to real things in the world or can miss them, and that beliefs about the world can be true or just believed, can drive people to test their analogies for fidelity to the causal structure of the world, and to prune away irrelevant features and zero in on the explanatory ones.

Needless to say, this combination of aptitudes does not endow any of us with a machine for churning out truths. Not only is a single mind limited in experience and ingenuity, but even a community of minds won't pool and winnow its inventions unless their social relationships are retuned for that purpose. Disagreements in everyday life can threaten our sense of face, which is why our polite interactions center on topics on which all reasonable people agree, like the weather, the ineptitude of bureaucracies, and the badness of airline or dormitory food. Communities that are supposed to evaluate knowledge, such as science, business, government, and journalism, have to find workarounds for this stifling desire for polite consensus. At a scientific conference, when a student points out a flaw in a presenter's experiment, it won't do to shut her up because the presenter is older and deserving of respect, or because he worked very hard on the experiment and the criticism would hurt his feelings. Yet these reactions would be perfectly legitimate in an everyday social interaction based on authority or communality.

In the sphere of social relations, no less than the sphere of pure ideas, we have to pry our mental models free of the domains they were designed for and apply them metaphorically and in new combinations to the business at hand. We know that the tools for doing so are available, because cultures differ in how they assign relationship types to commodities and communities. In science and other knowledge-driven cultures, the mindset of communality must be applied to the commodity of good ideas, which are treated as resources to be shared. This is a departure from the more natural mindset in which ideas are thought of as traits that reflect well on a person, or inherent wants that comrades must respect if they are to maintain their communal relationship. The evaluation of ideas also must be wrenched away from our intuitions of authority: department chairs can demand larger offices or higher salaries, but cannot demand that their colleagues acquiesce to their theories. These radically new rules for relationships are the basis for open debate and peer review in science, and for the checks and balances and accounting systems found in other formal institutions.

When all the pieces fall into alignment, people can grope their way toward the mouth of the cave. In elementary education, children can be taught to extend their number sense beyond "one, two, many" by sensing an analogy between an increase in rough magnitude and the order of number words in the counting sequence.[8] In higher education, people can be disabused of their fallacies in statistics or evolution by being encouraged to think of a population as a collection of individuals rather than as a holistic figure.[9] Or they can unlearn their faulty folk economics by thinking of money as something that can change in value as it is slid back and forth along a time line and of interest as the cost of pulling it forward. In science and engineering, people can dream up analogies to understand their subjects (a paintbrush is a pump, heat is a fluid, inheritance is a code) and to communicate them to others (sexual selection is a room with a heater and a cooler). Carefully interpreted, these analogies are not just alluring frames but actual theories, which make testable predictions and can prompt new discoveries. In the governance of institutions, openness and accountability can be reinforced by reminding people that the intuitions of truth they rely on in their private lives—their defense against being cheated or misinformed or deluded—also apply in the larger social arena. These reminders can militate against our natural inclinations toward taboo, polite consensus, and submission to authority.

None of this, of course, comes easily to us. Left to our own devices, we are apt to backslide to our instinctive conceptual ways. This underscores the place of education in a scientifically literate democracy, and even suggests a statement of purpose for it (a surprisingly elusive principle in higher education today). The goal of education is to make up for the shortcomings in our instinctive ways of thinking about the physical and social world. And education is likely to succeed not by trying to implant abstract statements in empty minds but by taking the mental models that are our standard equipment, applying them to new subjects in selective analogies, and assembling them into new and more sophisticated combinations.

The view from language shows us the cave we inhabit, and also the best way out of it. With the use of metaphor and combination, we can entertain new ideas and new ways of managing our affairs. We can do this even as our minds flicker with the agonists and antagonists, the points and lines and slabs, the activities and accomplishments, the gods and sex and effluvia, and the sympathy and deference and fairness that make up the stuff of thought.

NOTES

For complete publication data on works cited below, see the references.

I. WORDS AND WORLDS

1. The term *conceptual semantics* was coined by the linguist Ray Jackendoff; see Jackendoff, 1983; Jackendoff, 1990; Levin & Pinker, 1992; Pinker, 1989.
2. Lakoff, 2004; Lakoff, 2006; Nunberg, 2006. See S. Pinker, "Block That Metaphor," *New Republic*, October 9, 2006.
3. Jackendoff, 1990.
4. Pullum, 2003b.
5. Possibly apocryphal, or a paraphrase of sayings by Josh Billings; see Kim A. McDonald, "Many of Mark Twain's Famed Humorous Sayings Are Found to Have Been Misattributed to Him," *Chronicle of Higher Education*, September 4, 1991, A8.
6. Kripke, 1972/1980.
7. Wierzbicka, 1987.
8. Maureen Dowd, "Mel's Tequila Sunrise," *New York Times*, August 2, 2006.
9. "Popular Baby Names," Social Security Administration Web site; ssa.gov/cgi-bin/popularnames.cgi.
10. Personal communication, July 26, 2005.
11. Lieberson, 2000.
12. www.templetons.com/brad/spamterm.html.
13. Metcalf, 2002.
14. Harvey Silverglate, personal communication, June 5, 2006.
15. Kennedy, 2002.
16. www.snopes.com/travel/airline/obnox.htm.
17. Grice, 1975.

2. DOWN THE RABBIT HOLE

1. C. S. Lewis, *The Chronicles of Narnia*; Madeleine L'Engle, *A Wrinkle in Time*; Philip Pullman, *The Subtle Knife*; Dr. Seuss, *Horton Hears a Who*.
2. Some of the material in this chapter is taken from Pinker, 1989.
3. Pinker, 1997b, chap. 1.

4. Pinker, 1994b.

5. Pinker, 1984/1996; Pinker, 1994b, chap. 9.

6. From the writer Lloyd L. Brown's record of the speech of his daughter.

7. Pinker, 1979.

8. Goodman, 1983; Quine, 1969.

9. Chomsky, 1972a.

10. Gold, 1967; Nowak & Komarova, 2001; Pinker, 1979.

11. For examples of different approaches, see Bertolo, 2001; Goldberg, 2005; Pinker, 1984/1996; Tomasello, 2003; Yang, 2003.

12. For overviews of verb constructions and lexical semantics, see Borer, 2005a; Borer, 2005b; Croft, in press; Goldberg, 1995; Jackendoff, 1990; Levin, 1993; Levin & Pinker, 1992; Levin & Rappaport Hovav, 2005; Miller, 1991; Pinker, 1989; Pustejovsky, 1995; Van Valin, 2005; Wierzbicka, 1988b.

13. Attributed to the American ambassador Joseph Hodges Choate (1832–1917), who said it to a guest at an embassy function who mistook him for a doorman; see www .barrypopik.com/article/570/call-me-a-taxi—youre-a-taxi.

14. He continues the story: "Then these three white boys came up to me and said, Boy, we're givin' you fair warnin'. Anything you do to that chicken, we're gonna do to you. So I put down my knife and fork, I picked up that chicken and I kissed it."

15. Chomsky, 1972b; Fillmore, 1968.

16. See Schütze, 1996.

17. Berent, Pinker, & Ghavami, in press; Berent, Pinker, & Shimron, 1999; Berent, Pinker, & Shimron, 2002; Gropen et al., 1991a; Gropen et al., 1991b; Gropen et al., 1989; Kim et al., 1994; Kim et al., 1991; Marcus et al., 1995; Pinker & Birdsong, 1979; Prasada & Pinker, 1993; Senghas, Kim, & Pinker, 2004; Ullman, 1999. See also Bresnan, 2005, for dative constructions, and Wolff, 2003, for causative constructions.

18. Gropen et al., 1991b.

19. Gropen et al., 1991b.

20. For example, Baker, 1979, and Fodor & Crain, 1987.

21. These constructions are slight variants of container-locative construction; see Pinker, 1989, pp. 124–130.

22. Bowerman, 1982b.

23. Gropen et al., 1991b; Pinker, 1989.

24. Berko, 1958.

25. Gropen et al., 1991a.

26. Bowerman, 1988; Brown & Hanlon, 1970; Marcus, 1993; Morgan, Bonamo, & Travis, 1995; Morgan & Travis, 1989.

27. Francis & Kucera, 1982.

28. Called "Baker's paradox" in Pinker, 1989, based on a paper by the linguist C. Lee Baker, 1979. Baker himself attributed it to the psycholinguist Martin Braine, 1971.

29. Lederer, 1990.

30. Pinker, 1999.

31. Marcus et al., 1992; Ullman, 1999.

32. Baker, 1979; Gold, 1967; Osherson, Stob, & Weinstein, 1985; Pinker, 1979; Valiant, 1994.

33. Whorf, 1956.

34. Levin, 1985. See also Levin, 1993; Rappaport & Levin, 1985; Rappaport & Levin, 1988.

35. Chomsky, 1965; Chomsky, 1981.

36. Anderson, 1971.

37. Jackendoff, 1978; Jackendoff, 1983; Jackendoff, 1987; Jackendoff, 1990. First noticed by the linguist Jeffrey Gruber, 1965.

38. Lakoff & Johnson, 1980.

39. Pinker, 1990.
40. Talmy, 1983. See also Jackendoff, 1990; Jackendoff, 1991; Landau & Jackendoff, 1993.
41. Gropen et al., 1991a.
42. Levin, 1993; Pinker, 1989.
43. Gropen et al., 1991b; Pinker, 1989.
44. Bowerman, 1988; Gropen et al., 1989; Mazurkewich & White, 1984.
45. Gropen et al., 1989.
46. Green, 1974; Oehrle, 1976.
47. From the linguist Joan Bresnan.
48. Green, 1974; Oehrle, 1976.
49. Counterexamples are discussed in Bresnan & Nikitina, 2003, and Pinker, 1989, pp. 82–84, 154–160.
50. See Rappaport Hovav & Levin, 2007, and a similar analysis in Pinker, 1989, pp. 82–84, for discussions of the subtleties of this effect.
51. From "Show Me," a song in *My Fair Lady*.
52. Reddy, 1993.
53. Bresnan, 2005; Gropen et al., 1989; Pinker, 1989.
54. Comrie, 1976; Dowty, 1979–1991; Sasse, 2002; Tenny, 1992; Vendler, 1957.
55. Bresnan, 2005; Gropen et al., 1989; Pinker, 1989.
56. From Joni Mitchell's song "Big Yellow Taxi."
57. Title of a song by Arthur Hamilton sung by Ella Fitzgerald, Joe Cocker, and others.
58. From Bob Dylan's song "Highway 61 Revisited."
59. Title of a song by Ralph Blane and Hugh Martin, sung by Judy Garland in *Meet Me in St. Louis*.
60. From Paul Simon's song "Late in the Evening."
61. From Bill Fries and Chip Davis's song "Convoy."
62. Jespersen, 1938/1982; Randall, 1989.
63. Green, 1974.
64. Levin, 1993.
65. Bowerman, 1982a; Pinker, 1989.
66. Unpublished experiment with Jess Gropen and Thomas Roeper, described in Pinker, 1989, pp. 27, 296, 318.
67. Fodor, 1970.
68. Fodor, 1970.
69. Wolff, 2003.
70. Bertolt Brecht, *Poems, 1913–1956*, translated by M. Hamburger. New York: Methuen, 1979.
71. Levin, 1985.
72. Haspelmath, 1993.
73. Pullum, 2003a.
74. Gropen et al., 1991a; Gropen et al., 1991b; Gropen et al., 1989; Wolff, 2003.
75. Gladwell, 2000.
76. Bresnan et al., in press.
77. "As Israel Goes for Withdrawal, Its Enemies Go Berserk," July 16, 2006.
78. Many of these examples were noticed and recorded by Ilavenil Subbiah.
79. From CIA director George Tenet's speech on July 11, 2003, accepting the blame for the notorious sixteen words in George W. Bush's State of the Union address.
80. See Pinker, 1989, chaps. 5 and 6.
81. See Pinker, 1989, pp. 94–97, and Comrie, 1985a; Fareh & Hamdan, 2000; Foley & Van Valin, 1985; Fukui, Miyagawa, & Tenny, 1985; Guerssel, 1986; Hirschbühler, 2003; Hirschbühler & Mchombo, 2006; Kim, 1999; Kordoni, 2003; Mateu, 2001; Nwachukwu, 1987.

82. They include Ojibwa and Cree, Nez Perce, Tzotzil (Mayan), Huichol and Yaqui (Uto-Aztecan), Palauan (Micronesian), Chi-Mwi:ni (Bantu), Khasi (an Austroasiatic language spoken in India), Lahu (a Lolo-Burmese language), Kokborok (a Sino-Tibetan language spoken in Assam), Kham (West Tibetan), Nengone (an Austronesian language spoken in New Caledonia), Bahasa (an Indonesian language), Fongbe (spoken in Benin), and Acooli and Lango (Nilotic languages spoken in Uganda). See Chung & Gordon, 1998; Comrie, 1985a; Dowty, 1979; Dryer, 1986; Foley & Van Valin, 1985; Guerrero Valenzuela, 2002; Haspelmath, 2005; Lefebvre, 1994; Levin, 2004.
83. Comrie, 1985a; Dixon, 2000; Haspelmath, 1993; Nedyalkov & Silnitsky, 1973; Shibatani, 1976; Wierzbicka, 1998.
84. Comrie, 1985a; Foley & Van Valin, 1985; Nwachukwu, 1987.
85. Dixon, 2000; Nedyalkov & Silnitsky, 1973.
86. Croft & The Manchester Cognitive Collective, 2001; Levin, 2004.
87. Haspelmath, 1993.
88. Dixon, 2000; Haspelmath, 1993; Levin, 2004; Levin & Rappaport Hovav, 1995.
89. Dixon, 2000; Gergely & Bever, 1986; McCawley, 1968; Nedyalkov & Silnitsky, 1973; Pinker, 1989; Shibatani, 1976.
90. Allan, 1977; Bybee, 1985; Carter, 1976; Carter, 1988; Denny, 1976; Pinker, 1989, chap. 5; Talmy, 1985.
91. Talmy, 1985.
92. Kemmerer, 2000a.
93. Breedin & Saffran, 1999; Druks & Masterson, 2003; Kemmerer, 2000b; Kemmerer, 2003; Kemmerer et al., 2007; Kemmerer & Wright, 2002; Marshall et al., 1996.
94. Etcoff, 2008; Gilbert, 2006; Myers & Diener, 1995.
95. Schacter, 2001.
96. Blakemore & Frith, 2005; Bok, 2006.
97. Lessig, 2001.
98. Halpern, 2000.
99. S. Pinker, "Gender: Science Promises Honest Investigation," *Nature,* 442, 2006. See also S. Pinker, "Sex Ed: The Science of Difference," *New Republic,* February 14, 2005.
100. A. Schaffer, "A President Felled by an Assassin and 1880's Medical Care," *New York Times,* July 25, 2006.

3. FIFTY THOUSAND INNATE CONCEPTS (AND OTHER RADICAL THEORIES OF LANGUAGE AND THOUGHT)

1. Dennett, 1997.
2. Simonson & Tversky, 1992.
3. Pinker, 2002.
4. Fodor, 1981a; Fodor, 1998.
5. Fodor, 1968; Fodor, 1981b.
6. Quoted in the editor's introduction to Loewer & Rey, 1991, p. xi.
7. Bates, 1976; Cole, 1981; Nunberg, 1979; Sadock, 1984; Sperber & Wilson, 1986.
8. McClelland & Kawamoto, 1986.
9. Gentner & Goldin-Meadow, 2003; Gordon, 2004; Whorf, 1956.
10. Fodor, 1975.
11. Leibniz, 1768/1949, bk. II, chap. I, p. 111.
12. Fodor, 1970.
13. Fodor et al., 1980.
14. Fodor, 1998.
15. Fodor, 1975.
16. Fodor, 1981a, p. 284.

17. Fodor, 2000. See Pinker, 2005, for a reply.
18. Fodor, 1998; Fodor, 2001.
19. Williams, 1966.
20. Piatelli-Palmarini, 1986; Piatelli-Palmarini, 1989.
21. Chomsky, 2000, pp. 65-66; see also Chomsky, 1988, p. 32.
22. Goodman, 1983; Quine, 1960.
23. Markman, 1989.
24. Fodor originally suggested relocating such information in "meaning postulates," but seems to have dropped the idea; it is not mentioned in Fodor, 1981a, Fodor, 1998, or his other recent writings.
25. Fodor, 1998, p. 137.
26. Pinker, 1999.
27. Fodor & Pylyshyn, 1988.
28. See Pinker, 1989, chap. 7.
29. See Pinker, 1989, chap. 7.
30. Gentner, 1975.
31. Thanks to Katya Rice for the example. See Pinker, 1989, chap. 7, for many more.
32. Fodor, 1981a, p. 286.
33. Fodor, 1981a, p. 287.
34. Fodor, 1981a, p. 288.
35. Fodor, 1981a, p. 288.
36. Levin, 1985; Levin, 1993. See Pinker, 1989, pp. 104-109.
37. Fillmore, 1967; Kemmerer, 2003.
38. See Bloom, 2003, for evidence that the distinction pervades human reasoning.
39. Wierzbicka, 1988b.
40. Keyser & Roeper, 1984.
41. James, 1890/1950, chap. 9. The original quote used the German word *Vorstellung,* a familiar term in psychology at the time; I have rendered it as "representation."
42. Cole, 1981; Nunberg, 1979; Sadock, 1984; Sperber & Wilson, 1986.
43. Nunberg, 2006.
44. Bates, 1976; Bates & MacWhinney, 1982; McClelland & Kawamoto, 1986; Smith & Thelen, 1993.
45. Apresjan, 1973; Jackendoff, 1990; Lehrer, 1990; Nunberg, 1979; Ostler & Atkins, 1992; Pustejovsky, 1995; Wierzbicka, 1988b.
46. From the Berkeley Unix 4.2 "Fortunes" program.
47. From the novel by Erich Maria Remarque, quoted in Niall Ferguson, 2006, *The War of the World.* New York: Penguin.
48. The examples are from Nunberg, 1979.
49. Nunberg, 1979.
50. See Pinker, 1989, pp. 107-109, 269-278, 372-373.
51. See Pinker, 1989, pp. 107-109, 269-278, 372-373.
52. Apresjan, 1973; Jackendoff, 1990; Klein & Murphy, 2001; Lehrer, 1990; Ostler & Atkins, 1992; Pustejovsky, 1995.
53. Baayen & Moscoso del Prado Martin, 2005.
54. Klein & Murphy, 2001.
55. Pylkkänen, Llinás, & Murphy, 2006.
56. Pustejovsky, 1995.
57. Apresjan, 1973; Jackendoff, 1990; Lehrer, 1990; Ostler & Atkins, 1992; Pustejovsky, 1995.
58. Ostler & Atkins, 1992.
59. Ostler & Atkins, 1992. Ostler and Atkins suggest an even more specific microclass: the pulses, or seeds from legume pods. I suspect it's simply aggregations, since a compote of cherries is called *cherries,* not *cherry.*

60. Burchfield, 1995.
61. Allan & Burridge, 1991.
62. From the film *You Can't Cheat an Honest Man.*
63. McClelland & Kawamoto, 1986.
64. Singer, 1984.
65. Gentner & Goldin-Meadow, 2003; Gumperz & Levinson, 1996. H. Fountain, "Proof Positive That People See Colors with the Tongue," *New York Times,* March 30, 1999; G. Cook, "Debate Opens Anew on Language and Its Effect on Cognition," *Boston Globe,* February 14, 2002; "Language Barriers: Can a Concept Exist Without Words to Describe It?" *Economist,* August 19, 2004.
66. *Newsweek,* July 22, 1991.
67. Pullum, 1991.
68. Tversky & Kahneman, 1981.
69. Bloom, 1999; Pinker, 1994a.
70. This is the problem, I believe, in Bickerton, 1990, and Spelke, 2003.
71. Brown, 1958; Kay & Kempton, 1984.
72. Newell & Simon, 1972.
73. Baddeley, 1986.
74. Carey, 2008; Dehaene, 1997; Hauser, MacNeilage, & Ware, 1996; Wynn, 1992.
75. Dehaene et al., 1999.
76. Varley et al., 2005.
77. Jackendoff, 1987.
78. Jackendoff, 1997b.
79. Slobin, 1996.
80. Gentner & Goldin-Meadow, 2003. See Gleitman & Papafragou, 2005, for a review.
81. Newman et al., 2003.
82. Kim Curtis, "Affair Cited as Husband's Motive in Calif. Murders," *Boston Globe,* February 25, 2004.
83. Whorf, 1956.
84. F. Nietzsche, 1901, *The Will to Power.* L. Wittgenstein, 1922, *Tractatus Logico-Philosophicus.* M. Heidegger, 1971, "Building Dwelling Thinking," in *Poetry, Language, Thought.* R. Barthes, 1972, "To Write: An Intransitive Verb?" in R. Macksey & E. Donato (Eds.), *The Languages of Criticism and the Science of Man: The Structuralist Controversy.*
85. J. J. Hayes-Rivas, "One World Scientific Language?" *Science, 304,* May 28, 2004, p. 1243.
86. Gordon, 2004, pp. 496, 498.
87. Bowerman & Levinson, 2001, p. 13; Majid et al., 2004.
88. Xu & Carey, 1996; Xu, Carey, & Welch, 1999.
89. Schaller, 1991.
90. Hauser, 2000.
91. Santos et al., 2002.
92. Gordon, 2004.
93. Butterworth, 1999; Dehaene, 1997; Devlin, 2000; Gelman & Gallistel, 1978; Wiese, 2003.
94. Casasanto, 2005. See also Gelman & Gallistel, 2004.
95. Everett, 2005.
96. Butterworth, 1999; Dehaene, 1997; Devlin, 2000; Gelman & Gallistel, 1978; Wiese, 2003.
97. Pica et al., 2004.
98. Gelman & Gallistel, 1978.
99. Levinson, 2003; Levinson, Kita, & Haun, 2002; Pederson et al., 1998.
100. Though it is claimed that Tzeltal speakers never extend left-right body terms to other objects, Linda Abarbanell, who also studied these speakers, notes that it is done by some speakers and some dialects (personal communication, March 28, 2006).

101. I am granting this claim, but Abarbanell (see the previous note) is skeptical: "From what I can tell, they primarily use a landmark-based system, with the landmark varying depending on the context. A Tenejapan (if they used teacups) would probably say the same thing as a Bostonian: 'Pass me the spoon by the cup—the one over there.' "
102. Gumperz & Levinson, 1996, p. 27.
103. Pederson et al., 1998, p. 586.
104. Pinker, 1997b, chap. 5; Corballis & Beale, 1976; Hinton & Parsons, 1981; Marr, 1982; Pinker, 1984; Pinker, 1988; Rock, 1983; Talmy, 1983.
105. "Wrong Again: Postage Stamp of Canyon Flipped," *Boston Globe*, February 3, 2000.
106. Corballis & Beale, 1976.
107. Rock, 1983.
108. Attneave, 1968.
109. Abarbanell, Li, & Papafragou, 2005; Li, Abarbanell, & Papafragou, 2005.
110. Abarbanell, Li, & Papafragou, 2005; Gleitman et al., 2005; Li & Gleitman, 2002. See also Levinson, Kita, & Haun, 2002, for a defense of the original conclusions.
111. Linda Abarbanell, personal communication, March 28, 2006.
112. Majid et al., 2004, did look for possible ecological and cultural influences, with equivocal results, but they tested global factors such as ecological zone and subsistence mode, rather than factors directly relevant to reference frames such as local topography, mobility, and literacy.
113. Gleitman et al., 2005.
114. Pinker, 1984/1996; Pinker, 1989; Roy & Pentland, 2002; Siskind, 1995.
115. Carey, 2008; Soja, Carey, & Spelke, 1991; Spelke, 1995.
116. Anderson, 1976; Anderson & Bower, 1973; Schacter, 1996.
117. Bransford & Franks, 1971. See also Anderson, 1976.
118. Adapted from Harlow, 1998, p. 13.
119. Pinker & Jackendoff, 2005.
120. Klein & Murphy, 2002.
121. From "Limitations."

4. CLEAVING THE AIR

1. Hume, 1748/1999.
2. The first part is commonly attributed to the physicist John Archibald Wheeler, but he wrote that he reproduced it from the wall of a men's room; Wheeler, 1994, n. 1.
3. Pinker, 1997b, chap. 4.
4. Berkeley, 1713/1929.
5. Cave et al., 1994.
6. Tegmark, 2003.
7. Kant, 1781/1998; Kant, 1783/1950; Körner, 1955; Walsh, 1967.
8. James, 1907/2005, pp. 76–77.
9. From the *Principia*; quoted in Körner, 1955, p. 33.
10. Körner, 1955; McCormick, 2005; Walsh, 1967.
11. Gardner, 1990; Randall, 2005.
12. Barbour, 2000; Goldstein, 2005; Hawking & Mlodinow, 2005.
13. Carlo Rovelli, "Rovelli's Two Principles: Space Does Not Exist; Time Does Not Exist," www.edge.org/q2004/page7.html#rovelli. The alphabet analogy comes from Brian Greene.
14. Kitcher, 1990.
15. Allison, 1973, pp. 135–136, quoted by Kitcher, 1990, pp. 15–16.
16. Chomsky, 1972a; Hirschfeld & Gelman, 1994; Pinker, 1997b; Tooby & Cosmides, 1992.
17. Pinker, 2002.

18. Pinker, 1997b, chap. 4. See also Kosslyn, 1980; Kubovy, 1981; Pinker, 1984; Pinker, 1988; Pinker, 1990; Robertson, 2003; Shepard, 1978.
19. See Cave et al., 1994, for experimental evidence that mental images always have a location in visual space.
20. Isenberg, Nissen, & Marchak, 1990; Kubovy, 1981; Robertson, 2003; Treisman & Gelade, 1980.
21. Tootell et al., 1982; Van Essen & Deyoe, 1995.
22. Bregman, 1990; Kubovy, 1981.
23. Gentner, 1981.
24. Wierzbicka, 1991.
25. Vise & Malseed, 2005.
26. Widely circulated on the Web.
27. Wierzbicka, 1988c.
28. Chierchia, 1998; Jackendoff, 1991; Rijkhoff, 2002; Semenza, 2005; Wierzbicka, 1988a; Winter, 2002.
29. Jackendoff, 1991.
30. Chierchia, 1998; Jackendoff, 1991; Rijkhoff, 2002; Winter, 2002.
31. Chierchia, 1998; Jackendoff, 1991; Rijkhoff, 2002; Winter, 2002.
32. Lederer, 1990.
33. From *Thoughts of a Biologist* (1939).
34. Soja, Carey, & Spelke, 1991.
35. Bach, 1986; Jackendoff, 1991.
36. Barner & Snedeker, 2005.
37. Bloom, 1994; Bloom, 1996; Giralt & Bloom, 2000.
38. Bloom, 1996.
39. Elizabeth Barrett Browning; an old Jewish saying; Paul Simon; Bob Dylan; Dorothy Parker.
40. Cushing, 1994.
41. Goodale & Milner, 2004.
42. Kemmerer, in press; Kosslyn, 1987; Kosslyn, 1994; Postma & Laeng, 2006.
43. Kemmerer, in press; Landau & Jackendoff, 1993.
44. Biederman, 1995.
45. Sinha & Kuteva, 1995.
46. Landau & Jackendoff, 1993; Talmy, 2000b; Tyler & Evans, 2003.
47. Bowerman & Levinson, 2001; Levinson, Meira, & The Language and Cognition Group, 2003; Levinson, 2003.
48. Brugman, 1988.
49. Levinson, Meira, & The Language and Cognition Group, 2003.
50. Levinson, Meira, & The Language and Cognition Group, 2003.
51. Talmy, 2000b.
52. Landau & Jackendoff, 1993; Talmy, 2000b.
53. Lloyd L. Brown, personal communication.
54. Jackendoff, 1991; Talmy, 2000b.
55. Landau & Jackendoff, 1993.
56. Sommers, 1963.
57. Casati, 2006; Pomerantz, 2003.
58. Biederman, 1995; Marr, 1982; Pinker, 1997b, chap. 4.
59. Tarr & Pinker, 1989.
60. Landau & Jackendoff, 1993.
61. Duncker, 1945; Glucksberg & Danks, 1968.
62. Rozin & Fallon, 1987.
63. Allan, 1977; Denny, 1976.

64. Talmy, 2000b.
65. Landau & Jackendoff, 1993.
66. Levinson, 2003.
67. Talmy, 2000b.
68. Francis & Kucera, 1982.
69. Eco, 1995; Pinker, 1999.
70. Pinker & Bloom, 1990.
71. *Boston Globe,* January 8, 1991.
72. Levinson & Wilkins, 2006; Pederson et al., 1998; Talmy, 2000b.
73. Pinker & Bloom, 1990.
74. Carlson & van der Zee, 2005; Coventry & Garrod, 2004. See also Feist & Gentner, 1998; Tyler & Evans, 2003.
75. Comrie, 1985b.
76. Brown, 1991; Malotki, 1983.
77. Bybee, Perkins, & Pagliuca, 1994; Bybee, 1985; Comrie, 1985b.
78. Comrie, 1985b.
79. James, 1890/1950, p. 21.
80. Pöppel, 2003.
81. Comrie, 1985b.
82. Comrie, 1985b.
83. Ornstein, 1975.
84. Lakoff & Johnson, 1980; Lakoff & Johnson, 2000.
85. Haspelmath, 1997.
86. Boroditsky, 2001.
87. Núñez & Sweetser, 2006.
88. Sasse, 2002.
89. Comrie, 1985b; Reichenbach, 1947.
90. Comrie, 1985b; Jackendoff, 1997a.
91. Comrie, 1985b.
92. Bybee, Perkins, & Pagliuca, 1994.
93. Brown & Levinson, 1987b.
94. Bach, 1986; Comrie, 1976; Dowty, 1979/1991; Dowty, 1982; Jackendoff, 1990; Jackendoff, 1991; Jackendoff, 1997a; Pustejovsky, 1995; Sasse, 2002; Van Valin, 2005; Vendler, 1957.
95. Tenny, 1992.
96. Pinker, 1989.
97. Bach, 1986; Jackendoff, 1991.
98. Dowty, 1979/1991; Dowty, 1982; Jackendoff, 1991; Pinker, 1989; Sasse, 2002; Vendler, 1957.
99. Bach, 1986.
100. Lederer, 1990.
101. Sasse, 2002.
102. Comrie, 1976; Comrie, 1985b; Sasse, 2002.
103. The bartender says to him, "What the hell happened to you?" The duck replies, "It started out as a wart on me ass."
104. Russell Smith, "CBC's Mixing of Verb Tenses Gives Me Unwanted Linguistic Tension," *Globe and Mail,* February 16, 2006.
105. From "Transcript of Testimony of William Jefferson Clinton, President of the United States, Before the Grand Jury Empaneled for Independent Counsel Kenneth Starr, August 17, 1998" (released September 21, 1998). Office of the Independent Counsel, Washington, D.C. I have cleaned up some minor dysfluencies in Clinton's testimony.
106. Dowty, 1979/1991.
107. Katz, 1987.

108. Hume, 1740/1955.
109. Hume, 1740/1955.
110. Katz, 1987; Lewis, 1973; Spellman & Mandel, 1999.
111. Lewis, 1973.
112. Katz, 1987.
113. Lagnado & Sloman, 2004; Spellman, 1996; Spellman & Mandel, 1999.
114. Dowty, 1979/1991; Katz, 1987; Spellman, 1997; Spellman & Mandel, 1999; Walsh & Sloman, 2005; Wolff, 2002; Wolff & Song, 2003.
115. Dowty, 1979/1991; Hilton, 1995.
116. Dowty, 1979/1991.
117. Katz, 1987, p. 234; Lewis, 1973.
118. Spellman & Kincannon, 2001.
119. Unpublished experiment by Barbara Spellman; see also Mandel, 2003.
120. Gopnik et al., 2004; Pearl, 2000.
121. Dowe, 2000; Shultz, 1982; Talmy, 1988; Talmy, 2000a; Walsh & Sloman, 2005; White, 1995; Wolff, 2002; Wolff & Song, 2003.
122. Chapman & Chapman, 1982; Redelmeier & Tversky, 1996.
123. Goldvarg & Johnson-Laird, 2001; Wolff, 2002.
124. The example is from the psychologist Phillip Wolff, 2007.
125. Michotte, 1963.
126. Carey, 2008; Cohen et al., 1998; Hauser, 2006; Leslie, 1995.
127. Premack & Premack, 2003.
128. Hauser & Spaulding, 2006.
129. Talmy, 1988; Talmy, 2000a.
130. Wolff, 2003.
131. Wolff, 2003.
132. Dixon, 2000; Gergely & Bever, 1986; McCawley, 1968; Nedyalkov & Silnitsky, 1973; Pinker, 1989; Shibatani, 1976.
133. Wolff, 2007.
134. Walsh & Sloman, 2005; White, 1995; Wolff, 2003; Wolff, 2007; Wolff & Song, 2003.
135. McCloskey, 1983; McCloskey, Caramazza, & Green, 1980.
136. Quoted in M. Kumar, "Quantum Reality," *Prometheus, 2* (1999), pp. 20–21.
137. Russell, 1913.
138. Pinker, 1997b; Tooby & DeVore, 1987, chap. 5.
139. From http://home.howstuffworks.com/toilet.htm.
140. Dershowitz, 2005.
141. Foot, 1978; Thomson, 1985.
142. Hauser et al., 2007.
143. Greene, 2002.
144. Greene et al., 2001.
145. A. Harris, "Benihana Chef's Playful Food Toss Blamed for Diner's Death," *New York Law Journal,* November 23, 2004, www.law.com/jsp/article.jsp?id=1101136512535. C. Kilgannon, "Jury to Decide If Flying Sizzling Shrimp Led to Man's Death," *New York Times,* February 9, 2006.
146. Katz, 1987.

5. THE METAPHOR METAPHOR

1. Noticed by Ilavenil Subbiah.
2. From radio host Mike McConnel; www.calvin.edu/academic/engl/lang/mixmet.htm.
3. From David Horowitz, "The Loafing Class," *Salon,* February 8, 1998.

4. From an NFL football broadcast.
5. AWFUL was coined by Roger Tobin; thanks to David Birdsong for bringing it to my attention.
6. Lakoff & Johnson, 1980.
7. "Generative metaphor" from Schön, 1993; "conceptual metaphor" from Lakoff & Johnson, 1980.
8. The terms *vehicle* and *tenor* were introduced by the literary scholar I. A. Richards, 1936/1965.
9. Lakoff, 1993, p. 206.
10. Lakoff, 1987; Lakoff, 1993; Lakoff, 1996; Lakoff & Johnson, 1980; Lakoff & Johnson, 2000; Lakoff & Núñez, 2000.
11. Kolodner, 1997; Mayer, 1993.
12. Gould, 1980.
13. Pinker, 1997b, chap. 5.
14. Hauser & Spaulding, 2006; Hauser, 1997; Hauser, Pearson, & Seelig, 2002.
15. Pinker, 1989, chap. 8. See also Jackendoff, 2002; Pinker, 1997b, chap. 5.
16. Pinker, 1997b, chap. 5.
17. Tversky & Kahneman, 1981.
18. Kahneman & Tversky, 1979.
19. Schön, 1993.
20. Boudin, 1986; see also Winter, 2001.
21. Lawley & Tompkins, 2000.
22. Fairhurst & Sarr, 1996; see also Clancy, 1989.
23. Harris, 1993.
24. Lakoff & Johnson, 1980, p. 3.
25. Lakoff, 1987; Lakoff & Johnson, 2000.
26. Lakoff & Johnson, 2000.
27. Lakoff & Núñez, 2000.
28. Lakoff, 1996. For an alternative, see Pinker, 2002, chap. 16, based on Sowell, 1987.
29. Lakoff, 2004.
30. Lakoff, 2003.
31. M. Bai, "The Framing Wars," *New York Times Magazine,* July 17, 2005.
32. Nagel, 1997; Searle, 1993b.
33. Dawkins, 1998.
34. Nagel, 1997; Searle, 1993b.
35. Keysar et al., 2000; see also Bowdle & Gentner, 2005.
36. Jackendoff, 1983; Jackendoff, 1992, chap. 3; Jackendoff, 2002; Jackendoff & Aaron, 1991; Maratsos, 1988; Murphy, 1996.
37. Kemmerer, 2005.
38. Gruber, 1965; Jackendoff, 1978; Jackendoff, 1983; Jackendoff, 1990; Jackendoff, 2002.
39. Jackendoff, 1992, chap. 3.
40. Gentner et al., 2001.
41. Schön, 1993.
42. Gentner, 1983; Gentner et al., 2001; Gentner & Jeziorski, 1989. See also Holyoak & Thagard, 1996.
43. Gentner & Jeziorski, 1989.
44. Gentner & Jeziorski, 1989.
45. Dawkins, 1986, pp. 210–211.
46. Boyd, 1993.
47. Putnam, 1975.
48. A similar idea was proposed by Schelling, 1978.

49. Pinker, 1997b, chap. 2.
50. M. Bai, "The Framing Wars," *New York Times Magazine,* July 17, 2005; K. S. Baer, "Word Games," *Washington Monthly,* January/February 2005; J. Green, "It Isn't the Message, Stupid," *Atlantic Monthly,* May 2005; M. Cooper, "Thinking of Jackasses: The Grand Delusions of the Democratic Party (Review of George Lakoff's *Don't Think of an Elephant),*" *Atlantic Monthly,* March 2005; W. Galston, "Chico Marxism," *Democracy: A Journal of Ideas,* Fall 2006.
51. Glucksberg & Keysar, 1993.
52. Lakoff & Turner, 1989.
53. Jackendoff & Aaron, 1991.
54. Jackendoff & Aaron, 1991; Searle, 1993a.
55. Jackendoff & Aaron, 1991.
56. Gentner & Jeziorski, 1989; Gentner et al., 2001.
57. Meier & Robinson, 2004.
58. Borregine & Kaschak, in press.
59. Boroditsky, 2000; Boroditsky & Ramscar, 2002.
60. McGlone & Harding, 1998.
61. Bowerman, 1983.
62. Unpublished experiments described in Pinker, 1989, p. 333.
63. Schank, 1982.
64. Proust, 1913/1982, p. 48.
65. "Ring" rather than "keys" in the original.
66. Catrambone & Holyoak, 1989; Gentner, Ratterman, & Forbus, 1993; Gick & Holyoak, 1980; Keane, 1987; Ross, 1984; Ross, 1987.
67. Newell & Simon, 1972.
68. Kotovsky, Hayes, & Simon, 1985. I have modified the example to make it clearer.
69. Duncker, 1945.
70. Gick & Holyoak, 1980.
71. Keane, 1987. See also Catrambone & Holyoak, 1989; Gentner, Ratterman, & Forbus, 1993; Ross, 1984; Ross, 1987.
72. Chi, Feltovich, & Glaser, 1981.
73. Dunbar, 2001.
74. Blanchette & Dunbar, 2000.
75. Blanchette & Dunbar, 2001.
76. Pinker, 1999.
77. From Rebecca Goldstein's *The Mind-Body Problem.* Goldstein, 1983, pp. 103–104.
78. From Ian McEwan's *Amsterdam.* McEwan, 1998, pp. 90–91.
79. From Vladimir Nabokov's *Lolita.* Nabokov, 1955, p. 11.
80. From Gustave Flaubert's *Madame Bovary.* Flaubert, 1857/1998, p. 173; composite of several translations.

6. WHAT'S IN A NAME?

1. Social Security Administration, 2006; Wattenberg, 2005.
2. Pinker & Rose, 1998.
3. www.ncseweb.org/resources/articles/3541_project_steve_2_16_2003.asp.
4. Scott, Matzke, Branch, & 284 scientists named "Steve," 2004.
5. Social Security Administration, 2006; Wattenberg, 2005.
6. Example from Fodor, 1994; see also Pinker, 1995.
7. Chierchia & McConnell-Ginet, 2000.
8. Kripke, 1972/1980; Marcus, 1961; Putnam, 1975.
9. Putnam himself identified the cats merely as robots, not as daleks.

10. Putnam, 1975, p. 161.
11. Block, 1986; Fodor, 1987; Fodor, 1994.
12. Fodor, 1994; Pinker, 1995.
13. Hull, 1989; Mayr, 1982.
14. Katz, 1987.
15. Ayer, 1936.
16. Kitcher, 1990, pp. 15-16, distinguishes three senses of *a priori* in Kant's writings, one of which is "universal and necessary."
17. Kant, 1781/1998; Körner, 1955; Walsh, 1967.
18. McGinn, 1993, p. 64.
19. McGinn, 1993. See also Pinker, 1997b, pp. 558-565.
20. Attributed to Mike Kellen by the Unix "fortunes" program.
21. www.macmillandictionary.com/2005/index.htm.
22. Pinker, 1999, pp. 80-82; Pinker & Birdsong 1979; Pinker, 1994b, chap. 6. See also www.trismegistos.com/IconicityInLanguage/.
23. Brown, 1958.
24. First noticed by Emmon Bach. See Beeman, 2001.
25. Ross, undated, quoted in Beeman, 2001.
26. Metcalf, 2002; Wallraff, 2006.
27. The insomnia example is from Donald Symons. Some of the others are from Wallraff, 2006.
28. Hall & Friends, 1984.
29. Adams & Lloyd, 1990.
30. Wallraff, 2006.
31. Metcalf, 2002.
32. Wallraff, 2006.
33. Pinker, 1989, chap. 5; Bloom, 1999; Carter, 1976; di Sciullo & Williams, 1987; Dowty, 1979/1991; Markman, 1989.
34. Kaplan & Bernays, 1997.
35. Lieberson, 2000; Ornstein, 2004; Social Security Administration, 2006; Wattenberg, 2005.
36. Brown, 1985.
37. Lieberson, 2000.
38. Lieberson, 2000, p. xi.
39. Ornstein, 2004; Wattenberg, 2005.
40. Wattenberg, 2005.
41. Wattenberg, 2005.
42. Lieberson, 2000.
43. Tenner, 1989, described in Lieberson, 2000.
44. Lieberson, 2000.
45. Bell, 1992; Lieberson, 2000.
46. Bell, 1992; Veblen, 1899/1994.
47. Kaplan & Bernays, 1997.
48. Ornstein, 2004.
49. Lieberson, 2000.
50. Social Security Administration, 2006.
51. Barry & Harper, 1993; Lieberson, Dumais, & Baumann, 2000; Wattenberg, 2005.
52. Lieberson, 2000; Ornstein, 2004; Wattenberg, 2005.
53. S. Pinker, "The Game of the Name," *New York Times*, April 3, 1994.
54. Gladwell, 2000.
55. Aunger, 2000; Blackmore, 1999; Dawkins, 1976/1989; Sperber, 1985.
56. Schelling, 1978, pp. 27-28.

7. THE SEVEN WORDS YOU CAN'T SAY ON TELEVISION

1. Allan & Burridge, 1991; Dooling, 1996; Hughes, 1991/1998; Jay, 2000; Wajnryb, 2005.
2. Bruce, 1965/1991.
3. Kennedy, 2002.
4. Denfeld, 1995; Dooling, 1996; Patai, 1998; Saporta, 1994.
5. Allan & Burridge, 1991; Jay, 2000; Wajnryb, 2005.
6. Allan & Burridge, 1991; Jay, 2000; Wajnryb, 2005.
7. Hughes, 1991/1998.
8. Hughes, 1991/1998, p. 3.
9. From the CD accompanying Collins & Skover, 2002.
10. Hughes, 1991/1998.
11. Allan & Burridge, 1991; Jay, 2000; Wajnryb, 2005.
12. Hughes, 1991/1998.
13. Allan & Burridge, 1991; Aman, 1987; Crystal, 1997; Jay, 2000; Wajnryb, 2005.
14. Wajnryb, 2005, p. 223.
15. Aman, 1987; Solt, 1987.
16. Allan & Burridge, 1991; Rosenblum & Pinker, 1983.
17. Osgood, Suci, & Tannenbaum, 1957.
18. Beeman, 2005.
19. LeDoux, 1996; Panksepp, 1998.
20. Isenberg et al., 1999; LaBar & Phelps, 1998; Lewis et al., in press.
21. Harris, Gleason, & Aycicegi, 2006; Jay, 2000; Matthew, Richards, & Eysenck, 1989.
22. Harris, Gleason, & Aycicegi, 2006.
23. Roelofs, in press.
24. MacKay et al., 2004.
25. Allan & Burridge, 1991; Hughes, 1991/1998.
26. S. Pinker, "Racist Language, Real and Imagined," *New York Times,* February 2, 1999.
27. Dronkers, Pinker, & Damasio, 1999.
28. Jay, 2000; Van Lancker & Cummings, 1999.
29. Van Lancker & Cummings, 1999.
30. Jay, 2000; Van Lancker & Cummings, 1999; Van Lancker & Sidtis, 2006.
31. Dronkers, Pinker, & Damasio, 1999; Pinker, 1997a; Pinker, 1999.
32. Etcoff, 1986.
33. Wise, Murray, & Gerfen, 1996.
34. Ullman et al., 1997.
35. Speedie, Wertman, & Heilman, 1993.
36. Singer, 2005.
37. Jay, 2000; Van Lancker & Cummings, 1999.
38. Van Lancker & Cummings, 1999.
39. Wegner, 1989.
40. Allan & Burridge, 1991; Hughes, 1991/1998.
41. Kiparsky, 1973.
42. Pinker, 1994b; Pinker, 1999.
43. Allan & Burridge, 1991; Crystal, 2003.
44. Hughes, 1991/1998, pp. 22–23.
45. Allan & Burridge, 1991; Hughes, 1991/1998.
46. Frank, 1988; Pinker, 1997b, chap. 6; Schelling, 1960.
47. Hughes, 1991/1998.
48. Allan & Burridge, 1991; Hughes, 1991/1998.
49. Pinker, 2002, chap. 15; Tetlock et al., 2000.
50. Hughes, 1991/1998, p. 12.

51. Hughes, 1991/1998.
52. Allan & Burridge, 1991.
53. Curtis & Biran, 2001, p. 21.
54. Allan & Burridge, 1991; Harris, 1989; Rozin & Fallon, 1987.
55. Rozin & Fallon, 1987.
56. Quoted by Curtis & Biran, 2001.
57. Curtis & Biran, 2001. See also Rozin & Fallon, 1987, and Pinker, 1997b, chap. 6.
58. Rozin & Fallon, 1987.
59. Curtis & Biran, 2001; Rozin & Fallon, 1987.
60. From the biographical movie *Lenny*.
61. Buss, 1994; Symons, 1979.
62. Buss, 1994; Symons, 1979.
63. Jay, 2000; Van Lancker & Cummings, 1999; Wajnryb, 2005.
64. Denfeld, 1995; Dooling, 1996; Patai, 1998.
65. Dworkin, 1979, p. 133; cited in Denfeld, 1995, p. 23.
66. Allan & Burridge, 1991.
67. Quoted in Hughes, 1991/1998.
68. Hughes, 1991/1998, p. 122.
69. Allan & Burridge, 1991; Aman, 1987.
70. Quoted in Wajnryb, 2005, p. 48.
71. Levin, 1985; Levin, 1993; Pinker, 1989.
72. The following examples are marginally acceptable to me (in their grammar, if not their practice): *Clarence fucked at the goat for three minutes but was interrupted by Old MacDonald* (conative); *He fucked her in the armpit / He fucked her armpit* (possessor-raising); *Goats fuck easily* (middle). The following are not: *John fucked a dildo into the goat* (contact locative; compare *John fucked the goat with a dildo,* which is fine); *At three o'clock, the goat fucked* (anticausative).
73. Quang Fuc Dong, 1971/1992b.
74. Sheidlower, 1995.
75. Zwicky et al., 1971/1992.
76. J. S. Farmer & W. E. Henley, *Slang and Its Analogues,* 1890–1904, quoted in Hughes, 1991/1998, p. 271.
77. Quang Fuc Dong, 1971/1992a.
78. Nunberg, 2004. Some adjectives, like *former* and *alleged,* also fail these tests, but they differ from *fucking* in other ways.
79. Shad, 1971/1992.
80. Nunberg, 2004.
81. The interview continues:
    INTERVIEWER: Who's your favorite Charlie's Angel?
    CLEESE: Noam Chomsky.
82. Thanks to Geoffrey Pullum for the laptop example.
83. Quang Fuc Dong, 1971/1992a; Potts, 2005.
84. Potts, 2005.
85. Bryson, 1990, p. 211.
86. Quang Fuc Dong, 1971/1992a.
87. Yeung, Botvinick, & Cohen, 2004.
88. Panksepp, 1998.
89. Dollard et al., 1939; Panksepp, 1998.
90. Panksepp, 1998.
91. Goffman, 1978.
92. Goffman, 1959.

93. Goffman, 1978.
94. Goffman, 1978, p. 814.
95. Code, in press; Darwin, 1874; Wray, 1998.
96. From a book review in a recent issue of a respected opinion magazine.
97. Larkin, 2003.
98. Harris, 1998, p. 350.

8. GAMES PEOPLE PLAY

1. *Two Gentlemen of Verona, The Taming of the Shrew, Twelfth Night, As You Like It, The Merchant of Venice, Cymbeline, The Winter's Tale, Measure for Measure.* Cohen, 2004.
2. Allan & Burridge, 1991; Brown & Levinson, 1987b.
3. See www.phrases.org.uk/bulletin_board/12/messages/1223.html.
4. Grice, 1975. See also Brown & Levinson, 1987b; Green, 1996; Holtgraves, 2002; Potts, 2005; Smith, 1982; Sperber & Wilson, 1986.
5. www.laughlab.co.uk/.
6. Dews, Kaplan, & Winner, 1995; Winner & Gardner, 1993.
7. Brown & Levinson, 1987b. See also Brown, 1987; Brown & Gilman, 1972; Fraser, 1990; Green, 1996; Holtgraves, 2002.
8. Goffman, 1967.
9. Fiske, 1992; Fiske, 2004; Haslam, 2004; Holtgraves, 2002.
10. Often mistakenly attributed to Mark Twain.
11. See Isaacs & Clark, 1990.
12. J. Gorman, "Like, Uptalk?" *New York Times Magazine,* 1993; reprinted in Hirschberg & Hirschberg, 1999.
13. Brown & Gilman, 1972.
14. Brown & Levinson, 1987b; Brown & Gilman, 1972.
15. Kaplan & Bernays, 1997.
16. Potter, 1950/1971, p. 190.
17. C. Murphy, "Feeling Entitled?" *Atlantic Monthly,* March 2005.
18. Though see also Fraser, 2005, for a skeptical appraisal.
19. J. Tierney, "The Big City: You Could Look It Up," *New York Times,* September 24, 1995.
20. Nisbett & Cohen, 1996.
21. Clark, 1996; Clark & Schunk, 1980; Francik & Clark, 1985; Gibbs, 1986; Searle, 1975.
22. Brown & Levinson, 1987b; Grice, 1975; Potts, 2005.
23. Brown & Levinson, 1987a; Clark & Schunk, 1980; Fraser, 1990; Holtgraves, 2002.
24. Holtgraves, 2002.
25. Holtgraves, 2002.
26. Pinker, 2002.
27. Kasher, 1977; Sampson, 1982.
28. Schelling, 1960, pp. 139–142.
29. Dawkins & Krebs, 1978.
30. Alan Dershowitz, presentation at a seminar on Indirect Speech at the Program for Evolutionary Dynamics, Harvard University, May 16, 2006; Harvey Silverglate, comments at the same seminar.
31. "Woman Is Found Guilty of Bribery to Win a Vote for Rights Proposal," *New York Times,* August 23, 1980; "Woman Convicted of Vote Bribe Is Ordered to Do Public Service," *New York Times,* November 8, 1980.
32. M. Langan, "The Language of Diplomacy," *Boston Globe,* April 19, 2001.
33. See http://en.wikipedia.org/wiki/UN_Security_Council_Resolution_242#Semantic_dispute.

34. Max Bazerman, personal communication, April 11, 2006.
35. B. Feiler, "Pocketful of Dough," *Gourmet,* October 2000.
36. Fiske, 1992; Fiske, 2004; Haslam, 2004. For a similar theory designed to explain the development of an individual's personality, see Judith Harris's *No Two Alike: Human Nature and Human Individuality,* 2006.
37. Dawkins, 1976/1989.
38. Tooby & Cosmides, 1996.
39. Daly, Salmon, & Wilson, 1997.
40. Ramachandran & Blakeslee, 1998; Wegner, 2002.
41. Fiske, 2004, p. 88.
42. Fiske & Tetlock, 1997; McGraw & Tetlock, 2005; Tetlock et al., 2000.
43. Dawkins, 1976/1989; Maynard Smith, 1988.
44. Schelling, 1960.
45. Provine, 1996.
46. Pinker, 1997b, chap. 8.
47. Cosmides & Tooby, 1992.
48. Sowell, 1980.
49. Fiske & Tetlock, 1997; McGraw & Tetlock, 2005; Tetlock et al., 2000.
50. Buss, 1994; Symons, 1979.
51. Rosovsky, 1990.
52. Brown, 1996, p. 8.
53. Brown & Levinson, 1987b. See also Clark, 1996; Isaacs & Clark, 1990.
54. Schelling, 1960.
55. Schelling, 1960, p. 67.
56. Schelling, 1960, p. 71.
57. Clark & Brennan, 1991; Clark & Marshall, 1991; Lewis, 1969; Schelling, 1960; Schiffer, 1972; Smith, 1982; Stalnaker, 1978; Vanderschraaf & Sillari, 2005. The terminology is confusing: some writers use "mutual knowledge" to refer to identical individual knowledge (which I think should be called "shared knowledge"), and reserve "common knowledge" for the case in which people *know* that their knowledge is shared (what I call "mutual knowledge"). Others use the terms interchangeably. Since the meaning of the word *mutual* captures the key concept better than the meaning of the word *common,* and since *common knowledge* is, confusingly, a vernacular term for shared individual knowledge, I will stick with *mutual knowledge.*
58. Vanderschraaf & Sillari, 2005.
59. Clark, 1996.
60. Sperber & Wilson, 1986.
61. Gigerenzer, 2004; Schelling, 1960. The term "rational ignorance" comes from the philosopher Sylvain Bromberger (Bromberger, 1992), though in reference to a different concept.
62. Gigerenzer, 2004.
63. Schelling, 1960.

9. ESCAPING THE CAVE

1. From *Great Dialogues of Plato: Complete Texts of the "Republic," "Apology," "Crito," "Phaido," "Ion," and "Meno,"* Vol. 1, ed. Warmington and Rouse (New York: Signet Classics, 1999), p. 316.
2. Shtulman, 2006.
3. Sowell, 1980; Sowell, 1996.
4. Chua, 2003.

5. Pinker, 2007.
6. Tetlock, 1999.
7. www.overheardinnewyork.com; www.overheardintheoffice.com.
8. Carey, 2008.
9. Cosmides & Tooby, 1996; Gigerenzer, 1991; Shtulman, 2006.

REFERENCES

Abarbanell, L., Li, P., & Papafragou, A. 2005. Spatial language and reasoning in Tseltal. Paper presented at the Annual Meeting of the Linguistics Society of America.

Adams, D., & Lloyd, J. 1990. *The deeper meaning of Liff*. New York: Harmony Books.

Allan, K. 1977. Classifiers. *Language, 53,* 285–311.

Allan, K., & Burridge, K. 1991. *Euphemism and dysphemism: Language used as shield and weapon*. New York: Oxford University Press.

Allison, H. E. 1973. *The Kant-Eberhard controversy: An English translation, together with supplementary materials*. Baltimore: Johns Hopkins University Press.

Aman, R. 1987. *The best of Maledicta: The International Journal of Verbal Aggression*. Philadelphia: Running Press.

Anderson, J. R. 1976. *Language, memory, and thought*. Mahwah, N.J.: Erlbaum.

Anderson, J. R., & Bower, G. H. 1973. *Human associative memory*. New York: Wiley.

Anderson, S. R. 1971. On the role of deep structure in semantic interpretation. *Foundations of Language, 6,* 197–219.

Apresjan, J. D. 1973. Regular polysemy. *Linguistics, 142,* 5–32.

Attneave, F. 1968. Triangles as ambiguous figures. *American Journal of Psychology, 81,* 447–453.

Aunger, R. 2000. *Darwinizing culture: The status of memetics as a science*. New York: Oxford University Press.

Ayer, A. J. 1936. *Language, truth, and logic*. New York: Oxford University Press.

Baayen, R. H., & Moscoso del Prado Martin, F. 2005. Semantic density and past-tense formation in three Germanic languages. *Language, 81,* 666–698.

Bach, E. 1986. The algebra of events. *Linguistics and Philosophy, 9,* 5–16.

Baddeley, A. D. 1986. *Working memory*. New York: Oxford University Press.

Baker, C. L. 1979. Syntactic theory and the projection problem. *Linguistic Inquiry, 10,* 533–581.

Barbour, J. B. 2000. *The end of time: The next revolution in physics*. New York: Oxford University Press.

Barner, D., & Snedeker, J. 2005. Quantity judgments and individuation: Evidence that mass nouns count. *Cognition, 97,* 41–66.

Barry, H., & Harper, A. 1993. Feminization of unisex names from 1960 to 1990. *Names, 41,* 228–238.

Barthes, R. 1972. To write: An intransitive verb? In R. Macksey & E. Donato (Eds.), *The languages of criticism and the science of man: The structuralist controversy.* Baltimore: Johns Hopkins University Press.

Bates, E. 1976. *Language and context: The acquisition of pragmatics.* New York: Academic Press.

Bates, E., & MacWhinney, B. 1982. Functionalist approaches to grammar. In E. Wanner & L. R. Gleitman (Eds.), *Language acquisition: The state of the art.* New York: Cambridge University Press.

Beeman, J.-J. 2005. Bilateral brain processes for comprehending natural language. *Trends in Cognitive Science, 9,* 512–518.

Beeman, W. O. 2001. The elusive butterfly. *Iconicity in Language.* http://www.trismegistos.com/IconicityInLanguage/.

Bell, Q. 1992. *On human finery.* London: Allison & Busby.

Berent, I., Pinker, S., & Ghavami, G. In press. The dislike of regular plurals in compounds: Phonological familiarity or morphological constraint? *The Mental Lexicon.*

Berent, I., Pinker, S., & Shimron, J. 1999. Default nominal inflection in Hebrew: Evidence for mental variables. *Cognition, 72,* 1–44.

Berent, I., Pinker, S., & Shimron, J. 2002. The nature of regularity and irregularity: Evidence from Hebrew nominal inflection. *Journal of Psycholinguistic Research, 31,* 459–502.

Berkeley, G. 1713/1929. Three dialogues between Hylas and Philonous. In M. W. Calkins (Ed.), *Berkeley selections.* New York: Scribners.

Berko, J. 1958. The child's learning of English morphology. *Word, 14,* 150–177.

Bertolo, S. (Ed.). 2001. *Language acquisition and learnability.* New York: Cambridge University Press.

Bickerton, D. 1990. *Language and species.* Chicago: University of Chicago Press.

Biederman, I. 1995. Visual object recognition. In S. M. Kosslyn & D. N. Osherson (Eds.), *An invitation to cognitive science,* Vol. 2: *Visual cognition and action.* Cambridge, Mass.: MIT Press.

Blackmore, S. J. 1999. *The meme machine.* New York: Oxford University Press.

Blakemore, S.-J., & Frith, U. 2005. *The learning brain: Lessons for education.* Malden, Mass.: Blackwell.

Blanchette, I., & Dunbar, K. 2000. How analogies are generated: The roles of structural and superficial similarity. *Memory & Cognition, 28,* 108–124.

Blanchette, I., & Dunbar, K. 2001. Analogy use in naturalistic settings: The influence of audience, emotion, and goals. *Memory & Cognition, 29,* 730–735.

Block, N. 1986. Advertisement for a semantics for psychology. In P. A. French, T. E. Uehling, & H. K. Wettstein (Eds.), *Midwest studies in philosophy: Studies in the philosophy of mind* (Vol. 10). Minneapolis: University of Minnesota Press.

Bloom, P. 1994. Syntax-semantics mappings as an explanation for some transitions in language development. In Y. Levy (Ed.), *Other children, other languages: Theoretical issues in language development.* Mahwah, N.J.: Erlbaum.

Bloom, P. 1996. Possible individuals in language and cognition. *Current Directions in Psychological Science, 5,* 90–94.

Bloom, P. 1999. *How children learn the meanings of words.* Cambridge, Mass.: MIT Press.

Bloom, P. 2003. *Descartes' baby: How the science of child development explains what makes us human.* New York: Basic Books.

Bok, D. C. 2006. *Our underachieving colleges: A candid look at how much students learn and why they should be learning more.* Princeton, N.J.: Princeton University Press.

Borer, H. 2005a. *Structuring sense,* Vol. 1: *In name only.* New York: Oxford University Press.

Borer, H. 2005b. *Structuring sense,* Vol. 2: *The normal course of events.* New York: Oxford University Press.

Boroditsky, L. 2000. Metaphoric structuring: Understanding time through spatial metaphors. *Cognition, 75,* 1–28.

Boroditsky, L. 2001. Does language shape thought? Mandarin and English speakers' conceptions of time. *Cognitive Psychology, 43,* 1–22.

Boroditsky, L., & Ramscar, M. 2002. The roles of body and mind in abstract thought. *Psychological Science, 13,* 185–188.

Borregine, K. L., & Kaschak, M. P. In press. The action-sentence compatibility effect: It's all in the timing. *Cognitive Science.*

Boudin, M. 1986. Antitrust doctrine and the sway of metaphor. *Georgetown Law Journal, 75,* 395–422.

Bowdle, B., & Gentner, D. 2005. The career of metaphor. *Psychological Review, 112,* 193–216.

Bowerman, M. 1982a. Evaluating competing linguistic models with language acquisition data: Implications of developmental errors with causative verbs. *Quaderni di Semantica, 3,* 5–66.

Bowerman, M. 1982b. Reorganizational processes in lexical and syntactic development. In E. Wanner & L. R. Gleitman (Eds.), *Language acquisition: The state of the art.* New York: Cambridge University Press.

Bowerman, M. 1983. Hidden meanings: The role of covert conceptual structures in children's development of language. In D. R. Rogers & J. A. Sloboda (Eds.), *The acquisition of symbolic skills.* New York: Plenum.

Bowerman, M. 1988. The "no negative evidence" problem: How do children avoid constructing an overly general grammar? In J. A. Hawkins (Ed.), *Explaining language universals.* Malden, Mass.: Blackwell.

Bowerman, M., & Levinson, S. C. 2001. Introduction. In M. Bowerman & S. Levinson (Eds.), *Language acquisition and conceptual development.* New York: Cambridge University Press.

Boyd, R. 1993. Metaphor and theory change: What is "metaphor" a metaphor for? In A. Ortony (Ed.), *Metaphor and thought* (2nd ed.). New York: Cambridge University Press.

Braine, M. D. S. 1971. On two types of models of the internalization of grammars. In D. I. Slobin (Ed.), *The ontogenesis of grammar: A theoretical symposium.* New York: Academic Press.

Bransford, J. D., & Franks, J. J. 1971. The abstraction of linguistic ideas. *Cognitive Psychology, 2,* 331–350.

Breedin, S. D., & Saffran, E. M. 1999. Sentence processing in the face of semantic loss: A case study. *Journal of Experimental Psychology: General, 128,* 547–562.

Bregman, A. S. 1990. *Auditory scene analysis: The perceptual organization of sound.* Cambridge, Mass.: MIT Press.

Bresnan, J. 2005. Is knowledge of syntax probabilistic? Experiments with the English dative alternation. In S. Kepser & M. Reis (Eds.), *Linguistic evidence: Empirical, theoretical, and computational perspectives.* New York: Mouton de Gruyter.

Bresnan, J., Cueni, A., Nikitina, T., & Baayen, R. H. In press. Predicting the dative alternation. In G. Bourne, I. Kraemer, & J. Zwarts (Eds.), *Cognitive foundations of interpretation.* Amsterdam: Royal Netherlands Academy of Science.

Bresnan, J., & Nikitina, T. 2003. On the gradience of the dative alternation. Unpublished manuscript, Dept. of Linguistics, Stanford University.

Bromberger, S. 1992. *On what we know we don't know: Explanation, theory, linguistics, and how questions shape them.* Chicago: University of Chicago Press.

Brown, D. E. 1991. *Human universals.* New York: McGraw-Hill.

Brown, P., & Levinson, S. C. 1987a. Introduction to the reissue: A review of recent work. In *Politeness: Some universals in language use.* New York: Cambridge University Press.

Brown, P., & Levinson, S. C. 1987b. *Politeness: Some universals in language usage.* New York: Cambridge University Press.

Brown, R. 1958. *Words and things*. New York: Free Press.

Brown, R. 1985. *Social psychology* (2nd ed.). New York: Free Press.

Brown, R. 1987. *Theory of politeness: An exemplary case*. Paper presented at the Society of Experimental Social Psychologists.

Brown, R. 1996. *Against my better judgment: An intimate memoir of an eminent gay psychologist*. Binghamton, N.Y.: Haworth Press.

Brown, R., & Gilman, A. 1972. The pronouns of power and solidarity. In *Psycholinguistics: Selected papers by Roger Brown*. New York: Free Press.

Brown, R., & Hanlon, C. 1970. Derivational complexity and order of acquisition in child speech. In J. R. Hayes (Ed.), *Cognition and the development of language*. New York: Wiley.

Bruce, L. 1965/1991. *How to talk dirty and influence people: An autobiography*. New York: Simon & Schuster.

Brugman, C. 1988. *The story of* over: *Polysemy, semantics, and the structure of the lexicon*. New York: Garland.

Bryson, B. 1990. *The mother tongue: English and how it got that way*. New York: Morrow.

Burchfield, R. 1995. *The English language*. New York: Oxford University Press.

Buss, D. M. 1994. *The evolution of desire*. New York: Basic Books.

Butterworth, B. 1999. *The mathematical brain*. London: Macmillan.

Bybee, J. L. 1985. *Morphology: A study of the relation between meaning and form*. Philadelphia: Benjamins.

Bybee, J. L., Perkins, R., & Pagliuca, W. 1994. *The evolution of grammar: Tense, aspect, and modality in the languages of the world*. Chicago: University of Chicago Press.

Carey, S. 2008. *Origins of concepts*. Cambridge, Mass.: MIT Press.

Carlson, L., & van der Zee, E. (Eds.). 2005. *Functional features in language and space*. New York: Oxford University Press.

Carter, R. J. 1976. Some constraints on possible words. *Semantikos, 1*, 27–66.

Carter, R. J. 1988. Some linking regularities. In B. Levin & C. Tenny (Eds.), *On linking: Papers by Richard Carter* (Lexicon Project Working Paper #25). Cambridge, Mass.: MIT Center for Cognitive Science.

Casasanto, D. 2005. Crying "Whorf" (letter). *Science, 307*, 1721–1722.

Casati, R. 2006. The cognitive science of holes and cast shadows. *Trends in Cognitive Science, 10*, 54–55.

Catrambone, R., & Holyoak, K. J. 1989. Overcoming contextual limitations on problem-solving transfer. *Journal of Experimental Psychology: Learning, Memory, and Cognition, 15*, 1147–1156.

Cave, K. R., Pinker, S., Giorgi, L., Thomas, C, Heller, L., Wolfe, J. M., & Lin, H. 1994. The representation of location in visual images. *Cognitive Psychology, 26*, 1–32.

Chapman, L. J., & Chapman, J. 1982. Test results are what you think they are. In D. Kahneman, P. Slovic, & A. Tversky (Eds.), *Judgment under uncertainty: Heuristics and biases*. New York: Cambridge University Press.

Chi, M. T. H., Feltovich, P., & Glaser, R. 1981. Categorization and representation of physics problems by experts and novices. *Cognitive Science, 5*, 121–152.

Chierchia, G. 1998. *Plurality of mass nouns and the notion of "semantic parameter."* Boston: Kluwer.

Chierchia, G., & McConnell-Ginet, S. 2000. *Meaning and grammar: An introduction to semantics* (2nd ed.). Cambridge, Mass.: MIT Press.

Chomsky, N. 1965. *Aspects of the theory of syntax*. Cambridge, Mass.: MIT Press.

Chomsky, N. 1972a. *Language and mind* (extended edition). New York: Harcourt Brace.

Chomsky, N. 1972b. *Studies on semantics in generative grammar*. The Hague: Mouton.

Chomsky, N. 1981. *Lectures on government and binding*. Dordrecht, Netherlands: Foris.

Chomsky, N. 1988. *Language and problems of knowledge: The Managua lectures*. Cambridge, Mass.: MIT Press.

Chomsky, N. 2000. *New horizons in the study of language and mind.* New York: Cambridge University Press.

Chua, A. 2003. *World on fire: How exporting free market democracy breeds ethnic hatred and global instability.* New York: Doubleday.

Chung, T. T. R., & Gordon, P. 1998. The acquisition of Chinese dative constructions. Paper presented at the Boston University Conference on Language Development, Boston.

Clancy, J. J. 1989. *The invisible powers: The language of business.* Lexington, Mass.: Lexington Books.

Clark, H. H. 1996. *Using language.* New York: Cambridge University Press.

Clark, H. H., & Brennan, S. E. 1991. Grounding in communication. In L. B. Resnick, J. M. Levine, & S. D. Teasley (Eds.), *Perspectives on socially shared cognition.* Washington, D.C.: American Psychological Association.

Clark, H. H., & Marshall, C. R. 1991. Definite reference and mutual knowledge. In A. K. Joshi, B. L. Webber, & I. A. Sag (Eds.), *Elements of discourse understanding.* New York: Cambridge University Press.

Clark, H. H., & Schunk, D. 1980. Polite responses to polite requests. *Cognition, 8,* 111–143.

Code, C. In press. First in, last out? The evolution of aphasic lexical speech automatisms to agrammatism and the evolution of human communication. *Interaction Studies.*

Cohen, D. B. 2004. Plays of genius: A psychological exploration of Shakespeare's ten great themes. Unpublished manuscript, Department of Psychology, University of Texas.

Cohen, L. B., Amsel, G., Redford, M. A., & Casasola, M. 1998. The development of infant causal perception. In A. Slater (Ed.), *Perceptual development: Visual, auditory, and speech perception in infancy.* East Sussex, UK: Psychology Press.

Cole, P. (Ed.). 1981. *Radical pragmatics.* New York: Academic Press.

Collins, R. K. L., & Skover, D. M. 2002. *The trials of Lenny Bruce: The fall and rise of an American icon.* Naperville, Ill.: Sourcebooks.

Comrie, B. 1976. *Aspect.* New York: Cambridge University Press.

Comrie, B. 1985a. Causative verb formation and other verb-deriving morphology. In T. Shopen (Ed.), *Language typology and syntactic description III: Grammatical categories and the lexicon.* New York: Cambridge University Press.

Comrie, B. 1985b. *Tense.* New York: Cambridge University Press.

Corballis, M. C., & Beale, I. L. 1976. *The psychology of left and right.* Mahwah, N.J.: Erlbaum.

Cosmides, L., & Tooby, J. 1992. Cognitive adaptations for social exchange. In J. H. Barkow, L. Cosmides, & J. Tooby (Eds.), *The adapted mind: Evolutionary psychology and the generation of culture.* New York: Oxford University Press.

Cosmides, L., & Tooby, J. 1996. Are humans good intuitive statisticians after all? Rethinking some conclusions from the literature on judgment under uncertainty. *Cognition, 58,* 1–73.

Coventry, K. R., & Garrod, S. C. 2004. *Saying, seeing, and acting: The psychological semantics of spatial prepositions.* New York: Psychology Press.

Croft, W. In press. *Verbs: Aspect and argument structure.* New York: Oxford University Press.

Croft, W., & The Manchester Cognitive Collective. 2001. Discriminating verb meanings: The case of transfer verbs. Paper presented at the conference Language Acquisition in Great Britain.

Crystal, D. 1997. *The Cambridge Encyclopedia of Language* (2nd ed.). New York: Cambridge University Press.

Crystal, D. 2003. *The Cambridge Encyclopedia of the English Language* (2nd ed.). New York: Cambridge University Press.

Curtis, V., & Biran, A. 2001. Dirt, disgust, and disease: Is hygiene in our genes? *Perspectives in Biology and Medicine, 44,* 17–31.

Cushing, S. 1994. *Fatal words: Communication clashes and aircraft crashes.* Chicago: University of Chicago Press.

Daly, M., Salmon, C., & Wilson, M. 1997. Kinship: The conceptual hole in psychological studies of social cognition and close relationships. In J. Simpson & D. Kenrick (Eds.), *Evolutionary social psychology.* Mahwah, N.J.: Erlbaum.

Darwin, C. 1874. *The descent of man, and selection in relation to sex* (2nd ed.). New York: Hurst & Company.

Dawkins, R. 1976/1989. *The selfish gene* (new ed.). New York: Oxford University Press.

Dawkins, R. 1986. *The blind watchmaker: Why the evidence of evolution reveals a universe without design.* New York: Norton.

Dawkins, R. 1998. *Unweaving the rainbow: Science, delusion, and the appetite for wonder.* Boston: Houghton Mifflin.

Dawkins, R., & Krebs, J. R. 1978. Animal signals: Information or manipulation? In J. R. Krebs & N. B. Davies (Eds.), *Behavioral ecology.* Malden, Mass.: Blackwell.

Dehaene, S. 1997. *The number sense: How the mind creates mathematics.* New York: Oxford University Press.

Dehaene, S., Spelke, L., Pinel, P., Stanescu, R., & Tsivkin, S. 1999. Sources of mathematical thinking: Behavioral and brain-imaging evidence. *Science, 284,* 970–974.

Denfeld, R. 1995. *The new Victorians: A young woman's challenge to the old feminist order.* New York: Warner Books.

Dennett, D. C. 1997. Darwinian fundamentalism: An exchange. *New York Review of Books, 44.*

Denny, J. P. 1976. What are noun classifiers good for? *Papers from the Twelfth Regional Meeting of the Chicago Linguistics Society.*

Dershowitz, A. 2005. The marketplace of ideas: Know who you are listening to or reading: The Norman Finkelstein Top Ten Lists. http://www.law.harvard.edu/faculty/dershowitz/currentlist.html.

Devlin, K. 2000. *The Math Gene: How mathematical thinking evolved and why numbers are like gossip.* New York: Basic Books.

Dews, S., Kaplan, J., & Winner, E. 1995. Why not say it directly? The social functions of irony. *Discourse Processes, 19,* 347–367.

di Sciullo, A. M., & Williams, E. 1987. *On the definition of word.* Cambridge, Mass.: MIT Press.

Dixon, R. M. W. 2000. A typology of causatives: Form, syntax, and meaning. In R. M. W. Dixon & A. Y. Aihenvald (Eds.), *Changing valency.* New York: Cambridge University Press.

Dollard, J., Miller, N. E., Doob, L. W., Mowrer, O. H., & Sears, R. R. 1939. *Frustration and aggression.* New Haven, Conn.: Yale University Press.

Dooling, R. 1996. *Blue streak: Swearing, free speech, and sexual harassment.* New York: Random House.

Dowe, P. 2000. *Physical causation.* New York: Cambridge University Press.

Dowty, D. R. 1979. Dative "movement" and Thomason's extensions of Montague Grammar. In S. Davis & M. Mithun (Eds.), *Linguistics, philosophy, and Montague Grammar.* Austin: University of Texas Press.

Dowty, D. R. 1979/1991. *Word meaning and Montague Grammar: The semantics of verbs and times in generative semantics and in Montague's PTQ.* Boston: Kluwer.

Dowty, D. R. 1982. Tenses, time adverbs, and compositional semantic theory. *Linguistics and Philosophy, 5,* 23–55.

Dronkers, N., Pinker, S., & Damasio, A. R. 1999. Language and the aphasias. In E. R. Kandel, J. H. Schwartz, & T. M. Jessell (Eds.), *Principles of neural science* (4th ed.). Norwalk, Conn.: Appleton & Lange.

Druks, J., & Masterson, J. 2003. The neural basis of verbs. *Journal of Neurolinguistics, 16* (Special Issue).

Dryer, M. S. 1986. Primary objects, secondary objects, and antidative. *Language, 62,* 808–845.

Dunbar, K. 2001. The analogical paradox: Why analogy is so easy in naturalistic settings yet so difficult in the psychological laboratory. In D. Gentner, K. J. Holyoak, & B. N. Kokinov (Eds.), *The analogical mind: Perspectives from cognitive science.* Cambridge, Mass.: MIT Press.

Duncker, K. 1945. On problem solving. *Psychological Monographs, 58.*

Dworkin, A. 1979. *Pornography: Men possessing women.* New York: Penguin.

Eco, U. 1995. *The search for the perfect language.* Malden, Mass.: Blackwell.

Etcoff, N. L. 1986. The neuropsychology of emotional expression. In G. Goldstein & R. E. Tarter (Eds.), *Advances in clinical neuropsychology* (Vol. 3). New York: Plenum.

Etcoff, N. L. 2008. *Liking, wanting, having, being: The science of happiness.* New York: Farrar, Straus & Giroux.

Everett, D. 2005. Cultural constraints on grammar and cognition in Pirahã: Another look at the design features of human language. *Current Anthropology, 46,* 621–646.

Fairhurst, G. T., & Sarr, R. A. 1996. *The art of framing: Managing the language of leadership.* San Francisco: Jossey-Bass.

Fareh, S., & Hamdan, J. 2000. Locative alternation in English and Jordanian spoken Arabic. *Papers and Studies in Contrastive Linguistics, 36,* 71–93.

Feist, M. I., & Gentner, D. 1998. On plates, bowls, and dishes: Factors in the use of English IN and ON. Paper presented at the Twentieth Annual Conference of the Cognitive Science Society.

Fillmore, C. 1967. The grammar of hitting and breaking. In R. Jacobs & P. Rosenbaum (Eds.), *Readings in English transformational grammar.* Waltham, Mass.: Ginn.

Fillmore, C. 1968. The case for case. In E. Bach & R. J. Harms (Eds.), *Universals in linguistic theory.* New York: Holt, Rinehart & Winston.

Fiske, A. P. 1992. The four elementary forms of sociality: Framework for a unified theory of social relations. *Psychological Review, 99,* 689–723.

Fiske, A. P. 2004. Four modes of constituting relationships: Consubstantial assimilation; space, magnitude, time, and force; concrete procedures; abstract symbolism. In N. Haslam (Ed.), *Relational models theory: A contemporary overview.* Mahwah, N.J.: Erlbaum.

Fiske, A. P., & Tetlock, P. E. 1997. Taboo trade-offs: Reactions to transactions that transgress the spheres of justice. *Political Psychology, 18,* 255–297.

Flaubert, G. 1857/1998. *Madame Bovary: Life in a country town* (G. Hopkins, Trans.). New York: Oxford University Press.

Fodor, J. A. 1968. *Psychological explanation: An introduction to the philosophy of psychology.* New York: Random House.

Fodor, J. A. 1970. Three reasons for not deriving "kill" from "cause to die." *Linguistic Inquiry, 1,* 429–438.

Fodor, J. A. 1975. *The language of thought.* New York: Crowell.

Fodor, J. A. 1981a. The present status of the innateness controversy. In J. A. Fodor (Ed.), *RePresentations.* Cambridge, Mass.: MIT Press.

Fodor, J. A. 1981b. *RePresentations: Philosophical essays on the foundations of cognitive science.* Cambridge, Mass.: MIT Press.

Fodor, J. A. 1987. *Psychosemantics: The problem of meaning in the philosophy of mind.* Cambridge, Mass.: MIT Press.

Fodor, J. A. 1994. *The elm and the expert: Mentalese and its semantics.* Cambridge, Mass.: MIT Press.

Fodor, J. A. 1998. *Concepts: Where cognitive science went wrong.* New York: Oxford University Press.

Fodor, J. A. 2000. *The mind doesn't work that way: The scope and limits of computational psychology*. Cambridge, Mass.: MIT Press.

Fodor, J. A. 2001. Doing without what's within: Fiona Cowie's critique of nativism. *Mind, 110*, 99–148.

Fodor, J. A., Garrett, M. F., Walker, E. C. T., & Parkes, S. 1980. Against definitions. *Cognition, 8*, 263–367.

Fodor, J. A., & Pylyshyn, Z. 1988. Connectionism and cognitive architecture: A critical analysis. *Cognition, 28*, 3–71.

Fodor, J. D., & Crain, S. 1987. Simplicity and generality of rules in language acquisition. In B. MacWhinney (Ed.), *Mechanisms of language acquisition*. Mahwah, N.J.: Erlbaum.

Foley, W. A., & Van Valin, R. D. 1985. Information packaging in the clause. In T. Shopen (Ed.), *Language typology and syntactic description I: Clause structure*. New York: Cambridge University Press.

Foot, P. 1978. *Virtues and vices and other essays in moral philosophy*. Berkeley: University of California Press.

Francik, E. P., & Clark, H. H. 1985. How to make requests that overcome obstacles to compliance. *Journal of Memory and Language, 24*, 560–568.

Francis, N., & Kucera, H. 1982. *Frequency analysis of English usage: Lexicon and grammar*. Boston: Houghton Mifflin.

Frank, R. H. 1988. *Passions within reason: The strategic role of the emotions*. New York: Norton.

Fraser, B. 1990. Perspectives on politeness. *Journal of Pragmatics, 14*, 219–236.

Fraser, B. 2005. Whither politeness. In R. T. Lakoff & S. Ide (Eds.), *Broadening the horizon of linguistic politeness*. Philadelphia: John Benjamins.

Fukui, N., Miyagawa, S., & Tenny, C. 1985. *Verb classes in English and Japanese: A case study in the interaction of syntax, morphology, and semantics*. Lexicon Project Working Paper #3 Cambridge, Mass.: MIT Center for Cognitive Science.

Gardner, M. 1990. *The new ambidextrous universe*. New York: W. H. Freeman.

Gelman, R., & Gallistel, C. R. 1978. *The child's understanding of number*. Cambridge, Mass.: Harvard University Press.

Gelman, R., & Gallistel, C. R. 2004. Language and the origin of numerical concepts. *Science, 306*, 441–443.

Gentner, D. 1975. Evidence for the psychological reality of semantic components: The verbs of possession. In D. A. Norman & D. E. Rumelhart (Eds.), *Explorations in cognition*. San Francisco: W. H. Freeman.

Gentner, D. 1981. Some interesting differences between verbs and nouns. *Cognition and Brain Theory, 4*, 161–178.

Gentner, D. 1983. Structure-mapping: A theoretical framework for analogy. *Cognitive Science, 7*, 155–170.

Gentner, D., Bowdle, B., Wolff, P., & Boronat, C. 2001. Metaphor is like analogy. In D. Gentner, K. J. Holyoak, & B. N. Kokinov (Eds.), *The analogical mind: Perspectives from cognitive science*. Cambridge, Mass.: MIT Press.

Gentner, D., & Goldin-Meadow, S. (Eds.). 2003. *Language in mind: Advances in the study of language and thought*. Cambridge, Mass.: MIT Press.

Gentner, D., & Jeziorski, M. 1989. Historical shifts in the use of analogy in science. In B. Gholson, W. R. Shadish, R. A. Beimeyer, & A. Houts (Eds.), *The psychology of science: Contributions to metascience*. New York: Cambridge University Press.

Gentner, D., Ratterman, M. J., & Forbus, K. D. 1993. The roles of similarity in transfer: Separating retrievability from inferential soundness. *Cognitive Psychology, 25*, 524–575.

Gergely, G., & Bever, T. G. 1986. Relatedness intuitions and mental representation of causative verbs. *Cognition, 23*, 211–277.

Gibbs, R. 1986. What makes some speech acts conventional? *Journal of Memory and Language, 25*, 181–196.

Gick, M., & Holyoak, K. J. 1980. Analogical problem solving. *Cognitive Psychology, 12,* 306–355.

Gigerenzer, G. 1991. How to make cognitive illusions disappear: Beyond heuristics and biases. *European Review of Social Psychology, 2,* 83–115.

Gigerenzer, G. 2004. Gigerenzer's law of indispensable ignorance. *Edge.* http://www.edge .org/q2004/page2.html#gigerenzer.

Gilbert, D. 2006. *Stumbling on happiness.* New York: Knopf.

Giralt, N., & Bloom, P. 2000. How special are objects? Children's reasoning about objects, parts, and wholes. *Psychological Science, 11,* 497–501.

Gladwell, M. 2000. *The tipping point: How little things make big differences.* Boston: Little, Brown.

Gleitman, L. R., Li, P., Papafragou, A., Gallistel, C. R., & Abarbanell, L. 2005. Spatial reasoning and cognition: Cross-linguistic studies. Unpublished presentation slides, Dept. of Psychology, University of Pennsylvania.

Gleitman, L. R., & Papafragou, A. 2005. Language and thought. In K. Holyoak & B. Morrison (Eds.), *Cambridge handbook of thinking and reasoning.* New York: Cambridge University Press.

Glucksberg, S., & Danks, J. 1968. Effects of discriminative labels and of nonsense labels upon availability of novel function. *Journal of Verbal Learning and Verbal Behavior, 7,* 72–76.

Glucksberg, S., & Keysar, B. 1993. How metaphors work. In A. Ortony (Ed.), *Metaphor and thought* (2nd ed.). New York: Cambridge University Press.

Goffman, E. 1959. *The presentation of self in everyday life.* New York: Doubleday.

Goffman, E. 1967. On face-work: An analysis of ritual elements in social interaction. In *Interaction ritual: Essays on face-to-face behavior.* New York: Random House.

Goffman, E. 1978. Response cries. *Language, 54,* 787–815.

Gold, E. M. 1967. Language identification in the limit. *Information and Control, 16,* 447–474.

Goldberg, A. 1995. *Constructions: A construction grammar approach to argument structure.* Chicago: University of Chicago Press.

Goldberg, A. 2005. *Constructions at work: The nature of generalization in language.* New York: Oxford University Press.

Goldstein, R. 1983. *The mind-body problem: A novel.* New York: Random House.

Goldstein, R. 2005. *Incompleteness: The proof and paradox of Kurt Gödel.* New York: Norton.

Goldvarg, E., & Johnson-Laird, P. N. 2001. Naive causality: A mental model theory of causal meaning and reasoning. *Cognitive Science, 25,* 565–610.

Goodale, M. A., & Milner, A. D. 2004. *Sight unseen: An exploration of conscious and unconscious vision.* New York: Oxford University Press.

Goodman, N. 1983. *Fact, fiction, and forecast.* Cambridge, Mass.: Harvard University Press.

Gopnik, A., Glymour, C., Sobel, D. M., Schulz, L. E., Kushnir, T., & Danks, D. 2004. A theory of causal learning in children: Causal maps and Bayes nets. *Psychological Review, 11,* 3–32.

Gordon, P. 2004. Numerical cognition without words: Evidence from Amazonia. *Science, 306,* 496–499.

Gould, S. J. 1980. Natural selection and the human brain: Darwin vs. Wallace. In *The panda's thumb.* New York: Norton.

Green, G. M. 1974. *Semantics and syntactic regularity.* Bloomington: Indiana University Press.

Green, G. M. 1996. *Pragmatics and natural language understanding* (2nd ed.). Mahwah, N.J.: Erlbaum.

Greene, J. D. 2002. The terrible, horrible, no good, very bad truth about morality and what to do about it. Unpublished Ph.D. dissertation, Princeton University.

Greene, J. D., Sommerville, R. B., Nystrom, L. E., Darley, J. M., & Cohen, J. D. 2001. An fMRI investigation of emotional engagement in moral judgment. *Science, 293,* 2105–2108.

Grice, H. P. 1975. Logic and conversation. In P. Cole & J. L. Morgan (Eds.), *Syntax & semantics 3: Speech acts.* New York: Academic Press.

Gropen, J., Pinker, S., Hollander, M., & Goldberg, R. 1991a. Affectedness and direct objects: The role of lexical semantics in the acquisition of verb argument structure. *Cognition, 41,* 153–195.

Gropen, J., Pinker, S., Hollander, M., & Goldberg, R. 1991b. Syntax and semantics in the acquisition of locative verbs. *Journal of Child Language, 18,* 115–151.

Gropen, J., Pinker, S., Hollander, M., Goldberg, R., & Wilson, R. 1989. The learnability and acquisition of the dative alternation in English. *Language, 65,* 203–257.

Gruber, J. 1965. Studies in lexical relations. Unpublished Ph.D. dissertation, MIT. Reprinted, 1976, as *Lexical structures in syntax and semantics.* Amsterdam: North Holland.

Guerrero Valenzuela, L. 2002. Macroroles and double-object constructions in Yaqui. Unpublished manuscript, Dept. of Linguistics, State University of New York at Buffalo.

Guerssel, M. 1986. *On Berber verbs of change: A study of transitivity alternations* (Lexicon Project Working Paper #9). Cambridge, Mass.: MIT Center for Cognitive Science.

Gumperz, J. J., & Levinson, S. C. (Eds.). 1996. *Rethinking linguistic relativity.* New York: Cambridge University Press.

Hall, R., & Friends. 1984. *Sniglets (Snig'lit: Any word that doesn't appear in the dictionary, but should).* New York: Collier.

Halpern, D. 2000. *Sex differences in cognitive abilities* (3rd ed.). Mahwah, N.J.: Erlbaum.

Harlow, R. 1998. Some languages are just not good enough. In L. Bauer & P. Trudgill (Eds.), *Language myths.* New York: Penguin.

Harris, C. L., Gleason, J. B., & Aycicegi, A. 2006. When is a first language more emotional? Psychophysiological evidence from bilingual speakers. In A. Pavlenko (Ed.), *Bilingual minds: Emotional experience, expression, and representation.* Clevedon, U.K.: Multilingual Matters.

Harris, J. R. 1998. *The nurture assumption: Why children turn out the way they do.* New York: Free Press.

Harris, J. R. 2006. *No two alike: Human nature and human individuality.* New York: Norton.

Harris, M. 1989. *Our kind: The evolution of human life and culture.* New York: Harper-Collins.

Harris, R. A. 1993. *The linguistics wars.* New York: Oxford University Press.

Haslam, N. (Ed.). 2004. *Relational models theory: A contemporary overview.* Mahwah, N.J.: Erlbaum.

Haspelmath, M. 1993. More on the typology of inchoative/causative verb alternations. In B. Comrie & M. Polinsky (Eds.), *Causatives and transitivity.* Philadelphia: John Benjamins.

Haspelmath, M. 1997. *From space to time: Temporal adverbials in the world's languages.* Newcastle, U.K.: Lincom Europa.

Haspelmath, M. 2005. Argument marking in ditransitive alignment types. *Linguistic Discovery, 3,* 1–21.

Hauser, M. D. 1997. Artifactual kinds and functional design features: What a primate understands without language. *Cognition, 64,* 285–308.

Hauser, M. D. 2000. *Wild minds: What animals really think.* New York: Henry Holt.

Hauser, M. D. 2006. *Moral minds.* New York: Ecco.

Hauser, M. D., Cushman, F., Young, L., Kang-Xing, J., & Mikhail, J. 2007. A dissociation between moral judgments and justifications. *Mind and Language, 22,* 1–21.

Hauser, M. D., MacNeilage, P., & Ware, M. 1996. Numerical representations in primates: Perceptual or arithmetic? *Proceedings of the National Academy of Sciences, 93,* 1514–1517.

Hauser, M. D., Pearson, H. E., & Seelig, D. 2002. Ontogeny of tool use in cotton-top tamarins (*Saguinus oedipus*): Recognition of functionally relevant features in the absence of experience. *Animal Behavior, 64,* 299–311.

Hauser, M. D., & Spaulding, B. 2006. Monkeys generate causal inferences about possible and impossible physical transformations. *Proceedings of the National Academy of Sciences, 103,* 7181–7185.

Hawking, S. W., & Mlodinow, L. 2005. *A briefer history of time.* London: Bantam.

Hilton, D. J. 1995. Logic and language in causal explanation. In D. Sperber, D. Premack, & A. Premack (Eds.), *Causal cognition: A multidisciplinary debate.* New York: Oxford University Press.

Hinton, G. E., & Parsons, L. M. 1981. Frames of reference and mental imagery. In J. Long & A. Baddeley (Eds.), *Attention and Performance IX.* Mahwah, N.J.: Erlbaum.

Hirschberg, S., & Hirschberg, T. (Eds.). 1999. *Reflections on language.* New York: Oxford University Press.

Hirschbühler, P. 2003. Cross-linguistic variation patterns in the locative alternation. Paper presented at the 13th Colloquium on Generative Grammar, Ciudad Real, Madrid.

Hirschbühler, P., & Mchombo, S. 2006. The location object construction in Romance and Bantu: Applicatives or not? *The Bantu-Romance Connection.* http://www.modern.lang.leeds .ac.uk/BantuRom/index.php?option=com_content&task=view&id=35&Itemid=40.

Hirschfeld, L. A., & Gelman, S. A. 1994. Toward a topography of mind: An introduction to domain specificity. In L. A. Hirschfeld & S. A. Gelman (Eds.), *Mapping the mind: Domain-specificity in cognition and culture.* New York: Cambridge University Press.

Holtgraves, T. M. 2002. *Language as social action.* Mahwah, N.J.: Erlbaum.

Holyoak, K. J., & Thagard, P. 1996. *Mental leaps: Analogy in creative thought.* Cambridge, Mass.: MIT Press.

Hughes, G. 1991/1998. *Swearing: A social history of foul language, oaths, and profanity in English.* New York: Penguin.

Hull, D. L. 1989. *The metaphysics of evolution.* Albany: State University of New York Press.

Hume, D. 1740/1955. An Abstract of A Treatise of Human Nature. In *An inquiry concerning human understanding: With a supplement, An abstract of A Treatise of Human Nature.* Indianapolis: Bobbs-Merrill.

Hume, D. 1748/1999. *An enquiry concerning human understanding.* New York: Oxford University Press.

Isaacs, E. A., & Clark, H. H. 1990. Ostensible invitations. *Language in Society, 19,* 493–509.

Isenberg, L., Nissen, M. J., & Marchak, L. C. 1990. Attentional processing and the independence of color and orientation. *Journal of Experimental Psychology: Human Perception and Performance, 16,* 869–878.

Isenberg, N., Silbersweig, D., Engelien, A., Emmerich, K., Malavade, K., Beati, B., et al. 1999. Linguistic threat activates the human amygdala. *Proceedings of the National Academy of Sciences, 96,* 10456–10459.

Jackendoff, R. 1978. Grammar as evidence for conceptual structure. In M. Halle, J. Bresnan, & G. A. Miller (Eds.), *Linguistic theory and psychological reality.* Cambridge, Mass.: MIT Press.

Jackendoff, R. 1983. *Semantics and cognition.* Cambridge, Mass.: MIT Press.

Jackendoff, R. 1987. *Consciousness and the computational mind.* Cambridge, Mass.: MIT Press.

Jackendoff, R. 1990. *Semantic structures.* Cambridge, Mass.: MIT Press.

Jackendoff, R. 1991. Parts and boundaries. *Cognition, 41,* 9–45.

Jackendoff, R. 1992. *Languages of the mind.* Cambridge, Mass.: MIT Press.

Jackendoff, R. 1997a. *The architecture of the language faculty.* Cambridge, Mass.: MIT Press.

Jackendoff, R. 1997b. How language helps us think. In *The architecture of the language faculty.* Cambridge, Mass.: MIT Press.

Jackendoff, R. 2002. *Foundations of language: Brain, meaning, grammar, evolution.* New York: Oxford University Press.

Jackendoff, R., & Aaron, D. 1991. Review of Lakoff & Turner's "More than cool reason: A field guide to poetic metaphor." *Language, 67,* 320–339.

James, W. 1890/1950. *The principles of psychology.* New York: Dover.

James, W. 1907/2005. *Pragmatism and the meaning of truth.* Cambridge, Mass.: Harvard University Press.

Jay, T. 2000. *Why we curse: A neuro-psycho-social theory of speech.* Philadelphia: John Benjamins.

Jespersen, O. 1938/1982. *Growth and structure of the English language.* Chicago: University of Chicago Press.

Kahneman, D., & Tversky, A. 1979. Prospect theory: An analysis of decisions under risk. *Econometrica, 47,* 313–327.

Kant, I. 1781/1998. *The critique of pure reason* (P. Guyer & A. W. Wood, Trans.). New York: Cambridge University Press.

Kant, I. 1783/1950. *Prolegomena to any future metaphysics.* Indianapolis: Bobbs-Merrill.

Kaplan, J., & Bernays, A. 1997. *The language of names.* New York: Simon & Schuster.

Kasher, A. 1977. Foundations of philosophical pragmatics. In R. E. Butts & J. Hintikka (Eds.), *Basic problems in methodology and linguistics.* Dordrecht, Netherlands: Reidel.

Katz, L. 1987. *Bad acts and guilty minds: Conundrums of criminal law.* Chicago: University of Chicago Press.

Kay, P., & Kempton, W. 1984. What is the Sapir-Whorf hypothesis? *American Anthropologist, 86,* 65–79.

Keane, M. 1987. On retrieving analogues when solving problems. *Quarterly Journal of Experimental Psychology: Human Experimental Psychology, 39,* 29–41.

Kemmerer, D. 2000a. Grammatical relevant and grammatical irrelevant features of word meaning can be independently impaired. *Aphasiology, 14,* 997–1020.

Kemmerer, D. 2000b. Selective impairment of knowledge underlying prenominal adjective order: Evidence for the autonomy of grammatical semantics. *Journal of Neurolinguistics, 13,* 57–82.

Kemmerer, D. 2003. Why can you hit someone on the arm but not break someone on the arm? A neuropsychological investigation of the English body-part possessor ascension construction. *Journal of Neurolinguistics, 16,* 13–36.

Kemmerer, D. 2005. The spatial and temporal meanings of English prepositions can be independently impaired. *Neuropsychologia, 43,* 797–806.

Kemmerer, D. In press. The semantics of space: Integrating linguistic typology and cognitive neuroscience. *Neuropsychologia.*

Kemmerer, D., Weber-Fox, C., Price, K., Zdanczyk, C., & Way, H. 2007. "Big brown dog" or "brown big dog"? An electrophysiological study of semantic constraints on prenominal adjective order. *Brain and Language, 100,* 238–256.

Kemmerer, D., & Wright, S. K. 2002. Selective impairment of knowlege underlying un-prefixation: Further evidence for the autonomy of grammatical semantics. *Journal of Neurolinguistics, 15,* 403–432.

Kennedy, R. 2002. *Nigger: The strange career of a troublesome word.* New York: Pantheon.

Keysar, B., Shen, Y., Glucksberg, S., & Horton, W. S. 2000. Conventional language: How metaphorical is it? *Journal of Memory and Language, 43,* 576–593.

Keyser, S. J., & Roeper, T. 1984. On the middle and ergative constructions in English. *Linguistic Inquiry, 15,* 381–416.

Kim, J. J., Marcus, G. F., Pinker, S., Hollander, M., & Coppola, M. 1994. Sensitivity of children's inflection to morphological structure. *Journal of Child Language, 21,* 173–209.

Kim, J. J., Pinker, S., Prince, A., & Prasada, S. 1991. Why no mere mortal has ever flown out to center field. *Cognitive Science, 15,* 173–218.

Kim, M. 1999. A cross-linguistic perspective on the acquisition of locative verbs. Unpublished Ph.D. dissertation, University of Delaware, Newark.

Kiparsky, P. 1973. The role of linguistics in a theory of poetry. *Daedalus, 102,* 231–244.

Kitcher, P. 1990. *Kant's transcendental psychology.* New York: Oxford University Press.

Klein, D. E., & Murphy, G. 2002. Paper has been my ruin: Conceptual relations of polysemous senses. *Journal of Memory and Language, 47,* 548–570.

Klein, D. E., & Murphy, G. L. 2001. The representation of polysemous words. *Journal of Memory and Language, 45,* 259–282.

Kolodner, J. L. 1997. Educational implications of analogy. *American Psychologist, 52,* 57–66.

Kordoni, V. 2003. Locative alternation in Modern Greek: At the syntax-semantics interface. Paper presented at the Sixth International Conference of Greek Linguistics, Rethymno, Greece.

Körner, S. 1955. *Kant.* London: Penguin Books.

Kosslyn, S. M. 1980. *Image and mind.* Cambridge, Mass.: Harvard University Press.

Kosslyn, S. M. 1987. Seeing and imagining in the cerebral hemispheres: A computational approach. *Psychological Review, 94,* 184–175.

Kosslyn, S. M. 1994. *Image and brain: The resolution of the imagery debate.* Cambridge, Mass.: MIT Press.

Kotovsky, K., Hayes, J. R., & Simon, H. A. 1985. Why are some problems hard? Evidence from the Tower of Hanoi. *Cognitive Psychology, 17,* 248–294.

Kripke, S. 1972/1980. *Naming and necessity.* Cambridge, Mass.: Harvard University Press.

Kubovy, M. 1981. Concurrent-pitch segregation and the theory of indispensable attributes. In M. Kubovy & J. R. Pomerantz (Eds.), *Perceptual organization.* Mahwah, N.J.: Erlbaum.

LaBar, K. S., & Phelps, E. A. 1998. Arousal-mediated memory consolidation: Role of the medial temporal lobe in humans. *Psychological Science, 9,* 490–493.

Lagnado, D. A., & Sloman, S. 2004. The advantage of timely intervention. *Journal of Experimental Psychology: Learning, Memory, and Cognition, 30,* 856–876.

Lakoff, G. 1987. *Women, fire, and dangerous things: What categories reveal about the mind.* Chicago: University of Chicago Press.

Lakoff, G. 1993. The contemporary theory of metaphor. In A. Ortony (Ed.), *Metaphor and thought* (2nd ed.). New York: Cambridge University Press.

Lakoff, G. 1996. *Moral politics: What conservatives know that liberals don't.* Chicago: University of Chicago Press.

Lakoff, G. 2003. Lakoff's First Law. *Edge.* http://www.edge.org/q2004#shpage4.html#lakoff.

Lakoff, G. 2004. *Don't think of an elephant! Know your values and frame the debate: The essential guide for progressives.* White River Junction, Vt.: Chelsea Green.

Lakoff, G. 2006. *Whose freedom? The battle over America's most important idea.* New York: Farrar, Straus & Giroux.

Lakoff, G., & Johnson, M. 1980. *Metaphors we live by.* Chicago: University of Chicago Press.

Lakoff, G., & Johnson, M. 2000. *Philosophy in the flesh*. Cambridge, Mass.: MIT Press.

Lakoff, G., & Núñez, R. E. 2000. *Where mathematics comes from: How the embodied mind brings mathematics into being*. New York: Basic Books.

Lakoff, G., & Turner, M. 1989. *More than cool reason: A field guide to poetic metaphor*. Chicago: University of Chicago Press.

Landau, B., & Jackendoff, R. 1993. "What" and "where" in spatial language and spatial cognition. *Behavioral and Brain Sciences, 16*, 217–238.

Larkin, P. 2003. *Collected poems* (A. Thwaite, Ed.). London: Faber & Faber.

Lawley, J., & Tompkins, P. 2000. *Metaphors in mind: Transformation through symbolic modeling*. London: The Developing Company Press.

Lederer, R. 1990. *Crazy English*. New York: Pocket Books.

LeDoux, J. E. 1996. *The emotional brain: The mysterious underpinnings of emotional life*. New York: Simon & Schuster.

Lefebvre, C. 1994. New facts from Fongbe on the double object construction. *Lingua, 94*, 69–123.

Lehrer, A. 1990. Polysemy, conventionality, and the structure of the lexicon. *Cognitive Linguistics, 1–2*, 207–246.

Leibniz, G. W. 1768/1949. *New essays concerning human understanding* (C. I. Gerhardt, Trans. 3rd ed.). LaSalle, Ill.: Open Court.

Leslie, A. M. 1995. A theory of agency. In D. Sperber, D. Premack, & A. Premack (Eds.), *Causal cognition: A multidisciplinary debate*. New York: Oxford University Press.

Lessig, L. 2001. *The future of ideas: The fate of the commons in a connected world*. New York: Random House.

Levin, B. 1985. *Lexical semantics in review: An introduction* (Lexicon Project Working Paper #1). Cambridge, Mass.: MIT Center for Cognitive Science.

Levin, B. 1993. *English verb classes and alternations: A preliminary investigation*. Chicago: University of Chicago Press.

Levin, B. 2004. Verbs and constructions: Where next? Paper presented at the Western Conference on Linguistics, University of Southern California.

Levin, B., & Pinker, S. (Eds.). 1992. *Lexical and conceptual semantics*. Malden, Mass.: Blackwell.

Levin, B., & Rappaport Hovav, M. 1995. *Unaccusativity: At the syntax-lexical semantics interface*. Cambridge, Mass.: MIT Press.

Levin, B., & Rappaport Hovav, M. 2005. *Argument realization*. New York: Cambridge University Press.

Levinson, S. C. 2003. *Space in language and cognition*. New York: Cambridge University Press.

Levinson, S. C., Kita, S., & Haun, D. 2002. Returning the tables: Language affects spatial reasoning. *Cognition, 84*, 155–188.

Levinson, S. C., Meira, S., & The Language and Cognition Group. 2003. "Natural concepts" in the spatial topological domain—adpositional meanings in crosslinguistic perspective: An exercise in semantic typology. *Language, 79*, 485–514.

Levinson, S. C., & Wilkins, D. (Eds.). 2006. *Grammars of space*. New York: Cambridge University Press.

Lewis, D. K. 1969. *Convention: A philosophical study*. Cambridge, Mass.: Harvard University Press.

Lewis, D. K. 1973. *Counterfactuals*. Cambridge, Mass.: Harvard University Press.

Lewis, P. A., Critchley, H. D., Rothstein, P., & Dolan, R. J. In press. Neural correlates of processing valence and arousal in affective words. *Cerebral Cortex*.

Li, P., Abarbanell, L., & Papafragou, A. 2005. Spatial reasoning skills in Tenejapan Mayans. Paper presented at the Twenty-sixth Annual Conference of the Cognitive Science Society.

Li, P., & Gleitman, L. R. 2002. Turning the tables: Spatial language and spatial cognition. *Cognition, 83*, 265–294.

Lieberson, S. 2000. *A matter of taste: How names, fashions, and culture change*. New Haven: Yale University Press.

Lieberson, S., Dumais, S., & Baumann, S. 2000. The instability of androgynous names: The symbolic maintenance of gender boundaries. *American Journal of Sociology, 105,* 1249–1287.

Loewer, B., & Rey, B. 1991. *Meaning in mind: Fodor and his critics*. Malden, Mass.: Blackwell.

MacKay, D. G., Shafto, M., Taylor, J. K., Marian, D. E., Abrams, L., & Dyer, J. R. 2004. Relations between emotion, memory, and attention: Evidence from taboo Stroop, lexical decision, and immediate memory tasks. *Memory & Cognition, 32,* 474–488.

Majid, A., Bowerman, M., Kita, S., Haun, D., & Levinson, S. C. 2004. Can language restructure cognition? The case for space. *Trends in Cognitive Science, 8,* 108–114.

Malotki, E. 1983. *Hopi time: A linguistic analysis of temporal concepts in the Hopi language*. Berlin: Mouton.

Mandel, D. R. 2003. Judgment dissociation theory: An analysis of differences in causal, counterfactual, and covariational reasoning. *Journal of Experimental Psychology: General, 132,* 419–434.

Maratsos, M. P. 1988. Metaphors of language: Metaphors of the mind? *Contemporary Psychology, 34,* 5–7.

Marcus, G. F. 1993. Negative evidence in language acquisition. *Cognition, 46,* 53–85.

Marcus, G. F., Brinkmann, U., Clahsen, H., Wiese, R., & Pinker, S. 1995. German inflection: The exception that proves the rule. *Cognitive Psychology, 29,* 189–256.

Marcus, G. F., Pinker, S., Ullman, M., Hollander, M., Rosen, T. J., & Xu, F. 1992. Overregularization in language acquisition. *Monographs of the Society for Research in Child Development, 57.*

Marcus, R. B. 1961. Modalities and intensional languages. *Synthèse, 13,* 303–322.

Markman, E. 1989. *Categorization and naming in children: Problems of induction*. Cambridge, Mass.: MIT Press.

Marr, D. 1982. *Vision*. San Francisco: W. H. Freeman.

Marshall, J., Chiat, S., Robson, J., & Print, T. 1996. Calling a salad a federation: An investigation of semantic jargon, Part 2: Verbs. *Journal of Neurolinguistics, 9,* 251–260.

Mateu, J. 2001. Locative and locatum verbs revisited: Evidence from Romance. In Y. D'Hulst, J. Rooryck, & J. Schroten (Eds.), *Romance languages and linguistic theory 1999*. Philadelphia: John Benjamins.

Matthew, A., Richards, A., & Eysenck, M. 1989. Interpretation of homophones related to threat in anxiety states. *Journal of Abnormal Psychology, 98,* 31–34.

Mayer, R. E. 1993. The instructive metaphor: Metaphoric aids to students' understanding of science. In A. Ortony (Ed.), *Metaphor and thought* (2nd ed.). New York: Cambridge University Press.

Maynard Smith, J. 1988. *Games, sex, and evolution*. New York: Harvester Wheatsheaf.

Mayr, E. 1982. *The growth of biological thought*. Cambridge, Mass.: Harvard University Press.

Mazurkewich, I., & White, L. 1984. The acquisition of the dative alternation: Unlearning overgeneralizations. *Cognition, 16,* 261–283.

McCawley, J. D. 1968. The role of semantics in grammar. In E. Bach & R. T. Harris (Eds.), *Universals in linguistic theory*. New York: Holt, Rinehart & Winston.

McClelland, J. L., & Kawamoto, A. H. 1986. Mechanisms of sentence processing: Assigning roles to constituents of sentences. In J. L. McClelland & D. E. Rumelhart (Eds.), *Parallel distributed processing: Explorations in the microstructure of cognition,* Vol. 2: *Psychological and biological models*. Cambridge, Mass.: MIT Press.

McCloskey, M. 1983. Intuitive physics. *Scientific American, 248,* 122–130.

McCloskey, M., Caramazza, A., & Green, B. 1980. Curvilinear motion in the absence of external forces: Naive beliefs about the motion of objects. *Science, 210,* 1139–1141.

McCormick, M. 2005. Immanuel Kant: Metaphysics. *The Internet Encyclopedia of Philosophy.* http://www.iep.utm.edu/k/kantmeta.htm.

McEwan, I. 1998. *Amsterdam.* London: Jonathan Cape.

McGinn, C. 1993. *Problems in philosophy: The limits of inquiry.* Malden, Mass.: Blackwell.

McGlone, M. S., & Harding, J. L. 1998. Back (or forward?) to the future: The role of perspective in temporal language comprehension. *Journal of Experimental Psychology: Learning, Memory, and Cognition, 24,* 1211–1223.

McGraw, A. P., & Tetlock, P. E. 2005. Taboo trade-offs, relational framing, and the acceptability of exchanges. *Journal of Consumer Psychology, 15,* 2–15.

Meier, B. P., & Robinson, M. D. 2004. Why the sunny side is up: Associations between affect and vertical position. *Psychological Science, 15,* 243–247.

Metcalf, A. 2002. *Predicting new words: The secret of their success.* Boston: Houghton Mifflin.

Michotte, A. 1963. *The perception of causality.* London: Methuen.

Miller, G. A. 1991. *The science of words.* New York: W. H. Freeman.

Morgan, J. L., Bonamo, K., & Travis, L. L. 1995. Negative evidence on negative evidence. *Developmental Psychology, 31,* 180–197.

Morgan, J. L., & Travis, L. L. 1989. Limits on negative information on language learning. *Journal of Child Language, 16,* 531–552.

Murphy, G. 1996. On metaphoric representation. *Cognition, 60,* 173–204.

Myers, D. G., & Diener, E. 1995. Who is happy? *Psychological Science, 6,* 10–19.

Nabokov, V. V. 1955. *Lolita.* New York: Vintage.

Nagel, T. 1997. *The last word.* New York: Oxford University Press.

Nedyalkov, V. P., & Silnitsky, G. G. 1973. The typology of morphological and lexical causatives. In F. Kiefer (Ed.), *Trends in Soviet theoretical linguistics.* Dordrecht, Netherlands: Reidel.

Newell, A., & Simon, H. A. 1972. *Human problem solving.* Englewood Cliffs, N.J.: Prentice-Hall.

Newman, M. L., Pennebaker, J. W., Berry, D. S., & Richards, J. M. 2003. Lying words: Predicting deception from linguistic styles. *Personality and Social Psychology Bulletin, 29,* 665–675.

Nisbett, R. E., & Cohen, D. 1996. *Culture of honor: The psychology of violence in the South.* New York: HarperCollins.

Nowak, M. A., & Komarova, N. L. 2001. Towards an evolutionary theory of language. *Trends in Cognitive Sciences, 5,* 288–295.

Nunberg, G. 1979. The non-uniqueness of semantic solutions: Polysemy. *Linguistics and Philosophy, 3,* 143–184.

Nunberg, G. 2004. Imprecational categories. *The Language Log.* http://itre.cis.upenn.edu/~myl/languagelog/archives/000614.html.

Nunberg, G. 2006. *Talking Right: How conservatives turned liberalism into a tax-raising, latte-drinking, sushi-eating, Volvo-driving, New York Times-reading, body-piercing, Hollywood-loving, left-wing freak show.* New York: PublicAffairs.

Núñez, R. E., & Sweetser, E. 2006. With the future behind them: Convergent evidence from Aymara language and gesture in the crosslinguistic comparison of spatial construals of time. *Cognitive Science, 30,* 401–450.

Nwachukwu, P. A. 1987. *The argument structure of Igbo verbs* (Lexicon Project Working Paper #18). Cambridge, Mass.: MIT Center for Cognitive Science.

Oehrle, R. T. 1976. The grammatical status of the English dative alternation. Unpublished Ph.D. dissertation, MIT.

Ornstein, P. 2004. Where have all the Lisas gone? In S. Pinker (Ed.), *The best American science and nature writing 2004.* Boston: Houghton Mifflin.

Ornstein, R. 1975. *On the experience of time.* New York: Penguin.

Osgood, C. E., Suci, G., & Tannenbaum, P. 1957. *The measurement of meaning.* Urbana, Ill.: University of Illinois Press.

Osherson, D. N., Stob, M., & Weinstein, S. 1985. *Systems that learn.* Cambridge, Mass.: MIT Press.

Ostler, N., & Atkins, B. T. S. 1992. Predictable meaning shift: Some linguistic properties of lexical implication rules. In J. Pustejovsky & S. Bergler (Eds.), *Lexical semantics and knowledge representation.* Berlin: Springer-Verlag.

Panksepp, J. 1998. *Affective neuroscience: The foundations of human and animal emotions.* New York: Oxford University Press.

Patai, D. 1998. *Heterophobia: Sexual harassment and the future of feminism.* New York: Rowman & Littlefield.

Pearl, J. 2000. *Causality.* New York: Oxford University Press.

Pederson, E., Danziger, E., Wilkins, D., Levinson, S. C., Kita, S., & Senft, G. 1998. Semantic typology and spatial conceptualization. *Language, 74,* 557–589.

Piatelli-Palmarini, M. 1986. The rise of selective theories: A case study and some lessons from immunology. In W. Demopoulos & A. Marras (Eds.), *Language learning and concept acquisition.* Norwood, N.J.: Ablex.

Piatelli-Palmarini, M. 1989. Evolution, selection, and cognition: From "learning" to parameter setting in biology and the study of language. *Cognition, 31,* 1–44.

Pica, P., Lemer, C., Izard, V., & Dehaene, S. 2004. Exact and approximate arithmetic in an Amazonian indigene group. *Science, 306,* 499–503.

Pinker, S. 1979. Formal models of language learning. *Cognition, 7,* 217–283.

Pinker, S. 1984. Visual cognition: An introduction. *Cognition, 18,* 1–63.

Pinker, S. 1984/1996. *Language learnability and language development.* Cambridge, Mass.: Harvard University Press.

Pinker, S. 1988. A computational theory of the mental imagery medium. In M. Denis, J. Engelkamp, & J. T. E. Richardson (Eds.), *Cognitive and neuropsychological approaches to mental imagery.* Amsterdam: Martinus Nijhoff.

Pinker, S. 1989. *Learnability and cognition: The acquisition of argument structure.* Cambridge, Mass.: MIT Press.

Pinker, S. 1990. A theory of graph comprehension. In R. Friedle (Ed.), *Artificial intelligence and the future of testing.* Mahwah, N.J.: Erlbaum.

Pinker, S. 1994a. How could a child use verb syntax to learn verb semantics? *Lingua, 92,* 377–410.

Pinker, S. 1994b. *The language instinct.* New York: HarperCollins.

Pinker, S. 1995. Beyond folk psychology (Review of J. A. Fodor's *The elm and the expert*). *Nature, 373,* 205.

Pinker, S. 1997a. Words and rules in the human brain. *Nature, 387,* 547–548.

Pinker, S. 1997b. *How the mind works.* New York: Norton.

Pinker, S. 1999. *Words and rules: The ingredients of language.* New York: HarperCollins.

Pinker, S. 2002. *The blank slate: The modern denial of human nature.* New York: Viking.

Pinker, S. 2005. So how *does* the mind work? *Mind & Language, 20,* 1–24.

Pinker, S. 2007. Introduction. In J. Brockman (Ed.), *What is your dangerous idea?* New York: HarperCollins.

Pinker, S., & Birdsong, D. 1979. Speakers' sensitivity to rules of frozen word order. *Journal of Verbal Learning and Verbal Behavior, 18,* 497–508.

Pinker, S., & Bloom, P. 1990. Natural language and natural selection. *Behavioral and Brain Sciences, 13,* 707–784.

Pinker, S., & Jackendoff, R. 2005. The faculty of language: What's special about it? *Cognition, 95,* 201–236.

Pinker, S., & Rose, S. 1998. The two Steves: A debate. *Edge.* http://www.edge.org/3rd_culture/pinker_rose/pinker_rose_p1.html.

Pomerantz, J. R. 2003. Wholes, holes, and basic features in vision. *Trends in Cognitive Science, 7,* 471–473.

Pöppel, E. 2003. Pöppel's universal. *Edge.* http://www.edge.org/q2004/page5.html#poppel.

Postma, A., & Laeng, B. 2006. New insights in categorical and coordinate processing of spatial relations. *Neuropsychologia, 44* (Special Issue).

Potter, S. 1950/1971. *The complete upmanship.* New York: Holt, Rinehart & Winston.

Potts, C. 2005. *The logic of conventional implicatures.* New York: Oxford University Press.

Prasada, S., & Pinker, S. 1993. Generalizations of regular and irregular morphological patterns. *Language and Cognitive Processes, 8,* 1–56.

Premack, D., & Premack, A. 2003. *Original intelligence.* New York: McGraw-Hill.

Proust, M. 1913/1982. *Remembrance of things past.* New York: Vintage Books.

Provine, R. R. 1996. Laughter. *American Scientist, 84,* 38–45.

Pullum, G. K. 1991. *The great Eskimo vocabulary hoax and other irreverent essays on the study of language.* Chicago: University of Chicago Press.

Pullum, G. K. 2003a. Passive voice and bias in Reuter headlines about Israelis and Palestinians. *Language Log.* http://itre.cis.upenn.edu/~myl/languagelog/archives/000236.html.

Pullum, G. K. 2003b. Verb semantics and justifying war. *Language Log.* http://itre.cis.upenn.edu/~myl/languagelog/archives/000015.html.

Pustejovsky, J. 1995. *The generative lexicon.* Cambridge, Mass.: MIT Press.

Putnam, H. 1975. The meaning of "meaning." In K. Gunderson (Ed.), *Language, mind, and knowledge.* Minneapolis: University of Minnesota Press.

Pylkkänen, L., Llinás, R., & Murphy, G. 2006. The representation of polysemy: MEG evidence. *Journal of Cognitive Neuroscience, 18,* 97–109.

Quang Fuc Dong. 1971/1992a. English sentences without overt grammatical subject. In A. M. Zwicky, P. H. Salus, R. I. Binnick, & A. L. Vanek (Eds.), *Studies out in left field: Defamatory essays presented to James D. McCawley on the occasion of his 33rd or 34th birthday.* Philadelphia: John Benjamins.

Quang Fuc Dong. 1971/1992b. A note on conjoined noun phrases. In A. M. Zwicky, P. H. Salus, R. I. Binnick, & A. L. Vanek (Eds.), *Studies out in left field: Defamatory essays presented to James D. McCawley on the occasion of his 33rd or 34th birthday.* Philadelphia: John Benjamins.

Quine, W. V. O. 1960. *Word and object.* Cambridge, Mass.: MIT Press.

Quine, W. V. O. 1969. Natural kinds. In W. V. O. Quine (Ed.), *Ontological relativity and other essays.* New York: Columbia University Press.

Ramachandran, V. S., & Blakeslee, S. 1998. *Phantoms in the brain: Probing the mysteries of the human mind.* New York: Morrow.

Randall, D. B. J. 1989. X me no X's: Some examples (mainly from the Renaissance) of the Neologizing imperative retort. *American Speech, 64,* 223–243.

Randall, L. 2005. *Warped passages: Unraveling the mysteries of the universe's hidden dimensions.* New York: HarperCollins.

Rappaport Hovav, M., & Levin, B. 2007. All dative verbs are not created equal. Unpublished manuscript, Hebrew University of Jerusalem and Stanford University.

Rappaport, M., & Levin, B. 1985. A case study in lexical analysis: The locative alternation. Unpublished manuscript, Hebrew University of Jerusalem.

Rappaport, M., & Levin, B. 1988. What to do with theta-roles. In W. Wilkins (Ed.), *Syntax and semantics 21: Thematic relations.* New York: Academic Press.

Reddy, M. 1993. The conduit metaphor: A case of frame conflict in our language about language. In A. Ortony (Ed.), *Metaphor and thought* (2nd ed.). New York: Cambridge University Press.

Redelmeier, D. A., & Tversky, A. 1996. On the belief that arthritis pain is related to the weather. *Proceedings of the National Academy of Sciences, 93,* 2895–2896.

Reichenbach, H. 1947. *Elements of symbolic logic*. New York: Macmillan.

Richards, I. A. 1936/1965. *The philosophy of rhetoric*. New York: Oxford University Press.

Rijkhoff, J. 2002. *The noun phrase*. New York: Oxford University Press.

Robertson, L. C. 2003. *Space, objects, brains, and minds*. New York: Psychology Press.

Rock, I. 1983. *The logic of perception*. Cambridge, Mass.: MIT Press.

Roelofs, A. In press. The visual-auditory color-word Stroop asymmetry and its time course. *Memory & Cognition*.

Rosenblum, T., & Pinker, S. 1983. Word magic revisited: Monolingual and bilingual preschoolers' understanding of the word-object relationship. *Child Development, 54*, 773–780.

Rosovsky, H. 1990. *The university: An owner's manual*. New York: Norton.

Ross, B. H. 1984. Remindings and their effects on learning a cognitive skill. *Cognitive Psychology, 16*, 371–416.

Ross, B. H. 1987. This is like that: The use of earlier problems and the separation of similarity effects. *Journal of Experimental Psychology: Learning, Memory, and Cognition, 13*, 629–639.

Ross, J. R. Undated. Butterfly gazette. Unpublished manuscript, University of North Texas.

Roy, D., & Pentland, A. 2002. Learning words from sights and sounds: A computational model. *Cognitive Science, 26*, 113–146.

Rozin, P., & Fallon, A. 1987. A perspective on disgust. *Psychological Review, 94*, 23–41.

Russell, B. 1913. On the notion of cause. *Proceedings of the Aristotelian Society, 13*, 1–26.

Sadock, J. 1984. Whither radical pragmatics? In D. Schiffrin (Ed.), *Georgetown University Round Table on Languages and Linguistics*. Washington, D.C.: Georgetown University Press.

Salkoff, M. 1983. Bees are swarming in the garden. *Language, 59*, 288–346.

Sampson, G. 1982. The economics of conversation. In N. V. Smith (Ed.), *Mutual knowledge*. New York: Academic Press.

Santos, L. R., Sulkowski, G. M., Spaepen, G. M., & Hauser, M. D. 2002. Object individuation using property/kind information in rhesus macaques (*Macaca mulatta*). *Cognition, 83*, 241–264.

Saporta, S. 1994. *Society, language, and the university*. New York: Vantage.

Sasse, H.-J. 2002. Recent activity in the theory of aspect: Accomplishments, achievements, or just non-progressive state? *Linguistic Typology, 6*, 199–271.

Schacter, D. L. 1996. *Searching for memory: The brain, the mind, and the past*. New York: Basic Books.

Schacter, D. L. 2001. *The seven sins of memory: How the mind forgets and remembers*. Boston: Houghton Mifflin.

Schaller, S. 1991. *A man without words*. New York: Summit Books.

Schank, R. C. 1982. *Dynamic memory: A theory of reminding and learning in computers and people*. New York: Cambridge University Press.

Schelling, T. C. 1960. *The strategy of conflict*. Cambridge, Mass.: Harvard University Press.

Schelling, T. C. 1978. *Micromotives and macrobehavior*. New York: Norton.

Schiffer, S. R. 1972. *Meaning*. New York: Oxford University Press.

Schön, D. A. 1993. Generative metaphor: A perspective on problem-setting in social policy. In A. Ortony (Ed.), *Metaphor and thought* (2nd ed.). New York: Cambridge University Press.

Schütze, C. 1996. *The empirical basis of linguistics: Grammaticality judgments and linguistic methodology*. Chicago: University of Chicago Press.

Scott, E. C., Matzke, N. J., Branch, G., & 284 scientists named "Steve." 2004. The morphology of Steve. *Annals of Improbable Research*, 24–29.

Searle, J. R. 1975. Indirect speech acts. In P. Cole & J. Morgan (Eds.), *Syntax and semantics 3: Speech acts*. New York: Academic Press.

Searle, J. R. 1993a. Metaphor. In A. Ortony (Ed.), *Metaphor and thought* (2nd ed.). New York: Cambridge University Press.

Searle, J. R. 1993b. Rationality and realism: What is at stake? *Daedalus, 122,* 55–83.

Semenza, C. 2005. The (neuro)-psychology of mass and count nouns. *Brain and Language, 95,* 88–89.

Senghas, A., Kim, J. J., & Pinker, S. 2004. The plurals-in-compounds effect. Unpublished manuscript, Barnard College.

Shad, U. P., 1971/1992. Some unnatural habits. In A. M. Zwicky, P. H. Salus, R. I. Binnick, & A. L. Vanek (Eds.), *Studies out in left field: Defamatory essays presented to James D. McCawley on the occasion of of his 33rd or 34th birthday.* Philadelphia: John Benjamins.

Sheidlower, J. 1995. *The F-word.* New York: Random House.

Shepard, R. N. 1978. The mental image. *American Psychologist, 33,* 125–137.

Shibatani, M. 1976. The grammar of causative constructions: A conspectus. In M. Shibatani (Ed.), *Syntax and semantics 6: The grammar of causative constructions.* New York: Academic Press.

Shtulman, A. 2006. Qualitative differences between naive and scientific theories of evolution. *Cognitive Psychology, 52,* 170–194.

Shultz, T. R. 1982. Rules of causal attribution. *Monographs of the Society for Research in Child Developments, 47.*

Simonson, I., & Tversky, A. 1992. Choice in context: Tradeoff contrast and extremeness aversion. *Journal of Marketing Research, 29,* 281–295.

Singer, H. S. 2005. Tourette syndrome: From behavior to biology. *Lancet Neurology, 4,* 149–159.

Singer, I. B. 1984. *Stories for children.* New York: Farrar, Straus & Giroux.

Sinha, C., & Kuteva, T. 1995. Distributed spatial semantics. *Nordic Journal of Linguistics, 18,* 167–199.

Siskind, J. 1995. A computational study of lexical acquisition. *Cognition, 50,* 1–25.

Slobin, D. I. 1996. From "thought and language" to "thinking for speaking." In J. J. Gumperz & S. C. Levinson (Eds.), *Rethinking linguistic relativity.* New York: Cambridge University Press.

Smith, L., & Thelen, E. (Eds.). 1993. *A dynamic systems approach to development: Applications.* Cambridge, Mass.: MIT Press.

Smith, N. V. (Ed.). 1982. *Mutual knowledge.* New York: Academic Press.

Social Security Administration. 2006. Popular baby names. http://www.ssa.gov/OACT/babynames/.

Soja, N. N., Carey, S., & Spelke, E. S. 1991. Ontological categories guide young children's inductions of word meaning: Object terms and substance terms. *Cognition, 38,* 179–211.

Solt, J. 1987. Japanese sexual maledicta. In R. Aman (Ed.), *The best of Maledicta: The International Journal of Verbal Aggression.* Philadelphia: Running Press.

Sommers, F. 1963. Types and ontology. *Philosophical Review, 72,* 327–363.

Sowell, T. 1980. *Knowledge and decisions.* New York: Basic Books.

Sowell, T. 1987. *A conflict of visions: Ideological origins of political struggles.* New York: Quill.

Sowell, T. 1996. *Migrations and cultures: A world view.* New York: Basic Books.

Speedie, L. J., Wertman, J. T., & Heilman, K. M. 1993. Disruption of automatic speech following a right basal ganglia lesion. *Neurology, 43,* 1768–1774.

Spelke, E. 1995. Initial knowledge: Six suggestions. *Cognition, 50,* 433–447.

Spelke, E. 2003. What makes us smart? Core knowledge and natural language. In D. Gentner & S. Goldin-Meadow (Eds.), *Language in mind: Advances in the study of language and thought.* Cambridge, Mass.: MIT Press.

Spellman, B. A. 1996. Acting as intuitive scientists: Contingency judgments are made while controlling for alternative potential causes. *Psychological Science, 7,* 337–343.

Spellman, B. A. 1997. Crediting causality. *Journal of Experimental Psychology: General, 126,* 323–349.

Spellman, B. A., & Kincannon, A. 2001. The relation between counterfactual ("but for") and causal reasoning and implications for jurors' decisions. *Law and Contemporary Problems* (special issue on Causation in Law and Science), *64,* 241–264.

Spellman, B. A., & Mandel, D. R. 1999. When possibility informs reality: Counterfactual thinking as a cue to causality. *Trends in Cognitive Science, 8,* 120–123.

Sperber, D. 1985. Anthropology and psychology: Towards an epidemiology of representations. *Man, 20,* 73–89.

Sperber, D., & Wilson, D. 1986. *Relevance: Communication and cognition.* Cambridge, Mass.: Harvard University Press.

Stalnaker, R. C. 1978. Assertion. In P. Cole (Ed.), *Syntax and semantics 9: Pragmatics.* New York: Academic Press.

Symons, D. 1979. *The evolution of human sexuality.* New York: Oxford University Press.

Talmy, L. 1983. How language structures space. In H. Pick & L. Acredolo (Eds.), *Spatial orientation: Theory, research, and application.* New York: Plenum.

Talmy, L. 1985. Lexicalization patterns: Semantic structure in lexical forms. In T. Shopen (Ed.), *Language typology and syntactic description III: Grammatical categories and the lexicon.* New York: Cambridge University Press.

Talmy, L. 1988. Force dynamics in language and cognition. *Cognitive Science, 12,* 49–100.

Talmy, L. 2000a. Force dynamics in language and cognition. In *Toward a cognitive semantics 1: Concept structuring systems.* Cambridge, Mass.: MIT Press.

Talmy, L. 2000b. How language structures space. In L. Talmy (Ed.), *Toward a cognitive semantics.* Cambridge, Mass.: MIT Press.

Tarr, M. J., & Pinker, S. 1989. Mental rotation and orientation-dependence in shape recognition. *Cognitive Psychology, 21,* 233–282.

Tegmark, M. 2003. Parallel universes. *Scientific American, 288,* 41–51.

Tenner, E. 1989. Talking through our hats. *Harvard Magazine, 91,* 21–36.

Tenny, C. 1992. The aspectual interface hypothesis. In I. A. Sag & A. Szabolcsi (Eds.), *Lexical matters.* Stanford: Center for the Study of Language and Information.

Tetlock, P. E. 1999. Coping with tradeoffs: Psychological constraints and political implications. In A. Lupia, M. McCubbins, & S. Popkin (Eds.), *Political reasoning and choice.* Berkeley: University of California Press.

Tetlock, P. E., Kristel, O. V., Elson, B., Green, M. C., & Lerner, J. 2000. The psychology of the unthinkable: Taboo tradeoffs, forbidden base rates, and heretical counterfactuals. *Journal of Personality and Social Psychology, 78,* 853–870.

Thomson, J. J. 1985. The trolley problem. *Yale Law Journal, 94,* 1395–1415.

Tomasello, M. 2003. *Constructing a language: A usage-based theory of language acquisition.* Cambridge, Mass.: Harvard University Press.

Tooby, J., & Cosmides, L. 1992. Psychological foundations of culture. In J. Barkow, L. Cosmides, & J. Tooby (Eds.), *The adapted mind: Evolutionary psychology and the generation of culture.* New York: Oxford University Press.

Tooby, J., & Cosmides, L. 1996. Friendship and the banker's paradox: Other pathways to the evolution of adaptations for altruism. *Proceedings of the British Academy, 88,* 119–143.

Tooby, J., & DeVore, I. 1987. The reconstruction of hominid evolution through strategic modeling. In W. G. Kinzey (Ed.), *The evolution of human behavior: Primate models.* Albany: State University of New York Press.

Tootell, R. B., Silverman, M. S., Switkes, E., & De Valois, R. L. 1982. Deoxyglucose analysis of retinotopic organization in primate striate cortex. *Science, 218,* 902–904.

Treisman, A., & Gelade, G. 1980. A feature-integration theory of attention. *Cognitive Psychology, 12,* 97–136.

Tversky, A., & Kahneman, D. 1981. The framing of decisions and the psychology of choice. *Science, 211,* 453–458.

Tyler, A., & Evans, V. 2003. *The semantics of English prepositions: Spatial scenes, embodied meaning, and cognition.* New York: Cambridge University Press.

Ullman, M. T. 1999. Acceptability ratings of regular and irregular past-tense forms: Evidence for a dual-system model of language from word frequency and phonological neighborhood effects. *Language and Cognitive Processes, 14,* 47–67.

Ullman, M. T., Corkin, S., Coppola, M., Hickok, G., Growdon, J. H., Koroshetz, W. J., & Pinker, S. 1997. A neural dissociation within language: Evidence that the mental dictionary is part of declarative memory, and that grammatical rules are processed by the procedural system. *Journal of Cognitive Neuroscience, 9,* 289–299.

Valiant, L. 1994. *Circuits of the mind.* New York: Oxford University Press.

Van Essen, D. C., & Deyoe, E. A. 1995. Concurrent processing in the primate visual cortex. In M. S. Gazzaniga (Ed.), *The cognitive neurosciences.* Cambridge, Mass.: MIT Press.

Van Lancker, D., & Cummings, J. L. 1999. Expletives: Neurolinguistic and neurobehavioral perspectives on swearing. *Brain Research Reviews, 31,* 83–104.

Van Lancker, D., & Sidtis, B. 2006. Formulaic expressions in spontaneous speech of left- and right-hemisphere-damaged subjects. *Aphasiology, 20,* 411–426.

Van Valin, R. D. 2005. *Exploring the syntax-semantics interface.* New York: Cambridge University Press.

Vanderschraaf, P., & Sillari, G. 2005. Common knowledge. *The Stanford Encyclopedia of Philosophy,* Winter 2005. http://plato.stanford.edu/archives/win2005/entries/common-knowledge/.

Varley, R. A., Klessinger, N. J. C., Romanowski, C. A. J., & Siegal, M. 2005. Agrammatic but numerate. *Proceedings of the National Academy of Sciences, 102,* 3519–3524.

Veblen, T. 1899/1994. *The theory of the leisure class.* New York: Penguin.

Vendler, Z. 1957. Verbs and times. *Philosophical Review, 66,* 143–160.

Vise, D., & Malseed, M. 2005. *The Google story.* New York: Delacorte Press.

Wajnryb, R. 2005. *Expletive deleted: A good look at bad language.* New York: Random House.

Wallraff, B. 2006. *Word fugitives.* New York: HarperCollins.

Walsh, C. R., & Sloman, S. A. 2005. The meaning of cause and prevent: The role of causal mechanism. Paper presented at the Proceedings of the Conference of the Cognitive Science Society, Stresa, Italy.

Walsh, W. H. 1967. Immanuel Kant. In P. Edwards (Ed.), *The encyclopedia of philosophy.* New York: Macmillan.

Wattenberg, L. 2005. The Baby Name Wizard's NameVoyager. www.babynamewizard.com/namevoyager/lnv0105.html.

Wegner, D. 1989. *White bears and other unwanted thoughts: Suppression, obsession, and the psychology of mental control.* New York: Guilford.

Wegner, D. 2002. *The illusion of conscious will.* Cambridge, Mass.: MIT Press.

Wheeler, J. A. 1994. Time today. In J. J. Halliwell, J. Pérez-Mercader, & W. H. Zurek (Eds.), *Physical origins of time asymmetry.* New York: Cambridge University Press.

White, P. A. 1995. Use of prior beliefs in the assignment of causal roles: Causal powers versus regularity-based accounts. *Memory & Cognition, 23,* 243–254.

Whorf, B. L. 1956. *Language, thought, and reality: Selected writings of Benjamin Lee Whorf.* Cambridge, Mass.: MIT Press.

Wierzbicka, A. 1987. *English speech act verbs: A semantic dictionary.* New York: Academic Press.

Wierzbicka, A. 1988a. Oats and wheat: Mass nouns, iconicity, and human categorization. In *The semantics of grammar.* Philadelphia: John Benjamins.

Wierzbicka, A. 1988b. *The semantics of grammar.* Philadelphia: John Benjamins.

Wierzbicka, A. 1988c. What's in a noun? (or: How do nouns differ in meaning from adjectives?). In *The semantics of grammar*. Philadelphia: John Benjamins.

Wierzbicka, A. 1991. *Cross-cultural pragmatics: The semantics of human interaction*. New York: Mouton de Gruyter.

Wierzbicka, A. 1998. The semantics of English causative constructions in a universal-typological perspective. In M. Tomasello (Ed.), *The new psychology of language*. Mahwah, N.J.: Erlbaum.

Wiese, H. 2003. *Numbers, language, and the human mind*. New York: Cambridge University Press.

Williams, G. C. 1966. *Adaptation and natural selection: A critique of some current evolutionary thought*. Princeton, N.J.: Princeton University Press.

Winner, E., & Gardner, H. 1993. Metaphor and irony: Two levels of understanding. In A. Ortony (Ed.), *Metaphor and thought* (2nd ed.). New York: Cambridge University Press.

Winter, S. L. 2001. *A clearing in the forest: Law, life, and mind*. Chicago: University of Chicago Press.

Winter, Y. 2002. Atoms and sets: A characterization of semantic number. *Linguistics Inquiry, 33*, 493–505.

Wise, S., Murray, E., & Gerfen, C. 1996. The frontal cortex-basal ganglia system in primates. *Critical Reviews in Neurobiology 10*, 317–356.

Wolff, P. 2002. A vector model of causal meaning. In W. D. Gray & C. D. Schunn (Eds.), *Proceedings of the 24th Annual Conference of the Cognitive Science Society*. Mahwah, N.J.: Erlbaum.

Wolff, P. 2003. Direct causation in the linguistic coding and individuation of causal events. *Cognition, 88*, 1–48.

Wolff, P. 2007. Representing causation. *Journal of Experimental Psychology: General*.

Wolff, P., & Song, G. 2003. Models of causation and the semantics of causal verbs. *Cognitive Psychology, 47*, 276–332.

Wray, A. 1998. Protolanguage as a holistic system for social interaction. *Language and Communication, 18*, 47–67.

Wynn, K. 1992. Addition and subtraction in human infants. *Nature, 358*, 749–750.

Xu, F., & Carey, S. 1996. Infants' metaphysics: The case of numerical identity. *Cognitive Psychology, 30*, 111–153.

Xu, F., Carey, S., & Welch, J. 1999. Infants' ability to use object kind information for object individuation. *Cognition, 70*, 137–166.

Yang, C. 2003. *Knowledge and learning in natural language*. New York: Oxford University Press.

Yeung, N., Botvinick, M. M., & Cohen, J. D. 2004. The neural basis of error detection: Conflict monitoring and the Error-Related Negativity. *Psychological Review, 11*, 931–959.

Zwicky, A. M., Salus, P. H., Binnick, R. I., & Vanek, A. L. (Eds.). 1971/1992. *Studies out in left field: Defamatory essays presented to James D. McCawley on the occasion of his 33rd or 34th birthday*. Philadelphia: John Benjamins.

INDEX

A *New York Times* bestseller

Finalist for the Pulitzer Prize

The Blank Slate

The Modern Denial of Human Nature

Steven Pinker, a leading expert on language and the mind, explores the idea of human nature and its moral, emotional, and political undertones. With characteristic wit, lucidity, and insight, Pinker tackles the theory that we are all born with a mind that is a blank slate—a doctrine held by many intellectuals during the past century—and alternatively proposes that humans inherit a common mental structure formed by the species' need for survival.

ISBN 978-0-14-200334-3